Advances in
Nutritional
Research
Volume 6

Advances in
Nutritional Research

A Continuation Order Plan is available for this series. A continuation order will bring delivery of each new volume immediately upon publication. Volumes are billed only upon actual shipment. For further information please contact the publisher.

Advances in
Nutritional
Research
Volume 6

Edited by Harold H. Draper

University of Guelph
Guelph, Ontario, Canada

Plenum Press · New York and London

The Library of Congress cataloged the first volume of this title as follows:

Advances in nutritional research. v.1-
    New York, Plenum Press, c1977-
    1 v. ill. 24 cm.
    Key title: Advances in nutritional research, ISSN 0149-9483

    1. Nutrition–Yearbooks.
QP141.A1A3   613.2′05   78-640645

ISBN-13: 978-1-4612-9805-2   e-ISBN-13: 978-1-4613-2801-8
DOI: 10.1007/978-1-4613-2801-8

© 1984 Plenum Press, New York
Softcover reprint of the hardcover 1st edition 1984
A Division of Plenum Publishing Corporation
233 Spring Street, New York, N.Y. 10013

Contributors

R. Anderson, The Immunology Section, Department of Medical Microbiology, Institute of Pathology, University of Pretoria, Republic of South Africa.

N.J. Benevenga, Departments of Meat Animal Science and Nutritional Sciences, University of Wisconsin, Madison, Wisconsin 53706.

G.G. Bruckner, Department of Clinical Nutrition, University of Kentucky, Medical Center Annex Two, Lexington, Kentucky 40536-0080.

J. Chen, Institute of Health, Chinese Academy of Medical Sciences, Beijing, People's Republic of China.

Xeucun Chen, Institute of Health, Chinese Academy of Medical Sciences, Beijing, People's Republic of China.

Xiaoshu Chen, Institute of Health, Chinese Academy of Medical Sciences, Beijing, People's Republic of China.

J.W. Coburn, Department of Pediatrics and Medicine, UCLA School of Medicine and the Medical and Research Services, VA Wadsworth Medical Center, Los Angeles, California 90073.

P.J. Garlick, London School of Hygiene and Tropical Medicine, Clinical Nutrition and Metabolism Unit, Hospital for Tropical Diseases, 4 St. Pancras Way, London, NWI.

K. Ge, Institute of Health, Chinese Academy of Medical Sciences, Beijing, People's Republic of China.

E. Gudmand-Høyer, Medical Department F (Gastroenterology), Copenhagen University Hospital in Gentofte, Denmark.

L.S. Hurley, Department of Nutrition, University of California, Davis, California 95616.

A.L. Jenkins, Department of Nutritional Sciences, Faculty of Medicine, University of Toronto, Toronto, Ontario M5S 1A8.

D.A.J. Jenkins, Department of Nutritional Sciences, Faculty of Medicine, University of Toronto, Toronto, Ontario M5S 1A8.

C.L. Keen, Department of Nutrition, University of California, Davis, California 95616.

G.L. Klein, Department of Pediatrics, City of Hope, National Medical Centre, Duarte, California 91010.

P.A. Krasilnikoff, Paediatric Department L, Copenhagen University in Gentofte, Denmark.

B. Lönnerdal, Department of Nutrition, University of California, Davis, California 95616.

C. Mettlin, Chief of Epidemiologic Research, Roswell Park Memorial Institute, Buffalo, New York 14263.

P.J. Reeds, Protein Biochemistry Department, Rowett Research Institute, Bucksburn, Aberdeen, Scotland AB2 9SB.

H. Skovbjerg, Medical Department F (Gastroenterology) Copenhagen University Hospital in Gentofte, Denmark.

J. Szabo, Department of Animal Hygiene, University of Veterinary Science, Budapest, Hungary.

Z. Wen, Institute of Health, Chinese Academy of Medical Sciences, Beijing, People's Republic of China.

G. Yang, Institute of Health, Chinese Academy of Medical Sciences, Beijing, People's Republic of China.

L. Zhu, Institute of Health, Chinese Academy of Medical Sciences, Beijing, People's Republic of China.

Preface

Volume 6 of *Advances in Nutritional Research* continues the theme of previous volumes in providing comprehensive reviews of the present state of knowledge on several topics of prime current interest in nutritional research. Some reviews relate to nutritional diseases of clinical or epidemiological interest, including the first detailed description of Keshan disease, a modern epidemic deficiency disease of man, and sucrase-isomaltase malabsorption, a racial-ethnic form of carbohydrase deficiency found among Arctic Eskimos. Others contain accounts of major new developments in knowledge of the metabolism of zinc and methionine. There is also a comprehensive summary of the nutritional and metabolic changes associated with the germfree state. The volume will be of value to graduate students and established investigators engaged in research in clinical or experimental nutrition and metabolism.

Contents

Chapter 1. Evidence for Alternative Pathways of Methionine
             Catabolism 1
             N.J. Benevenga

1. Introduction 1
2. Early Studies on Methionine Metabolism 1
3. Studies on the Oxidation of the Methionine Methyl Group 3
     3.1. Importance of Choline, Betaine and Sarcosine formation in
           Methionine Methyl Group Oxidation 3
     3.2. Importance of S-Adenosylmethionine Formation in
           Methionine Methyl Group Oxidation *In Vitro* 6
4. Importance of Transamination in Methionine Catabolism *In Vitro* 7
5. Decarboxylation of Methionine (α-Keto-γ-methiolbutyrate) *In
    Vitro* 9
6. Formation of 3-Methylthiopropionate From Methionine *In
    Vitro* 10
     6.1. Metabolism of 3-Methylthiopropionate *In Vitro* 12
7. Importance of the Transamination Pathway of Methionine
    Metabolism 13
8. Application of New Knowledge of Methionine Metabolism in
    Homocystinuria 15
9. Concluding Remarks 16
References 16

Chapter 2. The Immunostimulatory, Anti-Inflammatory and
             Anti-Allergic Properties of Ascorbate 19
             Ronald Anderson

1. Introduction 19

2.   Immunostimulatory Effects of Ascorbate   20
     2.1.   Leukocyte Motility   20
     2.2.   Ascorbate-Mediated Stimulation of Neutrophil Motility by
            an Antioxidant Mechanism   21
     2.3.   Importance of Extracellular Ascorbate in Stimulation of
            Leukocyte Motility   26
     2.4.   Ascorbate in the Treatment of Conditions Associated with
            Abnormal Neutrophil Motility   29
     2.5.   Ascorbate-Mediated Enhancement of Phagocyte
            Antimicrobial Activity   29
     2.6.   Effects of Ascorbate on Lymphocyte Proliferation and
            Migration   32
3.   Mechanisms of Ascorbate-Mediated Increased Lymphocyte
     Proliferation   32
     3.1.   Cyclic Nucleotide Effects   32
     3.2.   Antioxidant Mechanisms   33
4.   Anti-Allergic Properties of Ascorbate   33
     4.1.   Ascorbate-Mediated Neutralization of Histamine   35
     4.2.   Effects of Prostaglandins   35
     4.3.   Effects of Ascorbate on the Production of Leukotrienes   36
     4.4.   Ascorbate-Mediated Inhibition of Degranulation by an
            Antioxidant Mechanism   36
5.   Anti-Inflammatory Properties of Ascorbate   37
     5.1.   Prevention of Degradation of the Extracellular Matrix   37
     5.2.   Antioxidant Mechanisms   37
     5.3.   Other Mechanisms   40
6.   Conclusions   40
References   41

Chapter 3.   Epidemiologic Studies on Vitamin A and Cancer   47
             Curtis Mettlin

1.   Introduction   47
2.   Animal Studies   48
3.   Human Blood Level Studies   50
4.   Epidemiologic Studies   51
5.   Discussion   59
6.   Summary   64
References   64

Chapter 4.   Metabolic Bone Disease Associated with Total
             Parenteral Nutrition   67
             Gordon L. Klein and Jack W. Coburn

1. Introduction 67
2. Description of Index Case 68
3. Prospective Study of Biochemical Abnormalities Related to Bone
   in TPN Patients 70
4. Observations Made by Others 75
5. Bone Disease in Infants Treated with TPN 77
6. Controversy Regarding the Role of Vitamin D 78
   6.1. Alterations in Vitamin D Metabolism and the Possible Role
        of Vitamin D in TPN Bone Disease 79
   6.2. Causes of Low Serum Levels of $1,25(OH)_2$–vitamin D 81
   6.3. Do Low Serum Levels of $1,25 (OH)_2$–vitamin D Affect the
        Bone in TPN Patients? 82
   6.4. Does Vitamin D_2 Itself Contribute to the Bone Disease of
        TPN Patients? 82
7. Trace Elements in the Pathogenesis of TPN Bone Disease 84
   7.1. Aluminum Loading During Long-Term TPN as a Factor in
        Bone Abnormalities 84
8. Hypercalciuria in Patients Treated with Total Parenteral
   Nutrition 85
9. Is There a Circulating Inhibitor of Bone Mineralization? 86
10. Summary 87
11. Acknowledgments 88
References 88

Chapter 5. Nutrition and Protein Turnover in Man 93
            P.J. Reeds and P.J. Garlick

1. Introduction 93
2. Methods 95
   2.1. General Considerations 95
   2.2. Prescursor Methods 97
        2.2.1. Analysis of the Model 97
        2.2.2. Choice of Label and Amino Acid 98
        2.2.3. Method of Infusion 99
        2.2.4. Assessment of Labeled CO_2 Production 100
        2.2.5. The Precursor Pool Problem 102
        2.2.6. The Steady State 104
   2.3. End Product Methods 104
        2.3.1. Analysis of the Model 105
        2.3.2. Urinary End Products 106
        2.3.3. The Assumption of a Homogeneous Metabolic
               Pool 106
        2.3.4. The Assumption of a Steady State 108

2.4.	Comparison of Methods	110

3. Protein Synthesis and Turnover 112
   3.1. Protein Turnover in Man Related to That in Other
        Mammals 112
   3.2. Protein Synthesis During Development 116
4. Nutrition and Protein Turnover 119
   4.1. The Response to Food Intake 120
   4.2. Diet Composition 123
        4.2.1. Non-Protein Energy 123
        4.2.2. Dietary Protein 126
        4.2.3. Studies in Animals of the Effects of Food and
               Protein Intake 129
4.3. Protein Turnover in Severe Undernutrition and During Recovery . 131
5. Conclusion 133
6. Acknowledgment 134
References 134

Chapter 6. Zinc Binding Ligands and Complexes in Zinc
            Metabolism 139
            Bo Lönnderdal, Carl L. Keen, and Lucille S. Hurley

1. Introduction 139
2. Zinc Binding Ligands in Infant Nutrition 140
   2.1. Identification of Zinc Binding Ligands in Milk 140
        2.1.1. Background 140
        2.1.2. Citrate 141
        2.1.3. Picolinic Acid 144
        2.1.4. Prostaglandins 146
   2.2. Low Molecular Weight Zinc Complexes in Milk – A Role in
        Infant Nutrition 147
3. Genetic Abnormalities of Zinc Metabolism 147
   3.1. Acrodermatitis Enteropathica in Human Beings 148
        3.1.1. Genetic Acrodermatitis Enteropathica 148
        3.1.2. "Acquired Acrodermatitis Enteropathica" 149
   3.2. Animal Models 150
        3.2.1. Lethal Milk in Mice 150
        3.2.2. Adema Disease in Cattle 151
        3.2.3. Alaskan Malamute Chondrodysplasia in Dogs 152
4. Zinc Homeostasis 154
   4.1. Theories on Zinc Homeostasis 154
   4.2. Zinc Bind Ligands in Bile and Pancreatic Fluid 156
   4.3. Zinc Binding Ligands and Complexes in Duodenum 157
   4.4. Methodological Considerations 158

4.5.   A Proposed Model of Zinc Absorption 159
5. Concluding Remarks 160
References 161

Chapter 7.   The Clinical Implications of Dietary Fiber 169
          David J.A. Jenkins and Alexandra L. Jenkins

1. Introduction 169
2. Dietary Fiber Hypothesis 170
3. Diabetes 171
    3.1.   Purified Fiber and Diabetic Management 171
    3.2.   Fiber in Unrefined Foods and Diabetic Management 172
    3.3.   Mechanism of Action: Lente Carbohydrate 175
            3.3.1.   Reduced Postprandial Blood Glucose Response 175
            3.3.2.   Reduced Postprandial Endocrine Response 176
            3.3.3.   Reduced Rate of Absorption, Not Malabsorption 177
            3.3.4.   Gastric Emptying and Mouth to Cecum Transit
                      Time 178
            3.3.5.   Diversity of Response to Carbohydrate from
                      Different Foods 179
            3.3.6.   *In Vitro* Digestion – Lente Carbohydrate 180
            3.3.7.   Factors Other than Fiber 181
4. Dumping Syndrome 184
5. Dietary Fiber and Hyperlipidemia 185
    5.1.   Purified Fiber and Hypolipidemic Effects 186
    5.2.   High Fiber Foods and Hypolipidemic Effects 186
    5.3.   Mechanism of Action of Fiber on Lipid Metabolism 187
            5.3.1.   Purified Fiber 187
            5.3.2.   High Fiber Foods and Associated Hypolipidemic
                      Constituents 188
6. Weight Control 190
7. Constipation 191
8. Diverticular Disease 192
9. Crohn's Disease 193
10. Gallstones, Hiatus Hernia, and Varicose Veins 193
11. Fiber, Antinutrients, and Neoplastic Disease 194
12. Possible Adverse Effects on Fiber 194
13. Conclusion 195
14. Acknowledgments 195
References 195

Chapter 8.   The Role of Selenium in Keshan Disease 203
          Guangqi Yang, Junshi Chen, Zhimei Wen, Keyou Ge,
          Lianzhen Zhu, Xuecun Chen, and Xiaoshu Chen

1. Introduction 203
2. Brief Description of Keshan Disease 205
    2.1.   History and Epidemiology 205
    2.2.   Clinical Manifestations and Course 207
            2.2.1   Acute Type 207
            2.2.2   Chronic Type 207
            2.2.3. Subacute Type 208
            2.2.4. Insidious Type 209
    2.3.   Morphological Observations 209
    2.4.   Treatment 211
    2.5.   Etiology 211
            2.5.1. Infection 211
            2.5.2. Intoxication 212
            2.5.3. Nutrient Deficiency 212
3. Selenium Status and Keshan Disease 212
    3.1.   Selenium Status of Residents in Keshan Disease Areas 212
    3.2.   Concentrations of Selenium in Foods 215
    3.3.   Epidemiological Characteristics of Keshan Disease and Se   218
            3.3.1. Regional Distribution 219
            3.3.2. Population Susceptibility 219
            3.3.3. Seasonal Prevalence 220
4. The Efficacy of Sodium Selenite in the Prevention of Keshan
    Disease 220
    4.1.   Studies in Mianning County During 1974-1977 (Keshan
            Research Group, 1979) 220
    4.2.   Studies in Mianning County During 1976-1980 223
    4.3.   Studies in Five Counties During 1976-1980 223
5. Discusion 224
    5.1.   Role of Selenium in the Etiology of Keshan Disease 224
    5.2.   Possible Mechanisms of the Effects of Sodium Selenite in
            the Prevention of Keshan Disease 226
            5.2.1. Effects on Cell Antioxidant Potential 226
            5.2.2   Effects on Oxygen Metabolism 227
            5.2.3. The Anti-Infection Effect of Se 227
References 229

Chapter 9.   Sucrose – Isomaltose Malabsorption 233
             E. Gudmand-Høyer, P.A. Krasilnikoff, and H. Skovbjerg

1. Introduction 233
2. The Structure of the Brush Border Membrane 234
    2.1.   The Morphology of the Enterocyte 234
    2.2.   The Structure of the Membrane and Location of the
            Enzymes 235

3. The Digestive Functions of the Brush Border Enzymes 237
4. Distribution of Brush Border Enzymes 238
    4.1. The Distribution of Enzymes Along the Small Intestine 238
    4.2. The Distribution of Enzymes Along the Villus-Crypt Axis . 239
5. Biochemical Background of Sucrase-Isomaltase Deficiency 239
6. Pathophysiology 241
7. Symptoms of Sucrose-Isomaltose Malabsorption 243
8. Diagnosis 245
    8.1. Clinical Features 246
    8.2. Investigation of Stools 246
          8.2.1. Clinitest 247
          8.2.2. Chromatography 247
          8.2.3. pH 247
          8.2.4. Lactic Acid 247
    8.3. Barium-Sucrose Meal 247
    8.4. Oral Sucrose Tolerance Test 248
    8.5. Sucrose Hydrogen Breath Test 250
    8.6. Disaccharidase Assay 251
    8.7. Conclusion 251
9. Incidence and Genetic Conditions 251
10. S-I Malabsorption Diagnosis in Adults 254
11. S-I Malabsorption Combined With Other Specific Disaccharide
    Deficiencies 255
12. Unspecific (Secondary) S-I Malabsorption 256
13. Treatment 259
References 260

Chapter 10.   Nutrient Absorption in Gnotobiotic Animals 271
              Géza Bruckner and Jozsef Szabó

1. Introduction 271
2. The Normal Intestinal Microflora 272
3. Intestinal Morphology of Gnotobiotic Animals 274
    3.1. The Effects of Germfree Status on Various Intestinal
          Parameters 274
          3.1.1. Intestinal Length and Weight 274
          3.1.2. Intestinal Surface Area 274
          3.1.3. Intestinal Histology and Cell Constituents 275
          3.1.4. Cell Renewal and Mitotic Index 276
          3.2.1. Intestinal Histology and Cell Renewal Rate 276
          3.2.2. Cecal Reduction 277
4. Physical Characteristics of the Intestinal Contents of Gnotobiotic
    Animals 282

	4.1.	Dry Weight Percent	283
	4.2.	Relative Viscosity	284
	4.3.	Colloid Osmotic Pressure and Total Osmolality	284
	4.4.	pH	285
	4.5.	Redox Potential	285
5.	Biochemical Characteristics of the Intestinal Contents of Gnotobiotic Animals		285
6.	Water and Electrolyte Transport		287
7.	Minerals: Absorption and Utilization in Gnotobiotic Animals		294
	7.1.	Iron	294
	7.2.	Copper	296
	7.3.	Manganese	296
	7.4.	Zinc	298
	7.5.	Calcium	298
	7.6.	Magnesium	300
	7.7.	Phosphorus	300
8.	Vitamins		300
	8.1.	Thiamin	300
	8.2.	Riboflavin	301
	8.3.	Vitamin B_6, Nicotinic Acid, Pantothenic Acid, and Biotin	301
	8.4.	Vitamin B_{12}	301
	8.5.	Folic Acid	302
	8.6.	Ascorbic Acid	302
	8.7	Vitamin A	302
	8.8	Vitamins D and E	303
	8.9	Vitamin K	303
9.	Digestive Enzymes		303
	9.1.	Digestive Enzymes in Gnotobiotic Animals	303
10.	Bile Acids		308
11.	Fats and Fatty Acids		312
12.	Carbohydrates		315
13.	Proteins		316
14.	Acknowledgments		319
	References		319
	Index		333

Chapter 1

Evidence for Alternative Pathways of Methionine Catabolism

N. J. Benevenga

1. Introduction

The catabolism of methionine has been reviewed most recently by Green-berg (1975b), Mudd and Levy (1978), and Finkelstein (1975). These reviews propose that the major route of methionine degradation is via the transsulfuration pathway (Fig. 1). Following activation of methionine to S-adenosylmethionine, the methionine methyl group is transferred to methyl acceptors such as glycine and phosphatidylethanolamine. The sulfur atom of methionine is recovered in homocysteine, cystathionine, and cysteine. Carbon atoms 1 through 4 are recovered in homocysteine, cysta-thionine, and α-ketobutyrate.

2. Early Studies on Methionine Metabolism

The relationships shown in Fig. 1 for the metabolism of methionine by the transsulfuration pathway are in part based on the demonstrated transfer of sulfur from methionine to cystathionine and cysteine and the transfer of cystathionine sulfur to cysteine (Tarver and Schmidt, 1939; Reed et al., 1949; Rachele et al., 1950). Stetten (1942) showed that serine provided the carbon for cysteine synthesis. The transsulfuration pathway for methionine metabolism as outlined in Fig. 1 is consistent with the carbon trans-

N. J. Benevenga • Departments of Meat Animal Science and Nutritional Sciences, University of Wisconsin, Madison, Wisconsin 53706.

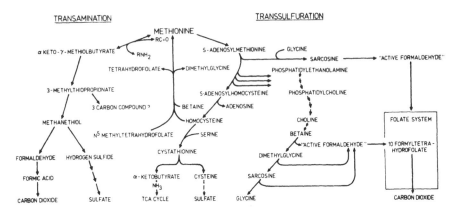

Fig. 1. The transsulfuration pathway of methionine metabolism (right) is compared with the proposed transamination pathway of methionine metabolism (left).

fers shown for the metabolism of specifically labeled methionine in the rat (Kisliuk *et al.*, 1956). In those studies rats were injected with DL-methionine labeled in the methyl, 4, 3, or 2 carbon and the distribution of labeled carbon in glycogen-glucose, serine, and aspartic acid was determined. The methyl carbon of methionine was recovered in the 3 carbon of serine and in carbons 1, 2, 5, and 6 of glucose. The 3 carbon of serine probably derived its label from the conversion of the methyl carbon of methionine to formate (Case and Benevenga, 1977). The latter is metabolized by the folate system. Glycine then combines with a one carbon unit to yield serine labeled in position 3. Serine labeled in the 3 position can also be obtained from methyl-labeled betaine, dimethylglycine and sarcosine derived from the metabolism of [14C]methyl-methionine via the transsulfuration pathway because these methyl groups are also metabolized by the folate system. The labeling of glucose could be by the conversion of serine-3[14C] to pyruvate-3[14C] followed by its conversion to 2,3-labeled oxaloacetate which yields glucose labeled mainly in carbons 1, 2, 5, and 6.

Methionine labeled in carbons 4 or 3 yielded glucose with label in positions 1, 2, 5, and 6. 2-[14C] methionine labeled glucose in carbons 3 and 4 and aspartate in carbons 1 and 4. These observations could be explained if carbons 4, 3, or 2 of methionine yield propionate labeled in the 3, 2, or 1 carbon, respectively. Propionate labeled in the 3 or 2 carbon yields 2,3-[14C] oxaloacetate and hence 1-, 2-, 5-, and 6-labeled glucose. Propionate labeled in the 1 carbon yields 1,4-[14C] oxaloacetate (i.e., aspartate) and thus glucose labeled in the 3 and 4 carbons.

The transaminative pathway of methionine metabolism (Fig. 1) outlined by us (Steele and Benevenga, 1978, 1979) leads to some of the same products as does the transsulfuration pathway, i.e., carbon dioxide from the methyl carbon and sulfate from the sulfur atom but should not yield carbon- or sulfur-labeled cystathionine. However, sulfur-labeled cysteine could be seen if the cysteine synthase (i.e., H_2S + serine $\longrightarrow$ cysteine) system (Greenberg, 1975a) was of importance in the rat. Thus, the metabolism of methionine via the transsulfuration and transamination pathways cannot be completely differentiated by examination of the end products. Additional experiments are necessary to demonstrate the possibility of additional pathways. Evidence for this will be reviewed below.

3.   Studies on the Oxidation of the Methionine Methyl Group

Early work showed that the rat could convert 6 to 25% of the methionine methyl carbon to carbon dioxide in 24 hr (Mackenzie *et al.*, 1950). The suggested route for this conversion was via the formation of S-adenosylmethionine and subsequently of choline (i.e., by *de novo* synthesis of phosphatidylcholine). Free choline would be oxidized to betaine and its methyl carbons oxidized to CO_2 via the folate system (Mackenzie, 1955). It should be noted that the metabolism of methionine, via formation of S-adenosylmethionine, provides methyl groups for approximately 40 different reactions; of these only the *de novo* synthesis of choline and sarcosine leads to a significant production of carbon dioxide (Mudd and Cantoni, 1964; Lombardini and Talalay, 1971). Thus, the metabolism of methionine appeared to be well understood. However, interest in the toxicity of methionine (Benevenga and Harper, 1967) soon led to experiments in which the metabolism of methyl- or carboxyl-labeled methionine was studied (Aguilar *et al.*, 1974; Benevenga and Harper, 1970).

The following observations raised questions regarding the proposed pathway for the conversion of methionine methyl carbon to CO_2: a) the conversion rate of methyl carbon to CO_2 is high (15% of the absorbed dose in 10 hr); b) this rate was increased when rats were adapted to a 10% casein diet containing 3% of L-methionine (i.e., to 21% of the absorbed dose); c) when this diet was supplemented with glycine or serine, methionine methyl oxidation was increased by 25% (Benevenga and Harper, 1970; Benevenga, 1974).

3.1.   Importance of Choline, Betaine and Sarcosine Formation in Methionine Methyl Group Oxidation

The importance of choline formation in the catabolism of the methionine methyl carbon was tested by substantially expanding the choline and betaine pools by feeding rats diets containing 10% casein, 1.5% methyl-la-

Table I.   Effect of Dietary Supplements of Choline Cl or Betaine
HCl on the Recovery of the Methionine Methyl Carbon in
Carbon Dioxide[a]

Diets[b]	Percent of fed dose recovered in CO_2 in 12 hours
10% casein + 1.5% L-Met[c]	31.0 A[d]
10% casein + 1.5% L-Met + 2.79% choline Cl	36.4 B
10% casein + 1.5% L-Met + 3.15% betaine HCl	39.9 B

[a]Modified from Benevenga (1974)

[b]Methionine 0.1 mole/kg diet.   Choline and betaine 0.2 mole/kg diet.

[c]L-$[^{14}CH_3]$methionine.

[d]Means followed by different letters are significantly different (P>0.01).

beled L-methionine, and 2.79% choline chloride or 3.15% betaine HCl.
The results of these experiments are shown in Table I. Addition of choline
or betaine (0.2 moles/kg) to a diet containing 0.1 mole of added methionine per kg diet did not reduce the recovery of the methionine methyl
group in CO_2. The interpretation of these results was equivocal until it
was shown that the amount of betaine used in these experiments did,
indeed, create a trapping pool (Case et al., 1976). Thus, the formation of
choline and its subsequent oxidation cannot account for the oxidative
catabolism of large quantities of the methionine methyl carbon in the rat
in short term (12 hr) experiments. These results do not imply that the
methionine methyl carbon does not flow into the phosphatidylcholine pool;
others (Lyman et al., 1969; Lyman et al., 1971) have unquestionably
shown substantial incorporation of the methionine methyl carbon into liver
phospholipids. Apparently, the half-life of methyl-labeled phosphatidylcholine is long enough to delay the release of labeled carbon derived from
the methyl group of methionine.

Experiments with rats fed diets containing 10% casein supplemented
with 0.3, 1.0, or 3.0 percent of methyl-labeled methionine and sufficient
sarcosine (10%) to exceed the kidney threshold for sarcosine showed that
the methionine methyl carbon was incorporated into sarcosine (Mitchell
and Benevenga, 1976). Calculations showed that from 10 to 19% of the
radioactive carbon dioxide produced from the methionine methyl group
was dependent on sarcosine formation. Thus, two presumably important
routes of methionine methyl group oxidation, the formation and oxidation
of choline and the formation and oxidation of sarcosine, account for less
than one-half of the oxidation of the methionine methyl carbon observed
in short-term experiments. Additional routes of methionine degradation
therefore must exist.

Table II. Comparison of the Rate of Conversion of the Methyl Carbons of Methionine and S-Adenosylmethionine to Carbon Dioxide, Serine and Sarcosine in Rat Liver Homogenates and Slices

Preparation	Substrate	Products		
		CO_2	Serine	Sarcosine
		--------µmole/g/hour--------		
Liver Homogenate [a]	L-[methyl-^{14}C]Met 10 mM	0.68	0.11	0.40
	" + glycine 10 mM	0.64	0.51	0.98
	L-[methyl]-^{14}C]AdoMet 0.5 mM	0.01	0.002	0.18
	" + glycine 10 mM	0.07	0.27	1.65
Liver Slice [b]	L-[methyl-^{14}C]Met 10 mM	0.88, 0.96	0.03, 0.08	0.03
	" + glycine 10 mM	0.85 ± 0.18	0.28 ± 0.08	0.14 ± 0.05
	L-[methyl-^{14}C]AdoMet 3 mM	0.31, 0.31	N.D.[c]	0.19, 0.25
	" + glycine 10 mM	0.58 ± 0.16	0.54 ± 0.09	3.33 ± 0.22

[a] Modified from Case and Benevenga (1976)

[b] Modified from Mitchell and Benevenga (1976)

[c] None detected.

3.2. Importance of S-Adenosylmethionine Formation in Methionine Methyl Group Oxidation *In Vitro*

Further evidence for alternative pathways of methionine degradation comes from direct comparison of the incorporation rates of the methionine and S-adenosylmethionine (AdoMet) methyl carbons into carbon dioxide, serine, and sarcosine in liver homogenates and slices (Table II). The results obtained with homogenates were inconsistent with the idea that methionine metabolism is solely dependent on the formation of AdoMet. It is clear that the methyl carbon of methionine was converted to carbon dioxide at a rate greater than that of AdoMet. The addition of glycine stimulated carbon dioxide production from the methyl carbon of AdoMet but not from that of methionine. Glycine should have stimulated carbon dioxide production from the methionine methyl carbon if formation of AdoMet was of major importance in the metabolism of methionine in the liver homogenate system. It is possible that homogenization of a tissue alters the relative importance of pathways available for the degradation of methionine; however, results obtained with rat liver slices were similar to those obtained with the homogenate. More carbon dioxide was recovered from the methyl carbon of methionine than from that of AdoMet. This could be due to a difference in the uptake of the two substances. Again the effect of supplemental glycine was to stimulate carbon dioxide production from AdoMet, but not from methionine. These results are not consistent with the hypothesis that methionine degradation is totally dependent on the formation of AdoMet.

The importance of AdoMet as an intermediate in the oxidative catabolism of methionine was tested directly in a liver homogenate system. The rate of conversion of the methyl carbon of AdoMet to carbon dioxide was at a maximum of 22 nmole/g/hr when the AdoMet concentration exceeded 0.6 mM (Case and Benevenga, 1976). Addition of AdoMet (2 mM) to a homogenate system containing methyl-labeled methionine at concentrations of 2, 10, 20, or 50mM suppressed the recovery of the methyl carbon of methionine in sarcosine to 20-25 % of that recovered in its absence but did not affect recovery of the methyl carbon of methionine in carbon dioxide at any of the concentrations of methionine used. In summary, these results show that oxidation of the methionine methyl carbon to carbon dioxide in a homogenate system is not dependent on the formation of AdoMet and that oxidation of the methyl carbon of methionine in a liver slice system is only partially dependent on formation of AdoMet.

Additional differences in the catabolism of the methyl groups of methionine and AdoMet were shown in experiments utilizing trapping pools of semicarbazide or sodium formate. The methyl carbon of methionine could be trapped as formaldehyde (Table III) or as free formate

Table III. Effect of Addition of Semicarbazide to a Liver Homogenate
System on Recovery of L-[Methyl-^{14}C]Methionine in Formaldehyde
Formate and Carbon Dioxide[a]

Semicarbazide	Formaldehyde	Formate Carbon	Carbon Dioxide	Total
mM	-----------------------	--------µmole/g/hour--------	--------------------------	
0	0.00	0.32	1.08	1.40
1	0.35	0.18	0.70	1.23
5	0.60	0.03	0.13	0.76
10	0.64	0.00	0.10	0.74

[a]Modified from Case and Benevenga (1976). The methionine concentration was
10 mM.

(Table IV). As the level of semicarbazide was increased, recovery of the methyl carbon of methionine as formate and carbon dioxide declined steadily while that recovered as formaldehyde increased. The semicarbazide treatment depressed the total oxidative catabolism of the methionine methyl carbon by about 50%. Addition of sodium formate to the liver homogenate system (Table IV) resulted in a substantial increase in the recovery of the methionine methyl carbon in formate and a concomitant reduction in carbon dioxide. The formate treatment suppressed the total oxidative catabolism of methionine by about 25%. The results in Table III suggest that formate is derived from formaldehyde. Whereas semicarbazide suppressed conversion of the methyl carbon of methionine to carbon dioxide, it did not affect the conversion of the methyl carbon of AdoMet to carbon dioxide in a rat liver homogenate system (Case and Benevenga, 1976). It thus became necessary to consider alternative pathways for methionine catabolism.

4.   Importance of Transamination in Methionine Catabolism *In Vitro*

Transamination is the initial step in the degradation of a number of amino acids and methionine has long been known to participate in transamination reactions (Meister and Tice, 1950; Meister *et al.,* 1952). For example, methionine is a better substrate than leucine for the leucine transaminase found in rat liver mitochondria (Ikeda *et al.,* 1976). Methionine, methionine sulfoxide, and ethionine will transaminate with α-ketoglutarate in the glutamine transaminase system isolated from rat liver (Cooper and Meister, 1972) and methionine transaminates with the histidine pyruvate transaminase purified from rat liver mitochondria (Noguchi *et al.,* 1976).

The importance of transamination in the metabolism of methionine

Table IV.   Effect of Addition of Formate to a Liver Homogenate System on Recovery of L-[Methyl-^{14}C]Methionine in Formate and Carbon Dioxide[a]

Formate	Formate	Carbon Dioxide	Total
nM	----------------------- μmole/g/hour ---------------------		
0	0.07	0.41	0.48
0.8	0.15	0.22	0.37
1.6	0.19	0.17	0.36
2.4	0.22	0.12	0.34
3.2	0.25	0.08	0.33
4.0	0.26	0.07	0.33
6.0	0.29	0.04	0.33
8.0	0.32	0.03	0.35

[a]Modified from Case and Benevenga (1976). The methionine concentration was 10 mM.

was tested directly in a liver homogenate system (Mitchell and Benevenga, 1978). Pyruvate (10mM) and α-keto-γ-methiolbutyrate (1mM) increased the production of carbon dioxide from methyl labeled methionine by 1.8- and 2.8-fold, respectively. These keto acids increased the recovery of methionine as a phenylhydrazone derivative by 10- to 40-fold. Later experiments with L-[1-^{14}C]methionine, which involved conversion of the phenylhydrazone derivative to an amino acid, showed that only methionine was recovered. This indicated that α-keto-γ-methiolbutyrate but not α-ketobutyrate (see Fig. 1) was being formed. These results were confirmed by Beliveau (1976). Livesey and Lund (1980) reported substantial increases in methionine transamination with a variety of keto acids in a rat liver cell preparation, but little increase in carbon dioxide production from carboxyl-labeled methionine was noted. The production of carbon dioxide and α-keto-γ-methiolbutyrate was suppressed by the transaminase inhibitor amino oxyacetate, suggesting that transamination of methionine was required before it could be further catabolized by a rat liver homogenate system (Mitchell and Benevenga, 1978).

Livesey and Lund (1980) suggest that, in the liver, cytosolic glutamine transaminase is responsible for transamination of methionine. However, other work (Beliveau, 1976) does not support this idea (Table V). Glutamine suppressed the recovery of the carboxyl carbon of methionine in carbon dioxide or α-keto-γ-methiolbutyrate (phenylhydrazone) only when pyruvate was the keto acid co-substrate. The results with pyruvate could be explained by a competition between methionine and glutamine

Table V.   Effect of Glutamine on the Catabolism of L-[1-^{14}C]Methionine in a Rat Liver Homogenate System[a]

| Glutamine | With 10 mM pyruvate | | With 10 mM α-keto-γ-methiolbutyrate | |
	CO_2	α-keto-γ-methiolbutyrate accumulation	CO_2	α-keto-γ-methiolbutyrate accumulation
nM	------------------------------μmole/g/hour[b]------------------------------			
0	5.8	4.7	6.8	59.8
1	4.2	1.3	6.8	50.8
2	3.0	.4	7.9	50.7
5	1.6	ND[c]	8.0	57.8
10	1.2	ND	----	----

[a]Taken from Beliveau (1976)

[b]Rate calculated from the specific activity of methionine added to the homogenate.

[c]Not detectable.

for transamination or by a rapid resynthesis of methionine from α-keto-γ-methiolbutyrate by donation of an amino group by glutamine. Resynthesis of methionine would decrease the concentration of α-keto-γ-methiolbutyrate and hence the rate of $^{14}CO_2$ production from carboxyl-labeled methionine. The use of α-keto-γ-methiolbutyrate as a co-substrate provided a trapping pool for the radioactive α-keto-γ-methiolbutyrate made from methionine. Although glutamine may increase the rate of conversion of α-keto-γ-methiolbutyrate to methionine, no detectable decrease in the rate of conversion of the carboxyl carbon of methionine to CO_2 was observed. These results suggest that a transaminase(s) other than glutamine transaminase is involved in the catabolism of methionine.

The tissue distribution of methionine transamination has been studied in the rat (Mitchell and Benevenga, 1978; Beliveau, 1976). The highest specific activities were noted in the liver and kidney. However, measurable transamination of methionine occurred in the heart, brain, spleen, skeletal muscle, and small intestine. Mitchell and Benevenga (1978) calculated that 48% of the transamination of methionine may occur in the muscle and another 40% may occur in the liver.

5.   Decarboxylation of Methionine (α-Keto-γ-methiolbutyrate) *In Vitro*

The results above and those below suggest that in the rat liver homogenate system, methionine must be transaminated and then decarboxylated

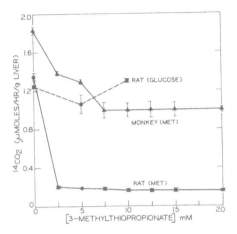

Fig. 2. Effect of 3-methylthiopropionate on the conversion of 10 mM
L-[methyl-^{14}C]methionine to CO_2 in rat (●—●) and monkey (▲—▲)
liver homogenate and of 10 mM [U-^{14}C]glucose to CO_2 in rat (●---●)
liver homogenates. Taken from Steele and Benevenga (1978).

before degradation of the methyl carbon can proceed. Mitchell and Benevenga (1978) showed that addition of α-ketobutyrate (10mM) to a liver
homogenate system inhibited the conversion of both the methyl and carboxyl carbons of methionine and α-keto-γ-methiolbutyrate to carbon dioxide by 80%. Dixon and Benevenga (1980) showed that in the liver 70% of
the decarboxylation of α-keto-γ-methiolbutyrate occurred in the mitochondria and that 0.4 mM α-ketobutyrate decreased the decarboxylation
of α-keto-γ-methiolbutyrate (2.5 mM) to 30% of the control value. The
decarboxylation of the α-keto-γ-methiolbutyrate by mitochondria was
inhibited 30-50% by pyruvate (1 mM) or the keto acid of tyrosine (1
mM), whereas the keto acids of the branched chain amino acids (1 mM)
caused an inhibition of 75-95% (Dixon and Benevenga, 1980; Livesey and
Lund, 1980). Dixon and Benevenga (1980) suggested that more than one
decarboxylase may play a role in methionine catabolism.

6.   Formation of 3-Methylthiopropionate From Methionine *In Vitro*

The observations above indicate that a new and unique metabolite of
methionine should be formed. The most likely product is 3-methylthiopro-
pionyl-CoA, which may be released as 3-methylthiopropionate. As predicted, addition of graded levels of 3-methylthiopropionate to rat or monkey liver homogenates reduced the conversion of the methyl carbon of
methionine to CO_2 to 15 or 55% of control values, respectively (Fig. 2)

Table VI. Percent of Radioactivity Recovered from the Gas Chromatograph
           Elution Profile Found in the 18-20 Minute Sample[a],[b]

Substrate	Rat	Monkey
	---------------------%----------------	
L-[methyl-^{14}C]Met	76.1 ± 0.5[c]	63.6 ± 2.0
L-[^{35}S]Met	85.7 ± 1.2	70.8 ± 2.3
L-[1-^{14}C]Met	17.4 ± 2.5	18.6 ± 1.1

[a]Authentic 3-methylthiopropionate has a retention time of 19.5 minutes.

[b]Modified from Steele and Benevenga (1978).

[c]Mean of 2 or 3 ± SEM.

(Steele and Benevenga, 1978). The conversion of the carboxyl carbon of methionine or the carboxyl carbon of α-keto-γ-methiolbutyrate to carbon dioxide was reduced to 30% of the control value when 3-methylthiopropionate (2.5mM) was included in a rat liver homogenate system. The inhibition of methionine metabolism by 3-methylthiopropionate is not due to a general effect on metabolism because glucose metabolism was unaffected (Fig. 2).

Studies to determine if methionine was converted to 3-methylthiopropionate in liver homogenates required the use of a trapping pool because 3-methylthiopropionate could not be recovered from tissue homogenates without it. When 3-methylthiopropionate was added to rat or monkey liver homogenates incubated with labeled methionine, radioactive 3-methylthiopropionate could be recovered by gas chromatography (Steele and Benevenga, 1978). The results obtained with methyl-, sulfur-, or carboxyl-labeled methionine indicated that the methyl carbon and sulfur atom of methionine can be recovered in 3-methylthiopropionate (Table VI). These results and the identification of the eluted material as 3-methylthiopropionate by gas chromatography-mass spectrometry (Steele and Benevenga, 1978) showed for the first time that 3-methylthiopropionate is produced from methionine by rat and monkey liver homogenates.

The evidence suggesting that 3-methylthiopropionyl-CoA and not 3-methylthiopropionate is the product of decarboxylation is circumstantial. Livesey and Lund (1980) state that "omission of either MgSO$_4$, thiamin pyrophosphate, coenzyme A or NAD decreased the rates of ^{14}CO$_2$ formation." Dixon, Harper, and Benevenga (1982) showed that the production of ^{14}CO$_2$ from 1-[^{14}C]-α-keto-γ-methiolbutyrate was increased twofold upon addition of NAD (1.0 mM) and coenzyme A (0.6 mM) but only when mitochondria were first treated with Ca^{++}. Activities obtained with Ca^{++}-treated mitochondria without supplemental NAD or coenzyme A were one-half of those with mitochondrial which had not been treated with Ca^{++}. Apparently, the swelling produced by the Ca^{++} treatment decreased the integrity of the mitochondrial membrane and allowed the cofactors and substrates to leak into the incubation medium. Thus, the increased activity noted upon addition of NAD or coenzyme A was due to the increased accessibility of the added substrates and cofactors to the mitochondrial enzymes.

6.1. Metabolism of 3-Methylthiopropionate *In Vitro*

Metabolism of 3-methylthiopropionate labeled in the methyl carbon or sulfur atom by rat liver homogenates gave rise to a gaseous product that contained radioactive carbon and sulfur (Steele and Benevenga, 1979). The gas produced could be composed of any or several of the following: carbon dioxide, methanethiol, dimethylsulfide, methyldisulfide, or hydrogen sulfide. The head space gas was collected in a sodium hydroxide trap, released and analyzed by gas chromatography. Both methanethiol and hydrogen sulfide were found. When these compounds were first removed by a mercuric acetate trap, the gas trapped in alkali was shown to be carbon dioxide derived from the methyl carbon of 3-methylthiopropionate. Ion exchange chromatography of the acid extract of the incubation medium showed that the sulfur atom of 3-methylthiopropionate had been converted to sulfate. These studies suggest that methanethiol, hydrogen sulfide, sulfate, and carbon dioxide are products of methionine catabolism via the transamination pathway.

The relative proportions of carbon dioxide, methanethiol, and hydrogen sulfide recovered varied with pH. No hydrogen sulfide was recovered at pH 7.8. An increase in hydrogen sulfide production was noted only below pH 7.2; nearly five times as much hydrogen sulfide as methanethiol was produced when the pH was 6.3. These results are apparently due to the narrow pH optimum for sulfide oxidation by rat tissues (Baxter *et al.,* 1958).

In the rat, the liver is the major site of metabolism of 3-methylthiopropionate because the rate of production of carbon dioxide from the methyl carbon and of hydrogen sulfide from the sulfur atom of 3-methylthiopropionate was 20 times higher in liver than in kidney homogenates (Steele and Benevenga, 1979). Little activity was found in other tissues tested (i.e., brain, muscle, and heart). The observation that both methanethiol and hydrogen sulfide can be formed from 3-methylthiopropionate and hence from methionine and its keto acid (Benevenga and Haas, 1982) may help explain the rather marked toxicity of methionine.

7. Importance of the Transamination Pathway of Methionine Metabolism

Because the methyl carbon of methionine has been shown to be a precursor of formate (Siekevitz and Greenberg, 1950; Weinhouse and Friedmann, 1952) and our observation that as little as 1 mM formate decreased the conversion of the methionine methyl carbon to carbon dioxide by 60% in rat liver homogenates (Case and Benevenga, 1976) we decided to determine the importance of formate production from methionine in the intact rat. Because formate is an important product of the metabolism of methionine via the transamination pathway, its production may be used as an estimate of the relative importance of this pathway in the intact animal as dietary methionine levels are increased. Addition of six percent sodium formate to the diet resulted in approximately one-half of the formate being excreted in the urine and one-half excreted as carbon dioxide (Case and Benevenga, 1977). Thus, if the methionine methyl carbon mixes with the free formate pool during its conversion to carbon dioxide, it too should be excreted as urinary formate. Determination of the amount of methyl carbon recovered as urinary formate and correction for the fraction of formate oxidized to carbon dioxide makes it possible to calculate the fraction of methionine methyl carbon converted to carbon dioxide via the intermediate formation of formate. From 65 to 75% of the methionine methyl carbon recovered in carbon dioxide traveled through a free formate pool regardless of the level (0.3 to 2.5%) of methionine added to the diet (Case and Benevenga, 1977).

The proportion of the methyl carbon flowing through a free formate pool did not change with dietary methionine level. It was the same in rats fed a diet containing methionine at the requirement (10% casein plus 0.3% of L-Met) or at levels of methionine which are four times the requirement (10% casein plus 2.5% of L-Met) and are toxic. Although these experiments may be criticized because a diluting pool was used, they do suggest that the transamination pathway of methionine metabolism is important and cannot be considered a pathway that operates only at high levels of methionine intake.

Other evidence supporting the significance of the transamination pathway of methionine degradation comes from recent studies with isolated rat hepatocytes (Engstrom and Benevenga, 1981). The rate of conversion of the methionine and S-adenosylmethionine methyl carbons to carbon dioxide was compared at methionine concentrations of 0.1 and 1.0 mM. When the specific activity of S-adenosylmethionine was used as the basis for the calculation of the rate of oxidation for both methionine and S-adenosylmethionine (the specific activity of methionine was one-fifth to

Table VII. Comparison of the Inhibitory Effects of 3-Methylthiopropionate and Formate on Conversion of the Methionine Methyl Carbon Dioxide in Rat, Monkey, Pig and Sheep Liver Homogenates

Animal	L-[methyl-^{14}C]Met	L-[methyl-^{14}C]Met + 5 mM 3-methylthiopropionate	L-[methyl-^{14}C]Met = 2 mM Formate	Reference
		---------- µmole/g/hr^a ----------		
Rat	1.4	0.2	0.4	Steele and Benevenga(1978) Case and Benevenga(1976)
Monkey	1.8	1.3	--	Steele and Benevenga(1978)
Pig	1.2	0.6	0.05	Benevenga and Haas (1979)
Sheep	2.3	0.3	0.2	Benevenga et al. (1982)

^{a}Values are taken from studies in which a range of these inhibitors were used.

one-half that of AdoMet), the rate of conversion of the methyl carbon of methionine to carbon dioxide was 4.7 times that of AdoMet at 0.1 mM methionine and 7.9 times that of AdoMet when the concentration of methionine was raised to 1 mM. These results suggest that pathways of conversion of the methyl carbon of methionine to carbon dioxide independent of AdoMet formation are important and that they operate at concentrations of methionine that are found in tissues.

The metabolism of methionine by the transamination pathway (i.e., pathway(s) independent of AdoMet formation) is not restricted to the rat. Based on the inhibition of conversion of the methyl carbon of methionine to carbon dioxide in a liver homogenate system by 3-methylthiopropionate and/or formate (Table VII) the monkey, pig, and sheep are also able to metabolize methionine by this pathway. It would seem, therefore, that pathways of methionine catabolism other than the well recognized trans-sulfuration pathway may be widely distributed and may be of significance in the catabolism of methionine.

8. Application of New Knowledge of Methionine Metabolism in Homocystinuria

The recognition of the existence of multiple pathways of methionine degradation makes possible new approaches to the treatment of disorders of sulfur amino acid metabolism in man. Approximately one-half of patients with the inherited disorder homocystinuria are helped by dietary supplements of vitamin B_6 (50-1000 mg/day) whereas the others do not respond (Mudd and Levy, 1978). Treatment of patients with the form of the disease that does not respond to supplemental vitamin B_6 has centered around diet modification so that methionine intake is kept low and cysteine intake is made high. Because these diets often involve the use of ill-tasting free amino acids or modified proteins, their acceptance by patients may be marginal and adherence to the dietary plan sporadic. Thus, a single dietary supplement that would affect homocysteine metabolism could be beneficial if it were well accepted by patients. Although a number of methyl donors have been used as dietary supplements (Komrower and Sardharwalla, 1971) they were abandoned because a marked elevation in plasma methionine occurred. This was of great concern because of the well known toxicity of methionine. A reinvestigation of the use of the methyl donor betaine (Smolin, Benevenga, and Berlow, 1981) showed, on the other hand, that this compound was beneficial. Two patients received daily supplements of 6 to 10 g of betaine over three years with no obvious detrimental effects although plasma methionine levels rose to 10 to 14 mg/dl (normal: 0.45 mg/dl). The circulating homocystine levels were reduced to 20-30% of those prior to treatment, appetite improved, hypertension was dramatically reduced and a generalized improvement in clinical status was evident. The explanation for the beneficial effect of betaine was that it depressed circulating homocystine levels by providing a methyl group for the conversion of homocysteine to methionine (see Fig. 1). The explanation for the lack of detrimental effects due to the elevated methionine may be that the alternative pathways for methionine degradation shown in the rat and monkey were operating in man and that the rate of production of methanethiol and hydrogen sulfide was such that they were rapidly metabolized and hence were harmless. Although this argument is intriguing, it does require documentation. It is, however, supported by the finding that the high levels of methionine observed in these patients are not as deleterious as were the high homocysteine concentrations observed before the use of betaine supplements.

9.   Concluding Remarks

Whereas it is obvious that considerable work remains before the suggestions made in this chapter for alternative pathways of methionine metabolism are fully documented and their significance evaluated, it seems clear that more than one degradation pathway for methionine exists and that a more complete knowledge of the pathways involved in methionine degradation will be of importance in nutrition and medicine.

References

Aguilar, T.S., Benevenga, N.J., and Harper, A.E., 1974, Effect of dietary methionine level on its metabolism in rats, *J. Nutr.* **104**:761.

Baxter, C.F., VanReen, R., Pearson, P.B., and Rosenberg, C., 1958, Sulfide oxidation in rat tissues, *Biochim. Biophys. Acta,* **27**:584.

Beliveau, G.P., 1976, A study of methionine transamination in rats, *M.S. Thesis,* University of Wisconsin-Madison.

Benevenga, N.J., 1974, Toxicities of methionine and other amino acids, *Agr. Food Chem.* **22**:2.

Benevenga, N.J., and Haas, L.G., 1979, Studies on methionine metabolism in pig, Abstract No. 41, *12th Meeting Midwest Section American Society of Animal Science,* June 14 and 15, St. Louis, Missouri.

Benevenga, N.J., and Haas, L.G., 1982, Unpublished results.

Benevenga, N.J., and Harper, A.E., 1967, Alleviation of methionine and homocystine toxicity in the rat, *J. Nutr.* **93**:44.

Benevenga, N.J., and Harper, A.E., 1970, Effect of glycine and serine on methionine metabolism in rats fed diets high in methionine, *J. Nutr.* **100**:1205.

Benevenga, N.J., Radcliffe, B.C., and Egan, A.R., 1982, Tissue metabolism of methionine in sheep, *Aust. J. Biol. Sci.,* submitted.

Case, G.L., and Benevenga, N.J., 1976, Evidence for S-adenosylmethionine independent catabolism of methionine, *J. Nutr.* **106**:1721.

Case, G.L., and Benevenga, N.J., 1977, Significance of formate as an intermediate in the oxidation of the methionine, S-methyl-L-cysteine and sarcosine methyl carbons to carbons to CO_2 in the rat. *J. Nutr.* **107**:1665.

Case, G.L., Mitchell, A.D., Harper, A.E., and Benevenga, N.J., 1976, Significance of choline synthesis in the oxidation of the methionine methyl group in rats, *J. Nutr.* **106**:735.

Cooper, J.L., and Meister, A., 1972, Isolation and properties of highly purified glutamine transaminase, *Biochemistry* **11**:661.

Dixon, J.L., and Benevenga, N.J., 1980, The decarboxylation of α-keto-γ-methiolbutyrate in rat liver mitochondria, *Biochem. Biophys. Res. Commun.* **97**:939.

Dixon, J.L., Harper, A. E., and Benevenga, N.J., 1982, Unpublished observations.

Engstrom, M.A., and Benevenga, N.J., 1981, Oxidation rates of the methionine and S-adenosylmethionine methyl carbons in isolated rat hepatocytes, *Fed. Proc.* **40**:841.

Finkelstein, J.D., 1975, Enzyme defects in sulfur amino acid metabolism in man, in: *Metabolic Pathways,* Third Edition, Vol. VII, Metabolism of Sulfur Compounds (David M. Greenberg, ed.), pp. 547-597, Academic Press, New York.

Greenberg, D.M., 1975a, Biosynthesis of cysteine and cystine, in: *Metabolic Pathways,* Third Edition, Vol. VII, Metabolism of Sulfur Compounds (David M. Greenberg, ed), pp. 505-528, Academic Press, New York.

Greenberg, D.M., 1975b, Utilization and dissimilation of methionine, in: *Metabolic Pathways,* Third Edition, Vol. VII, Metabolism of Sulfur Compounds (David M. Greenberg, ed), pp. 529-534, Academic Press, New York.

Ikeda, T., Konishi, Y., and Ichihara, A., 1976, Transaminase of branched-chain amino acids (XI), leucine (methionine) transaminase of rat liver mitochondria, *Biochim. Biophys. Acta.* **445**:622.

Kisliuk, R.L., Sakami, W., and Patwardhan, M.V., 1956, The metabolism of the methionine carbon chain in the intact rat, *J. Biol. Chem.* **221**:885.

Komrower, G.M., and Sardharwalla, I.B., 1971, The dietary treatment of homocystinuria, in: *Inherited Disorders of Sulfur Metabolism* (N.A. Carson and D.N. Raine, eds.), pp. 254, Churchill Livingston, London.

Livesey, G., and Lund, P., 1980, Methionine metabolism via the transamination pathway in rat liver, *Biochem. Soc. Trans.* **8**:540.

Lombardini, J.B., and Talalay, P., 1971, Formation, functions and regulatory importance of S-adenosyl-L-methionine, *Adv. Enz. Reg.* **9**:349.

Lyman, R.L., Hopkins, S.M., Sheehan, G., and Tinoco, J., 1969, Incorporation and distribution of (Me-[^{14}C]) methionine methyl into liver phosphatidyl choline fractions from control and essential fatty acid deficient rats, *Biochim. Biophys. Acta.* **176**:86.

Lyman, R.L., Sheehan, G., and Tinoco, J., 1971, Diet and $^{14}CH_3$-methionine incorporation into liver phosphatidylcholine fractions of male and female rats, *Can. J. Biochem.* **49**:71.

Mackenzie, C.G., 1955 Conversion of N-methyl glycine to active formaldehyde and serine, in: *A Symposium on Amino Acid Metabolism* (W.D. McElroy and H.B. Glass, eds.), pp. 684-726, Johns Hopkins Press, Baltimore, Maryland.

Mackenzie, C.G., Rachele, J.R., Gross, N., Chandler, J.P., and du Vigneaud, V., 1950, A study of the rate of oxidation of the methyl group of dietary methionine *J. Biol. Chem.* **183**:617.

Meister, A., and Tice, S.V., 1950, Transamination from glutamine to α-keto acids, *J. Biol. Chem.* **187**:173.

Meister, A., Sober, H.A., Tice, S.V., and Fraser, P.E., 1952, Transamination and associated deamidation of asparagine and glutamine, *J. Biol. Chem.* **197**:319.

Mitchell, A.D., and Benevenga, N.J., 1976, Importance of sarcosine formation in methionine methyl carbon oxidation in the rat, *J. Nutr.* **106**:1702.

Mitchell, A.D., and Benevenga, N.J., 1978, The role of transamination in methionine oxidation in the rat, *J. Nutr.* **108**:67.

Mudd, S.H., and Cantoni, G.L., 1964, Biological transmethylation, methyl-group neogenesis and other "one-carbon" metabolic reactions dependent on tetrahydrofolic acid, in: *Comprehensive Biochemistry*, Vol. 15 (M. Florkin and E.H. Stotz, eds.), pp. 1-47, Elsevier, Amsterdam.

Mudd, S.H., and Levy, H.L., 1978, Disorders of transsulfuration, in: *The Metabolic Basis of Inherited Disease* (J.B. Stanbury, J.B. Wyngaarden and D.S. Fredrickson, eds.), pp. 458-503, McGraw-Hill, New York.

Noguchi, T., Okuno, E., and Kido, R., 1976, Identity of isoenzyme of histidine-pyruvate aminotransferase with serine-pyruvate amino transferase, *Biochem. J.* **159**:607.

Rachele, J.R., Reed, L.J., Kidwai, A.R., Ferger, M.F., and du Vigneaud, V., 1950, Conversion of cystathionine labeled with ^{35}S to cystine *in vivo*, *J. Biol. Chem.* **185**:817.

Reed, L.J., Cavallini, D., Plum, F., Rachele, J.R., and du Vigneaud, V., 1949, Conversion of methionine to cystine in a human cystinuric, *J. Biol. Chem.* **180**:387.

Siekevitz, P., and Greenberg, D.M., 1950, The biological formation of formate from methyl compounds in liver slices, *J. Biol. Chem.* **186**:275.

Smolin, L.A., Benevenga, N.J., and Berlow, S., 1981, The use of betaine for the treatment of homocystinuria, *J. Pediat.* **99**:467.

Steele, R.D., and Benevenga, N.J., 1978, Identification of 3-methylthiopropionic acid as an intermediate in the mammalian methionine metabolism *in vitro*, *J. Biol. Chem.* **253**:7844.

Steele, R.D., and Benevenga, N.J., 1979, The metabolism of 3-methylthiopropionate in rat liver homogenates, *J. Biol. Chem.* **254**:8885.

Stetten, D., Jr., 1942, The fate of dietary serine in the body of the rat, *J. Biol. Chem.* **144**:501.

Tarver, H., and Schmidt, C.L.A., 1939, The conversion of methionine to cystine: experiments with radioactive sulfur, *J. Biol. Chem.* **130**:67.

Weinhouse, S., and Friedmann, B., 1952, Study of precursors of formate in the intact rat, *J. Biol. Chem.* **197**:733.

Chapter 2

The Immunostimulatory, Anti-Inflammatory and Anti-Allergic Properties of Ascorbate

Ronald Anderson

1. Introduction

Assessment of the effects of single nutrient supplements given to normal or immunocompromised individuals and of single nutrient deficiency states on cellular and humoral immune functions is a field of considerable interest which has led to the development of the science of nutritional immunology. Of the vitamins investigated, ascorbate has received most attention. Interest in ascorbate as an immunostimulatory agent is due largely to the wide publicity given to the theories and claims of Linus Pauling. However, the role of ascorbate as a potentially important immunomodulator remains controversial. Many investigators have attempted to evaluate the vitamin in the therapy and prevention of various diseases such as the common cold and in individuals with different types of cancer. In most investigations no attempt was made to identify which components of the immune system were compromised in individuals with the various diseases studied or which cellular and humoral immune functions are altered during ingestion of large doses of vitamin C. Furthermore, many studies have been performed using dosages of ascorbate ranging from milligram amounts to 70 g daily or greater in the absence of adequate data relating

Ronald Anderson • The Immunology Section, Department of Medical Microbiology, Institute of Pathology, University of Pretoria, Republic of South Africa.

to the *in vivo* requirements for enhancement of specific immune functions. Results obtained by different researchers often have been conflicting and difficult to compare. However, recent investigations using reliable immunological techniques have identified the cellular immune functions which are enhanced by exposure of leukocytes to ascorbate *in vitro* and by ingestion of the vitamin.

In this presentation the role of ascorbate as an immunostimulatory, anti-allergic, and anti-inflammatory agent has been reviewed. Emphasis has been placed on the mechanisms by which ascorbate mediates these activities. Recently acquired data on the ability of ascorbate to reverse oxidative loss of leukocyte function and to prevent auto-oxidation-induced degranulation by human blood neutrophils has been included.

2.   Immunostimulatory Effects of Ascorbate

There are now numerous reports relating to the enhancing effects of ascorbate on leukocyte motility (especially of neutrophils), the antimicrobial activity of phagocytes and the proliferative responses of lymphocytes induced by mitogens and antigens.

2.1.   Leukocyte Motility

Goetzl *et al.* (1974) reported that ascorbate at concentrations of $10^{-3}M$-$10^{-2}M$ caused stimulation of neutrophil random motility and migration to the leukoattractants kallikrein and C5a *in vitro*. Eosinophil and monocyte migration was likewise increased. These observations have been confirmed by other investigators. Goetzl *et al.* (1974) suggested that increased leukocyte motility may be related to ascorbate-mediated enhancement of hexose-monophosphate shunt (HMPS) activity, but this association has been disputed by Anderson (1979) who found that concentrations of ascorbate which cause increased neutrophil locomotion inhibit oxidative metabolism. Sandler *et al.* (1975) found that increased neutrophil and monocyte migration in the presence of ascorbate *in vitro* was associated with increased levels of intracellular cyclic 3,5-guanosine monophosphate (cGMP). There are a number of other reports which confirm the enchancing effect of ascorbate on leukocyte intracellular cGMP levels (Gallin *et al.*, 1978; Atkinson *et al.*, 1978; Anderson and Theron, 1979a; Panush *et al.*, 1981). However, the relationship between intracellular cGMP levels and increased leukocyte motility is questionable, since other researchers have reported that some agents such as acetylcholine and carbamylcholine, which elevate neutrophil cGMP levels, have no effect on motility (Wilkinson, 1976; Anderson and Van Rensburg, 1979).

Recent investigations have shown that the ascorbate-mediated stimulation of neutrophil motility *in vitro* and *in vivo* is a serum-dependent phenomenon (Anderson and Theron, 1979; Dallegri, *et al.*, 1980; Anderson, 1981a). The serum may be necessary to maintain the ascorbate in the unoxidized state. Ascorbate concentrations which stimulate neutrophil motility *in vitro* and *in vivo* have been shown to cause inhibition of the myeloperoxidase/hydrogen peroxide/halide system (MPO/H_2O_2/halide system) and this inverse relationship has been shown to occur both *in vitro* and *in vivo* (Anderson, 1981a). The MPO/H_2O_2/halide system is activated during phagocytosis (Klebanoff and Clark, 1978) and during exposure of neutrophils to leukoattractants (Becker *et al.*, 1974; Becker *et al.*, 1979) and makes an important contribution to antimicrobial activity within the phagocytic vacuole (Klebanoff and Clark, 1978). MPO is contained exclusively within the primary granules of the neutrophil, and H_2O_2 is derived from superoxide generated by activation of a membrane-linked oxidase thought to be NADH or NADPH oxidase following exposure to phagocytic stimuli or leukoattractants (Klebanoff and Clark, 1978). The participating halides are usually chloride or iodide, which are oxidized by MPO and H_2O_2 to the potent microbicidal oxidizing agents hypochlorous acid (Harrison and Schultz, 1976) and molecular iodine respectively (Thomas and Aune, 1978a). Both agents oxidize sulfhydryl groups, and hypochlorous acid also converts amino groups to toxic chloramines and chloramides (Johnson and Green, 1975; Thomas and Aune, 1978b).

2.2. Ascorbate-Mediated Stimulation of Neutrophil Motility by an Antioxidant Mechanism

Although the MPO/H_2O_2/halide system is an important component of the phagocyte intracellular defence system it is also released extracellularly by degranulation during phagocytosis and exposure to leukoattractants (Becker *et al.*, 1974). At least 17% of neutrophil MPO is released extracellularly during phagocytosis (Bradley and Rothstein, 1981). Although the MPO released may contribute to antimicrobial activity outside the cell, it can also cause decreased immune reactivity by mediating inhibition of neutrophil motility by autooxidation of the neutrophil membrane (Anderson and Grabow, 1980) and by oxidative inactivation of leukoattractants (Clark and Klebanoff, 1979; Anderson and Grabow, 1980).

Fifteen minutes exposure to an *in vitro* system comprised of horseradish peroxidase (HRP), H_2O_2 and sodium iodide, which simulates the MPO/H_2O_2/halide system, has been shown to cause human blood neutrophils to lose their ability to respond to leukoattractants (Table I). This loss of chemotactic responsiveness is accompanied by oxidation of the neutrophil membrane as assessed by membrane iodination and loss of

Table I. Effects of Ascorbate on the Inhibition of Neutrophil Motility and Oxidation of the Cell Membrane Mediated by the $HRP/H_2O_2/Iodide$ System

Test System	Neutrophil Migration to		Oxidation of the Neutrophil	
	EAS	f-met-leu-phe	Sulfhydryl content $\mu moles/10^7$ cells	Iodination nmoles ^{125}I deposited
Untreated neutrophils	162 ± 27	92 ± 15	4 ± 0.6	0
Neutrophils exposed to:				
a) HRP ± H_2O_2 ± NaI [a]	42 ± 6 (P<0.005)[b]	14 ± 5 (P<0.005)	0.2 ± 0.14	0.29 ± 0.04
c) HRP ± H_2O_2 ± NaI 5 x 10^{-2}M ascorbate	131 ± 15	131 ± 4	N.D.	0

[a] Human blood neutrophils (5 x 10^6/ml) were exposed to 0.25 units HRP, 10^{-6} M H_2O_2 and 10^{-3} M NaI for 15 min in the presence and absence of migration stimulatory concentrations of sodium ascorbate after which the neutrophils were washed to remove the oxidizing system and tested for migratory responsiveness to the leukoattractants autologous endotoxin activated serum (EAS) and the snythetic chemotactic tripeptide N-formyl-L-methionyl-L-leucyl-L-phenylalanine (f-met-leu-phe). Results are expressed as neutrophils which have traversed a 5 μm pore size millipore filter during a 2 hr incubation period as cells/microscope high powered field (mean value with standard error of 6 experiments). Membrane sulfhydryl content was measured as μmoles 10^7 cells and iodination by substituting NaI with Na^{125}I in the oxidizing system. Ascorbate protected the chemotactic responsiveness of the neutrophils and prevented membrane oxidation.

[b] t statistic for 2 means.

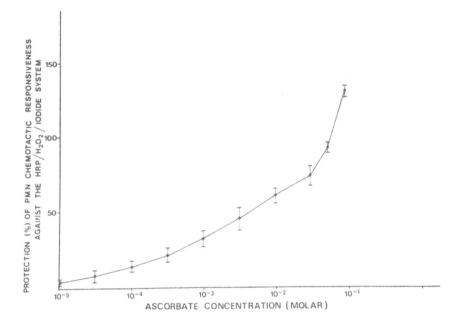

Fig. 1. Ascorbate mediated protection of neutrophil chemotactic responsiveness to endotoxin-activated autologous serum (EAS) from inactivation by the HRP/H_2O_2/iodide system in vitro. Neutrophils were exposed to 0.25 units HRP, 10^{-6}M H_2O_2 and 10^{-3}M NaI for 15 min in the presence and absence of sodium ascorbate and tested for chemotactic responsiveness to EAS. Ascorbate caused a dose-dependent protection of migratory responsiveness to EAS from inactivation by the HRP/H_2O_2/iodide system. Figure reproduced from Anderson (1981a).

sulfhydryl groups (Table I). However, exposure of neutrophils to the HRP/H_2O_2/iodide system in the presence of ascorbate at concentrations which stimulate neutrophil motility and inhibit activity of the MPO/H_2O_2/halide and HRP/H_2O_2/iodide systems *in vitro,* protects and even enhances neutrophil migratory activity (Fig. 1). This protection is accompanied by inhibition of oxidation of the neutrophil membrane as determined by prevention of iodination and loss of sulfhydryl groups (Table I). Although this mechanism of ascorbate-mediated stimulation of neutrophil motility is based on data obtained from *in vitro* experimentation, it probably reflects *in vivo* events (Anderson, 1981a). The association of increased neutrophil motility with ascorbate ingestion is well documented (Boxer *et al.,* 1976; Anderson *et al.,* 1980b; Rebora *et al.,* 1980; Weening *et al.,* 1981); it is related to inhibition of the

Table II. Assessment of the Effects of Exposure to Ascorbate of Oxidized Chemotactically Unresponsive Neutrophils

Test system	Neutrophil migration to	
	EAS	f-met-leu-phe
Untreated neutrophils	199 ± 31^{b}	123 ± 8
Oxidized neutrophils only[a]	75 ± 11 ($P<0.005$)	35 ± 3 ($P<0.005$)
Oxidized neutrophils:[b]		
a) 5×10^{-2} M ascorbate	253 ± 23	169 ± 10
b) 1×10^{-1} M ascorbate	310 ± 31	210 ± 18

[a]In these experiments human blood neutrophils were exposed to 0.25 units HRP, 10^{-6} M H_2O_2 and 10^{-3} M NaI in the absence of ascorbate. After washing to remove the HRP/H_2O_2/iodide system the cells were exposed to migration-stimulatory concentrations of ascorbate which reversed the oxidative loss of chemotactic responsiveness to both leukoattractants.
[b]Results are expressed as the mean value with standard error of 5 experiments as cells/microscope high-powered field.

Table III. Reversal of Neutrophil Membrane Oxidation (Iodination) by Exposure of Oxidized Cells to Ascorbate

Test system	Oxidation of the neutrophil membrane measured by uptake of ^{125}I
Oxidized neutrophils[a]	0.33 ± 0.09[b]
Oxidized neutrophils $\pm 10^{-1}$M ascorbate	0.22 ± 0.13 ($P<0.05$)

[a] In these experiments human blood neutrophils were exposed to 0.25 units HRP, 10^{-6}M H_2O_2 and 0.06 μCi Na ^{125}I for 30 min after which the cells were washed to remove the HRP/H_2O_2/iodide system and exposed to a 10^{-1}M concentration of sodium ascorbate. Exposure to ascorbate caused a dose-dependent reversal of membrane oxidation as measured by deiodination.
[b] Results are expressed as the mean value with standard error of 3 experiments in nmoles ^{125}I deposited.

$MPO/H_2O_2/halide$ system which accompanies ascorbate ingestion (Anderson, 1981a).

The nature and the targets of the auto-oxidative inhibition of neutrophil motility which can be partially or totally protected by ascorbate have yet to be identified. However, it has been shown that oxidative inactivation of neutrophil locomotion is a reversible process (Anderson and Jones, 1982). Neutrophils exposed to the $HRP/H_2O_2/iodide$ system recover their chemotactic responsiveness on exposure to ascorbate after the system is removed by washing. The recovery of motility is associated with a reversal of the cell membrane from the oxidized to the unoxidized state. These data are shown in Tables II and III. Furthermore, oxidative inhibition of neutrophil migration is not associated with any loss of neutrophil adherence, viability, glycolytic activity, activity of serine esterases or uptake of leukoattractants, all of which are essential for optimal leukotaxis (Anderson and Jones, 1982). The most likely targets for reversible oxidative inactivation are structural or contractile membrane proteins or glycoproteins, which may have to be maintained in the unoxidized state for optimal function. Oliver *et al.* (1978) have reported that critical minimal levels of reduced glutathione are necessary for microtubule assembly, since they protect tubulin from direct attack by oxidants. Furthermore, Boxer *et al.* (1979) have reported that ascorbate causes enhanced microtubule assembly, which is accompanied by increased chemotactic responsiveness. Oxidation of the thioether group of methionine to the sulfoxide form is a reversible reaction (Jori *et al.*, 1968) and maintenance of these groups in the reduced form also may be necessary for leukocyte motility. Alternatively, essential membrane lipids may be the target for oxidative inactivation of motility. However, we have observed no protective effects of the lipophilic antioxidant α-tocopherol acetate on the inhibition of neutrophil motility by the $HRP/H_2O_2/halide$ system, suggesting that inhibition of lipid oxidation is unlikely to be the mechanism of ascorbate protection (unpublished observations).

2.3. Importance of Extracellular Ascorbate in Stimulation of Leukocyte Motility

Although the targets of oxidant-mediated inhibition of neutrophil chemotactic responsiveness have not yet been identified, recent investigations have shown that it is extracellular ascorbate which is important in the protection of leukocytes from oxidation-mediated loss of function *in vivo* and *in vitro*. We have observed (Anderson, 1981a) that intravenous injection of 1 g of ascorbate in six normal adult volunteers was accompanied by a rapid increase in serum ascorbate levels but by insignificant changes in the concentration of leukocyte ascorbate. Intravenous injection of

Table IV. Assessment of the Extent of Uptake of Radiolabeled Ascorbate by Resting and Stimulated Human Blood Neutrophils

Test system	Uptake of radiolabeled [14C]ascorbate
Resting (unstimulated) neutrophils[a]	21.3 ± 3.6
Neutrophils + 10% EAS	17.0 ± 2.6
Neutrophils + 5 x 10^{-7}M f-met-leu-phe	14.1 ± 2.3
Neutrophils + opsonised Candida albicans	12.0 ± 2.0

[a]Neutrophils were incubated with radiolabeled ascorbate or with the chemotactic stimuli 10% EAS or 5 x 10^{-7}M f-met-leu-phe or opsonised Candida albicans at a ratio of 20:1 microorganisms/cell for 60 min. The concentrations of the leuko-attractants used were those which induce optimal migration in vitro. Exposure of the neutrophils to chemotactic or phagocytic stimuli was not associated with increased ascorbate uptake and values lower than those for unstimulated control cells were observed. Results expressed as the mean value with standard error of 5 experiments in nmoles L-1-[14C]ascorbate bound/2 x 10^6 neutrophils.

ascorbate was associated with increased neutrophil motility and lymphocyte transformation, and with decreased activity of the MPO/H_2O_2/halide systems. These effects on leukocyte function were associated entirely with the *serum ascorbate,* indicating that it is the extracellular ascorbate which is important in stimulating leukocyte functions. Further evidence has come from *in vitro* studies using radiolabelled ascorbate (Jones and Anderson, 1982). In these investigations neutrophils were exposed to the leukoattractants endotoxin-activated serum (EAS) and the synthetic chemotactic tripeptide N-formyl-L-methionyl-L-leucine-L-phenylalanine (f-met-leu-phe) or a phagocytic stimulus of opsonized *Candida albicans* for various time intervals in the presence of radiolabelled ascorbate. The results of these experiments, which are shown in Table IV, indicate no increased uptake of ascorbate above resting levels by neutrophils exposed to either chemotactic or phagocytic stimulation. Furthermore, the uptake of ascorbate was slightly decreased in the stimulated cells, suggesting that ascorbate may in fact be released during exposure of neutrophils to chemoattractants or opsonized antigens. This possibility requires further investigation. However, release of antioxidants extracellularly during locomotion or phagocytosis could be expected to protect by regulating membrane auto-oxidation during these functions. Although phagocytic cells have been reported to concentrate ascorbate to levels considerably higher than those found in serum (De Chatelet *et al.,* 1974) it seems that intracellular accumulation of the vitamin does not occur during this process. These findings are not in agreement with the observations of Loh and Wilson (1970), who reported active uptake of ascorbate by leukocytes *in vitro.*

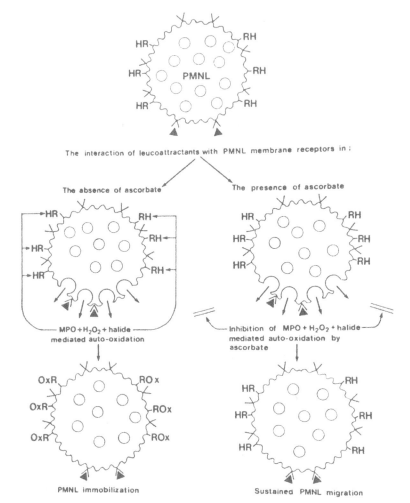

Fig. 2. This shows a schematic representation of the suggested
mechanism by which ascorbate causes increased leukocyte motility.
Interaction of leukoattractants (▲) with membrane receptors (∧)
on leukocytes (PMNL) causes degranulation with release of MPO and
H_2O_2. These agents, together with an appropriate halide in the
extracellular milieu, cause auto-oxidation of the leukocyte membrane (RH → ROx) with subsequent loss of chemotactic responsiveness. However, the presence of ascorbate in the extracellular
environment protects the cell membrane from auto-oxidative damage
mediated by the $MPO/H_2O_2/$halide system and migratory responsiveness is sustained.

A proposed mechanism of ascorbate-mediated stimulation of neutrophil motility by an antioxidant mechanism is shown in Fig. 2. Interaction of leukoattractants with receptors on the cell membrane causes stimulation of oxidative metabolism with extracellular release of H_2O_2 (Becker et al., 1979), degranulation (Becker et al., 1974) and release of MPO (Theron et al., 1981). These agents, together with an appropriate halide, cause auto-oxidation of the neutrophil membrane with an accompanying decrease in cell motility. However, the presence of an antioxidant such as ascorbate in the extracellular milieu protects the neutrophil from loss of chemotactic responsiveness resulting from membrane auto-oxidation by neutralizing oxidizing agents generated by the $MPO/H_2O_2/$halide system. The mechanism of ascorbate-mediated inhibition of auto-oxidative activity of the $MPO/H_2O_2/$halide system may be related to direct inhibition of MPO by ascorbate (McCall et al., 1971), neutralization of oxidants generated by this system, or to reversal of oxidation of essential groups on the cell membrane. Ascorbate is a scavenger of the reactive, unstable oxidants singlet molecular oxygen and hydroxyl radical (Halliwell, 1978; Bodaness and Chan, 1979).

2.4. Ascorbate in the Treatment of Conditions Associated with Abnormal Neutrophil Motility

There are a number of reports describing improved neutrophil motility with accompanying clinical improvement following ingestion of ascorbate by individuals with primary and acquired defects of leukocyte migratory responsiveness. Primary abnormalities of neutrophil locomotion completely or partially corrected following ascorbate therapy include those associated with the Chediak-Higashi syndrome (Boxer et al., 1976; Weening et al., 1981) and chronic granulomatous disease (Anderson, 1981b). Acquired defects of neutrophil motility responsive to ascorbate therapy include those found in some patients with chronic bacterial infections (Anderson and Theron, 1979b; Rebora et al., 1980) and hyperimmunoglobulinemia E (Friedenberg et al., 1979; Anderson et al., 1980b).

2.5. Ascorbate-Mediated Enhancement of Phagocyte Antimicrobial Activity

Although Greendyke et al. (1964) reported that ascorbate caused increased erythrophagocytosis in vitro, other researchers observed no enhancement of phagocytosis by human neutrophils in the presence of ascorbate in vitro. However, there are a number of reports on stimulatory effects of ascorbate ingestion on post-phagocytic antimicrobial activity in neutrophils from normal individuals (Anderson, 1981) and from individuals with primary and secondary abnormalities of intracellular antimicrobial activity associated with increased susceptibility to bacterial infection. Included in this latter group are individuals with the Chediak-Higashi syndrome (Boxer *et al.*, 1976; Weening *et al.*, 1981), chronic granuloma-tous disease (Anderson, 1981b) and primary defects of neutrophil bacteri-cidal activity (Rebora *et al.*, 1980).

The ability of ascorbate to potentiate neutrophil antimicrobial activity may be related to its ability to stimulate HMPS activity (De Chatelet *et al.*, 1972). These authors reported that ascorbate at a concentration of 0.01M caused stimulation of neutrophil HMPS activity. Dehydroascorbate had the same effects and was actually more effective than ascorbate. De Chatelet *et al.* (1972) described a series of reactions involving ascorbate, dehydroascorbate, reduced and oxidized glutathione and reduced nicoti-namide adenine dinucleotide phosphate linked to stimulation of HMPS activity and H_2O_2 production which may provide phagocytes with an alternative antimicrobial system by way of the interaction of ascorbate with H_2O_2. Ericsson and Lundbeck (1955) reported that oxidation of ascorbate *in vitro* using an ascorbate/H_2O_2 system generated toxic inter-mediates, possibly hydroxyl or hydroperoxyl free radicals, which were antimicrobial for a variety of bacteria and fungi and which inactivated viruses. Furthermore, Miller (1969) has reported that the antimicrobial activity of lysozyme is potentiated by ascorbate and H_2O_2 *in vitro*. Orr (1967) has reported that ascorbate causes inhibition of catalase activity *in vitro*. These findings have been confirmed by Anderson (1981b). They suggest that ascorbate may potentiate phagocyte antimicrobial activity by inactivating bacterial catalases, thus increasing the susceptibility of these microorganisms to the toxic effects of H_2O_2.

The antioxidant properties of ascorbate may also contribute to increased phagocyte antimicrobial activity. As previously described, the exposure of human neutrophils to chemotactic or phagocytic signals is accompanied by increased oxidative metabolism and degranulation with subsequent auto-oxidation of the cell membrane by the $MPO/H_2O_2/halide$ system. We have recently shown, using the $HRP/H_2O_2/sodium$ iodide system *in vitro,* that oxidation of the neutro-phil membrane causes the neutrophil to degranulate spontaneously (Anderson and Jones, 1982). This effect can be abrogated by inclusion of ascorbate in the oxidizing system. These data are shown in Table V, where it can be seen that ascorbate prevents the degranulation (as meas-ured by release of alkaline phosphatase) which accompanies membrane oxidation. These findings indicate a third mechanism by which ascorbate may enhance post-phagocytic antimicrobial activity. By preventing auto-oxidation of the cell membrane during exposure of neutrophils to leu-koattractants and phagocytic stimuli, extracellular degranulation is decreased. Consequently, more granules may be available to participate in

Table V. Assessment of the Effects of Ascorbate on the Increased Neutrophil Degranulation Mediated by the Peroxidase/H_2O_2/Halide System In Vitro

Test system	Neutrophil degranulation by measurement of alkaline phosphatase release	
	Resting neutrophils	Neutrophils stimulated with C.albicans
Untreated neutrophils	3.7 ± 1.0[b]	9.8 ± 0.9
Neutrophils + HRP + H_2O_2 + NaI[a]	6.1 ± 1.6	18.8 ± 4.4
Neutrophils + HRP + H_2O_2 + NaI[b]		
a) 5×10^{-2}M ascorbate	2.6 ± 0.7	7.4 ± 0.7
b) 1×10^{-2}M ascorbate	2.8 ± 0.9	8.0 ± 1.5
c) 1×10^{-3}M ascorbate	3.3 ± 1.3	8.8 ± 0.2

[a]Neutrophils were incubated with 0.25 units HRP, 10^{-6}M H_2O_2 and 10^{-3}M NaI for 15 min in the presence and absence of varying concentrations of sodium ascorbate after which they were washed to remove the oxidizing system and ascorbate and resuspended to 1 x 10^7 neutrophils/ml in phosphate-buffered saline (0.15M). The cells were then divided into two aliquots one of which was untreated and the other was exposed to opsonised C.albicans at a ratio of 1 cell: 20 microorganisms. After 30 min incubation the extent of alkaline phosphatase release in resting and stimulated cells was measured as previously described (Anderson and Jones, 1982). Exposure of neutrophils to the oxidizing system caused increased spontaneous release from resting cells and post-phagocytic release from stimulated cells.

[b]Results are expressed as the mean value with standard error in enzyme units/5 x 10^6 neutrophils of six experiments.

the intracellular destruction of ingested microorganisms. It may seem paradoxical that ascorbate causes enhanced neutrophil antimicrobial activity while inhibiting the activity of the $MPO/H_2O_2/halide$ system, which is one of the most important antimicrobial systems in the neutrophil. However, it should be remembered from the data shown in Table IV that ascorbate does not enter the neutrophil and it is therefore only the extracellular (auto-oxidative) activity and not the intracellular activity of the $MPO/H_2O_2/halide$ system which is affected.

2.6. Effects of Ascorbate on Lymphocyte Proliferation and Migration

Ascorbate has been reported to enhance the responsiveness of T-lymphocytes to mitogens *in vitro* (Munster *et al.,* 1977; Panush and Delafuente, 1979; Yonemoto, 1979). Manzella and Roberts (1979) have found that the vitamin prevents the inhibition of mitogen-induced lymphocyte proliferation induced by the influenza virus *in vitro*. Ingestion of ascorbate by mice (Siegel and Morton, 1977), normal human adult volunteers (Yonemoto, 1979; Anderson *et al.,* 1980b; Panush *et al.,* 1981) and individuals with abnormal lymphocyte transformation secondary to neoplasms (Yonemoto, 1979) or hyperimmunoglobulinemia E (Friedenberg *et al.,* 1979) is accompanied by increased T-lymphocyte reactivity to mitogens. Smörgorzewska *et al.* (1981) have reported increased migration of T-lymphocytes from normal individuals to a serum-derived leukoattractant in the presence of 0.05M ascorbate.

3.   Mechanisms of Ascorbate-Mediated Increased Lymphocyte Proliferation

3.1. Cyclic Nucleotide Effects

There has been much research recently on the effects of agents which modulate intracellular cyclic nucleotide levels on lymphocyte reactivity to antigens and mitogens. There is some evidence that increases in intracellular cGMP may be linked to lymphocyte activation (Coffey *et al.,* 1977). Ascorbate has been reported to increase intracellular cGMP in a variety of tissues (Sandler *et al.,* 1975; Goldberg and Haddox, 1977). Panush *et al.* (1981) have reported that ascorbate concentrations which promote increases in lymphocyte cGMP *in vitro* also enhance mitogen-induced proliferation. However, Atkinson *et al.* (1978) found that ascorbate at concentrations which increased intracellular cGMP had no effect on lymphocyte proliferation. The relationship between ascorbate-mediated increases in intracellular levels of cGMP and lymphocyte responsiveness to antigens and mitogens requires further investigation.

3.2. Antioxidant Mechanisms

Intravenous injection of 1 g of ascorbate has been reported to cause a serum-dependent stimulation of lymphocyte responsiveness to mitogens (Anderson, 1981b). This stimulation of lymphocyte proliferation was associated with a serum-dependent stimulation of neutrophil motility and inhibition of the $MPO/H_2O_2/halide$ system. These results suggest that oxidative products generated by the $MPO/H_2O_2/halide$ system may regulate lymphocyte proliferation. Confirmatory evidence was obtained from *in vitro* experiments in which lymphocytes were exposed to the $HRP/H_2O_2/sodium$ iodide system in the presence and absence of ascorbate and then tested for responsiveness to the mitogens phytohemagglutinin (PHA) and concanavalin A. Exposure of lymphocytes to the $HRP/H_2O_2/halide$ system in the absence of ascorbate caused considerable inhibition of reactivity to both mitogens. However, inclusion of ascorbate provided a dose-dependent protection of the lymphocyte response to both mitogens from inactivation by the oxidizing system. These results are shown in Fig. 3. The most probable explanation for these findings is that MPO and H_2O_2 released from neutrophils during exposure to leukoattractants or phagocytic stimuli may combine with a suitable halide in the extracellular medium to oxidize lymphocytes in the vicinity of the neutrophils with a resulting decrease in the lymphocyte proliferative response.

4.   Anti-Allergic Properties of Ascorbate

It has been reported that bronchoconstriction caused by inhalation of histamine aerosols (Zuskin *et al.*, 1973) textile dusts (Zuskin *et al.*, 1976), or metacholine aerosols (Ogilvy *et al.*, 1981) can be significantly reduced by prior administration of 500-1000 mg of ascorbate to adult volunteers. These findings, suggesting the possible value of ascorbate in the prophylaxis of bronchial asthma, are supported by the results of Anah *et al.* (1980). These investigators assessed the effects of ascorbate ingestion (1 g daily) on the severity and rate of asthmatic attacks in a group of 41 Nigerian children. The trial was performed on a double blind basis over a 14-week period. Exacerbations of the asthmatic condition in these children were precipitated by respiratory infection. After 14 weeks an assessment of the severity and frequency of attacks showed that children taking ascorbate suffered significantly fewer attacks than the control group. The

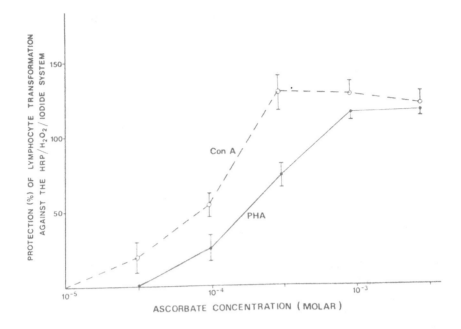

Fig. 3. Ascorbate-mediated protection of the lymphocyte proliferative response to the mitogens phytohemagglutinin (PHA) and concanavalin A (Con A) from inactivation by the HRP/H_2O_2/halide system.
Human blood lymphocytes exposed to the HRP/H_2O_2/iodide system lost
their responsiveness to both mitogens. However, in the presence
of ascorbate a dose-dependent protection and even enhancement of
mitogen-induced proliferation was observed. This figure is reproduced from Anderson (1981a).

severity of the attacks was also significantly reduced. Cessation of ascorbate ingestion was associated with increased attack rates. The authors did
not document the incidence of respiratory infections in the two groups,
but concluded that ascorbate is probably a good prophylaxis in some
bronchial asthmatics. Anderson *et al.* (1980a) reported that addition of
ascorbate (1 g daily) to a standard anti-asthmatic prophylactic regimen in
ten children with bronchial asthma was associated with improved neutrophil motility and lymphocyte transformation. However, the study was
uncontrolled and the effects of standard therapy only were not investigated. Other authors have been unable to detect any protective effects of
ascorbate on the intensity of histamine- or allergen-induced bronchoconstriction in mild asthmatics (Kreisman *et al.*, 1977; Kordansky *et al.*,
1979). Further research is therefore required to assess the potential beneficial effects of ascorbate as an adjunct to standard chemotherapy in the

prophylaxis of bronchial asthma. Nevertheless, it is interesting to consider the mechanisms by which ascorbate may mediate anti-allergic activity.

4.1. Ascorbate-Mediated Neutralization of Histamine

Subramanian *et al.* (1973) observed that in the presence of Cu^{2+} ascorbate causes degradation of histamine both *in vitro* and *in vivo*. These authors suggested that ascorbate thus may be important in the detoxification of histamine. Chatterjee *et al.* (1975) reported that the degradation of histamine is entirely dependent on the ongoing oxidation of ascorbate, since ascorbate alone or the combination of dehydroascorbate with hydrogen peroxide (the products of aerobic oxidation of the vitamin) are not effective in neutralizing histamine. They proposed that monodehydroascorbate, the intermediate of ascorbate auto-oxidation, may be the reactive species. These studies were extended by other researchers who reported that hydrogen peroxide and monodehydroascorbate in the presence of Cu^{2+} are capable of the breakdown of histamine (Yamamoto and Ohmori, 1981). The products of the oxidative inactivation of histamine are hydantoin-5-acetic acid and aspartate (Chatterjee *et al.*, 1975). These findings indicate ascorbate plays a role in allergic reactions by modulating the activity of histamine.

4.2. Effects on Prostaglandins

There have been a number of recent reports that ascorbate can alter prostaglandin (PG) formation in a variety of systems. Prostaglandins of the 2 series, *viz.* PGE_2 and $PGF_{2\alpha}$, have been reported to be involved in the modulation of bronchial smooth muscle. $PGF_{2\alpha}$ promotes contraction and PGE_2 promotes relaxation of bronchial smooth muscle (Mathé and Hedqvist, 1975). The reported inhibitory effects of ascorbate on histamine-induced bronchoconstriction may be mediated by a decrease in the synthesis of $PGF_{2\alpha}$ and an increase in the production of PGE_2. Pugh *et al.* (1975) found reduced formation of $PGF_{2\alpha}$ in homogenates of guinea pig uterus following ascorbate administration. Low dietary intakes of ascorbate decrease the rate of PGE_2 production with increased synthesis of $PGF_{2\alpha}$. Other researchers have reported that ascorbate modulates the basal tone and airway responses to histamine in guinea pigs by altering PG production (Brink *et al.*, 1978). In guinea pig tracheal tissue PGE_2 is the major bronchodilatory PG released upon exposure to histamine, and apparently mediates its bronchodilatory activity by elevating intracellular levels of cAMP which modulate the contractile process (Orehek *et al.*, 1973; Orehek *et al.*, 1975). The precise mechanism by which ascorbate affects the ratio of PGE_2: $PGF_{2\alpha}$ is not known, but it may be related to an inhibitory effect of the vitamin on the conversion of PGE_2 to $PGF_{2\alpha}$. A second possible mechanism by which ascorbate may inhibit the bronchoconstrictor effects of histamine is by promoting synthesis of PGE_1 from dihomo-gamma-linolenic acid (Manku *et al.*, 1978). Like PGE_2, PGE_1 causes an increase in intracellular cAMP (Zurier *et al.*, 1973) and it can therefore mimic the effects of PGE_2. A third possible mechanism is the inhibition of the arachidonic acid cascade by ascorbate via an antioxidant mechanism. In this pathway, in which the PG's of the 2 series are synthesized, the conversion of arachidonate to the unstable cyclic endoperoxides PGG_2 and PGH_2 is mediated by oxidative free radical intermediates (Simon and Mills, 1980). This step may be inhibited by ascorbate, which is a scavenger of oxygen radicals (Bodaness and Chan, 1979). A consequence of this inhibition may be an increased production of PGE_1 by the dihomo-gamma-linolenic acid pathway. However, these suggested mechanisms require considerable further investigation to establish or to disprove their existence.

4.3. Effects of Ascorbate on the Production of Leukotrienes

Interaction of leukocytes with leukoattractants, phagocytic stimuli or allergens (in the case of mast cells and basophils) causes generation of a potent group of arachidonic acid-derived leukoattractants, *viz.* the hydroxyeicosatetraenoic acids as well as the leukotrienes, especially leukotrienes C and D, which are the active components of slow-reacting substance of anaphylaxis (SRS-A). These important mediators of inflammatory and allergic reactions are generated from arachidonic acid by the lipoxygenase enzymes and are the subjects of two recent reviews (Samuelsson *et al.*, 1980; Goetzl, 1980). At the time of writing, this author is not aware of any reports on the effects of ascorbate on lipoxygenase activity. This would seem to be an important area for future research.

4.4. Ascorbate-Mediated Inhibition of Degranulation by an Antioxidant Mechanism

It has recently been demonstrated (Anderson and Jones, 1982) that oxidation of neutrophil membranes is accompanied by spontaneous degranulation and increased extracellular release of granule contents, as well as increased uptake of leukoattractants, during exposure to phagocytic stimuli. Ascorbate, as described in the following section, prevents membrane oxidation-induced degranulation in neutrophils by an antioxidant mechanism. Although this mechanism has not yet been described for mast cell degranulation, it is possible that ascorbate also interferes with mast cell and basophil degranulation by an antioxidant mechanism.

It is also possible that the immunostimulatory properties of ascorbate, which are related to an antioxidant mechanism, are important in preventing asthmatic attacks, which are associated with susceptibility to virus or bacterial infections (Anderson *et al.*, 1980a).

5.   Anti-Inflammatory Properties of Ascorbate

5.1.   Prevention of Degradation of the Extracellular Matrix

Ascorbate has been shown to stimulate the activity of the enzyme prolyl hydroxylase and to act as a co-factor in the hydroxylation reaction in collagen synthesis (Levene and Bates, 1976). Furthermore, ascorbate has been reported to stimulate the net biosynthesis of sulfated proteoglycans *in vitro* and to increase the deposition of these newly synthesized macromolecules in the cell layer in chondrocyte and cartilage cultures. This deposition is associated with inhibition of aryl sulfatase activity (Schwartz, 1979). Aryl sulfatases A and B, lysosomal enzymes which are involved in the breakdown of sulfated macromolecules, are present in elevated concentrations in zones of inflammation (Schwartz *et al.*, 1974). Apart from enhancing collagen synthesis by stimulating prolyl hydroxylase and preventing degradation of proteoglycans by inhibiting arylsulfatases, ascorbate also may stimulate proteoglycan synthesis by maintaining the sulfhydryl-containing amino acid cysteine in a reduced form (Schwartz, 1979). Cysteine residues are known components of the protein core of proteoglycans.

5.2.   Antioxidant Mechanisms

During the function of blood neutrophils auto-oxidation of the cell membrane occurs with subsequent immobilization and degranulation of the cell (Anderson and Jones, 1982). Ascorbate can prevent the oxidative induction of degranulation by neutrophils, thus preventing extracellular release of lysozomal enzymes such as elastase and collagenase which may cause destruction of the extracellular matrix. These effects are illustrated in Table V using alkaline phosphatase release as a measure of degranulation.

The oxidative inhibition of neutrophil motility is a reversible phenomenon. Exposure of oxidized cells which have completely lost their migratory responsiveness to concentrations of ascorbate which reverse membrane oxidation causes complete recovery of chemotactic responsiveness (Anderson and Jones, 1982). It is therefore possible that inflammatory cells which migrate into sites of inflammation undergo auto-oxidative

Table VI.   The Effects of Ascorbate on the Inhibition of Motility Caused by Exposure of Neutrophils to Aggregated IgG Complexes In Vitro

Test system	Neutrophil migration to	
	EAS	f-met-leu-phe
Untreated neutrophils	309 ± 23	227 ± 26
Neutrophils exposed to aggregated IgG only[a]	92 ± 33	77 ± 8
Neutrophils exposed to aggregated IgG[b]		
a) 1 x 10^{-2}M ascorbate	171 ± 44	192 ± 22
b) 2.5 x 10^{-2}M ascorbate	204 ± 39	206 ± 33
c) 5 x 10^{-2}M ascorbate	292 ± 25	245 ± 20

[a] Neutrophils (5x10^6/ml) were incubated with 2 mg/ml of heat aggregated human immunoglobin G for 30 min washed and tested in the presence and absence of varying concentrations of sodium ascorbate for chemotactic responsiveness to 10% autologous EAS and 5 x 10^{-7}M f-met-leu-phe.   Cells exposed to aggregated IgG lost their chemotactic responsiveness to both leukoattractants.   Ascorbate,
[b] however, prevented loss of motility during exposure of neutrophils to aggregated IgG. Results. as mean neutrophils per microscope high-powered field with standard error of 6 experiments.

immobilization with accompanying degranulation and release of proteolytic enzymes and toxic oxygen radicals which cause inflammation and tissue damage (Sacks *et al.*, 1978). However, antioxidants such as ascorbate may prevent immobilization of inflammatory cells and degranulation by inhibiting auto-oxidation of the cell membrane. These cells, especially neutrophils, therefore may enter the inflammatory zone, phagocytose and depart without contributing to the development of chronic inflammation.

To test this hypothesis, human blood neutrophils were exposed to 2 mg/ml of heat-aggregated IgG, which simulates immune complexes *in vitro*. Uptake of aggregated IgG was associated with increased oxidative metabolism as measured by chemiluminescence, increased degranulation and auto-oxidation of the neutrophil membrane as measured by loss of membrane sulfhydryl groups, and loss of neutrophil chemotactic responsiveness. Exposure of these cells to ascorbate was associated with recovery of chemotactic responsiveness and inhibition of degranulation. These results are shown in Table VI. They indicate the possible importance of antioxidants such as ascorbate in the prevention of auto-oxidative immobilization and degranulation of neutrophils.

A further anti-inflammatory property of antioxidants is the protection of anti-proteases such as alpha-l-antitrypsin from oxidative inactivation. It has been reported that oxidative factors generated by the MPO/H_2O_2/halide system cause inactivation of alpha-l-antitrypsin, which is the major elastase-inactivating protease (Carp and Janoff, 1981). Antioxidants such as sodium aurothiomalate and D-penicillamine, which inhibit the MPO/H_2O_2/halide system, protect alpha-l-antitrypsin from oxidative inactivation (Skosey and Chow, 1981). Although data are not yet available on the protective effects of ascorbate, it seems reasonable to assume that such evidence will be forthcoming, since ascorbate is a scavenger of oxidizing radicals generated by the MPO/H_2O_2/halide system. Oxidizing free radicals released from leukocytes during chemotaxis and exposure to immune complexes also have been reported to mediate breakdown of collagen (Greenwald and Moy, 1976) proteoglycan (Greenwald *et al.*, 1977) and hyaluronic acid (McCord, 1974). Since ascorbate is an antioxidant and scavenger of oxidative factors released from phagocytes, it also may protect collagen, proteoglycan and hyaluronic acid from oxidative degradation.

Should toxic oxidative factors generated by the extracellular activity of the MPO/H_2O_2/halide system be implicated in the pathogenesis of inflammation, the inhibitory effects of ascorbate on this system may be an important anti-inflammatory property of the vitamin.

5.3. Other Mechanisms

As mentioned in the previous section, a field of ascorbate research which requires further investigation is the effect of the vitamin on the activity of prostaglandin synthetase and lipoxygenase enzymes *in vitro* and *in vivo*. Since some of the arachidonate-derived products of these enzymes are pro-inflammatory, ascorbate-mediated inhibition of their generation could be an important anti-inflammatory mechanism.

6.   Conclusions

An attempt has been made to review the immunostimulatory anti-allergic and anti-inflammatory activities of ascorbate. While the enhancing effects of ascorbate on cellular immune responses are well documented, this review has been speculative in parts with regard to the possible anti-allergic and anti-inflammatory properties of the vitamin. Such an approach was deemed necessary to identify important areas for future research.

It is clear that the most important activities of the vitamin are related to its antioxidant properties. In addition to their effects on leukocytes, there is evidence indicating that antioxidants such as ascorbate inhibit the oxidative conversion of pre-carcinogens to carcinogens. This subject has not been considered in this presentation. A daily dosage of 1 g to 3 g is sufficient to achieve immuno-enhancing effects *in vivo* in adults (Anderson *et al.*, 1980b). This dosage is, of course, considerably in excess of the recommended daily allowance advocated by health authorities. Nutritionists recommend smaller daily dosages of ascorbate when used as a vitamin. For purposes of correcting abnormal cellular immune responses, however, it is used as an immunopharmacological agent. Evidence presently available shows that ascorbate stimulates leukocyte functions, especially neutrophil motility and antimicrobial activity, as well as lymphocyte responsiveness to antigens and mitogens, and that it is safe to use in the high doses required to achieve these effects. Ascorbate therefore may be considered as an immunopharmacological agent in the prophylaxis and therapy of immunocompromised individuals. The value of ingesting large daily supplements of ascorbate among immunologically normal individuals has long been controversial and seems likely to remain so. The acquisition of evidence showing that ascorbate can enhance and sustain cellular immune responsiveness and that it may function as a natural, safe anti-inflammatory and anti-allergic agent, strengthens the case for increasing the recommended daily intake of the vitamin. This does not mean that ascorbate should be used as a sole agent in the treatment of inflammatory and allergic disorders; however, it may be a useful adjunct to chemotherapy with standard anti-inflammatory and anti-allergic drugs.

It is unlikely that ascorbate is unique with regard to the properties documented here. Two other water soluble essential nutrients, viz. thiamine and cysteine, have been reported to possess immunostimulatory activity related to an antioxidant mechanism (Theron *et al.*, 1981; Anderson and Jones, 1982). Finally, an important point which has emerged from recent studies is that *extracellular ascorbate* may function in immunostimulation by protecting leukocyte membranes from oxidative damage. Arguments that the daily intake of the vitamin should be dictated by the amount required to achieve tissue saturation therefore may not be valid, since it is the circulating ascorbate level that is important in immunostimulation.

References

Anah, C.O., Jarike, L.N., and Baig, H.A., 1980, High dose ascorbic acid in Nigerian asthmatics, *Trop. Geogr. Med.* **32**:132.

Anderson, R., 1979, Effects of ascorbate on leucocytes, Part II. Effects of ascorbic acid and calcium and sodium ascorbate on neutrophil phagocytosis and post-phagocytic metabolic activity, *S. Afr. Med. J.* **56**:401.

Anderson, R., 1981a, Ascorbate mediated stimulation of neutrophil motility and lymphocyte transformation by inhibition of the peroxidase/H_2O_2/halide system *in vitro* and *in vivo*, *Am. J. Clin. Nutr.* **34**:1906.

Anderson, R., 1981b, Assessment of oral ascorbate in three children with chronic granulomatous disease and defective neutrophil motility over a 2-year period, *Clin. Exp. Immunol.* **43**:180.

Anderson, R., and Grabow, G., 1980, *In vitro* stimulation of neutrophil motility by metoprolol and sotalol related to inhibition of both H_2O_2 production and peroxidase mediated iodination of the cell and leukoattractant, *Int. J. Immunopharmac.* **2**:321.

Anderson, R., and Jones, P.T., 1982, Increased leukoattractant binding and reversible inhibition of neutrophil motility mediated by the peroxidase/H_2O_2/halide system: effects of ascorbate, cysteine, dithiothreitol, levamisole and thiamine, *Clin. Exp. Immunol.* **47**:487.

Anderson, R., and Theron, A., 1979a, Effects of ascorbate on leucocytes, Part I. Effects of ascorbate on neutrophil motility and intracellular cyclic nucleotide levels *in vitro*, *S. Afr. Med. J.* **56**:394.

Anderson, R., and Theron, A., 1979b, Effects of ascorbate on leucocytes, Part III. *In vitro* and *in vivo* stimulation of abnormal neutrophil motility by ascorbate, *S. Afr. Med. J.* **56**:429.

Anderson, R., and Van Rensburg, A.J., 1979, The *in vitro* effects of propranolol and atenolol on neutrophil motility and postphagocytic metabolic activity, *Immunology* **37**:15.

Anderson, R., Hay, I., Van Wyk, H., Oosthuizen, R., and Theron, A., 1980a, The effects of ascorbate on cellular and humoral immunity in ten selected asthmatic children, *S. Afr. Med. J.* **58**:974.

Anderson, R., Oosthuizen, R., Maritz, R., Theron, A., and Van Rensburg, A.J., 1980b, The effects of increasing weekly doses of ascorbate on certain cellular and humoral immune functions in volunteers, *Am. J. Clin. Nutr.* **33**:71

Atkinson, J.P., Kelly, J.P., Weiss, A., Wedner, H.J., and Parker, C.W., 1978, Enhanced intracellular cGMP concentrations and lectin-induced lymphocyte transformation, *J. Immunol.* **121**:2282.

Becker, E.L., Showell, H.J., Henson, P.M., and Hus, L.S., 1974, The ability of chemotactic factors to induce lysozomal enzyme release. I. The characteristics of the release, the importance of surfaces and the relation of enzyme release to chemotactic responsiveness, *J. Immunol.* **112**:2047.

Becker, E.L., Sigman, M., Oliver, J.M., 1979, Superoxide production induced in rabbit leukocytes by synthetic chemotactic peptides and A23187, *Am. J. Path.* **48**:288.

Bodaness, R.S., and Chan, P.G., 1979, Ascorbic acid as a scavenger of singlet oxygen, *FEBS Lett.* **105**:195.

Boxer, L.A., Watanabe, A.M., Rister, M., Besch, H.R., Allen, J., and Baehner, R.L., 1976, Correction of leukocyte function in Chediak-Higashi syndrome by ascorbate, *N. Engl. J. Med.* **295**:1041.

Bradley, P., and Rothstein, G., 1981, Release of myeloperoxidase during inflammation, *Clin. Res.* **29**:330A (abstract)

Brink, C., Ridgway, P., and Douglas, J.S., 1978, Regulation of guinea-pig airways *in vivo* by endogenous prostaglandins, *Pol. J. Pharmacol. Pharm.* **30**:157.

Carp, H., and Janoff, A., 1980, Potential mediator of inflammation: Phagocyte-derived oxidants suppress the elastase-inhibitory capacity of alpha-l-proteinase inhibitor *in vitro*, *J. Clin. Invest.* **66**:987.

Chatterjee, I.B., Majumder, A.K., Nandi, B.K., and Subramanian, N., 1975, Synthesis and some major functions of vitamin C, *Ann. N.Y. Acad. Sci.* **258**:24.

Clark, R. A., and Klebanoff, S.J., 1979, Chemotactic factor inactivation by the myeloperoxidase-hydrogen peroxide-halide system: an inflammatory control mechanism, *J. Clin. Invest.* **64**:913.

Coffey, R., Hadden, E.M., and Hadden, J.W., 1977, Evidence for cyclic GMP and calcium mediation of lymphocyte activation by mitogens, *J. Immunol.* **119**:1387.

Dallegri, F., Lanzi, G., and Patrone, F., 1980, Effects of ascorbic acid on neutrophil locomotion, *Int. Archs. Allergy Appl. Immun.* **61**:40.

De Chatelet, L.R., Cooper, M.R., and McCall, C.E., 1972, Stimulation of the hexose monophosphate shunt in human neutrophils by ascorbic acid. Mechanisms of action, *Antimicrob. Agents Chemother.* **1**:12.

De Chatelet, L.R., McCall, C.E., Cooper, M.R., and Shirley, P.S., 1974, Ascorbate levels in phagocytic cells, *Proc. Soc. Exp. Biol. Med.* **145**:1170.

Ericsson, Y., and Lundbeck, H., 1955, Antimicrobial effect *in vitro* of the ascorbic acid oxidation. II. Influence of various chemical and physical factors, *Acta. Pathol. Microbiol. Scand.* **37**:507.

Friedenberg, W.R., Marx, J.J., Hansen, R.L., and Haselby, R.C., 1979, Hyperimmunoglobulin E syndrome: Response to transfer factor and ascorbic acid therapy, *Clin. Immunol. Immunopathol.* **12**:132.

Gallin, J.I., Sandler, J.A., Clyman, R.I., Manganiello, V.C., and Vaughan, M., 1978, Agents that increase cyclic AMP inhibit accumulation of cGMP and depress human monocyte locomotion, *J. Immunol.* **120**:492.

Goetzl, E.J., 1980, Mediators of immediate hypersensitivity derived from arachidonic acid, *N. Engl. J. Med.* **303**:822.

Goetzl, E.J., Wasserman, S.I., Gigli, I., and Austen, K.F., 1974, Enhancement of random migration and chemotactic responses of human leukocytes by ascorbic acid, *J. Clin. Invest.* **53813**.

Goetzl, E.J., Wasserman, S.I., Gigli, I., and Austen, K.F., 1974, Enhancement of random migration and chemotactic responses of human leukocytes by ascorbic acid, *J. Clin. Invest.* **53**:813.

Goldberg, N.D., and Haddox, M.K., 1977, Cyclic GMP metabolism in biological regulation, *Ann. Rev. Biochem.* **46**:823.

Greendyke, R.M., Brierty, R.E., and Swisher, S.N., 1964, *In vitro* studies on erythrophayo-cytosis. II. Effects of incubating leokocytes with selected cell metabolites, *J. Lab. Clin. Med.* **63**:1016.

Greenwald, R.A., and Moy, W.W., 1979, Inhibition of collagen gelation by action of the superoxide radical, *Arthritis Rheum.* **22**:251.

Greenwald, R.A., Moy, W.W., and Lazarus, D., 1977, Degradation of cartilage proteogly-cans and collagen by superoxide radical, *Arthritis Rheum.* **19**:799.

Halliwell, B., 1978, Biochemical mechanisms accounting for the toxic action of oxygen on living organisms: The key role of superoxide dismutase, *Cell Biol. Internat. Reports* **2**:113.

Harrison, J.E., and Schultz, J., 1976, Studies on the chlorinating activity of myeloperoxi-dase, *J. Biol. Chem.* **251**:1371.

Johnson, R.A., and Green, F.D., 1975, Chlorination with N-chloro amides. I. Inter- and intra-molecular chlorination. *J. Org. Chem.* **40**:2186.

Jones, P.T., and Anderson, R., 1982, Oxidative inhibition of polymorphonuclear leukocyte motility mediated by the peroxidase/H_2O_2/halide system: Studies on the reversible nature of the inhibition and mechanisms of protection of migratory responsiveness by ascorbate, levamisole, thiamine and cysteine, *Clin. Exp. Immunol.* (in press).

Jori, G., Galiazzo, G., Marzotto, A., and Scoffone, E., 1968, Selective and reversible photo-oxidation of the methionyl residues in lysozyme, *J. Biol. Chem.* **243**:4272.

Klebanoff, S.J., and Clark, R.A., 1978, *The Neutrophil: Function and Clinical Disorders,* p. 409 North-Holland Pub. Co., Amsterdam.

Kordansky, D.V., Rosenthal, R.R., and Norman, P.S., 1979, The effects of vitamin C on antigen-induced bronchospasm, *J. Allergy Clin. Immunol.* **63**:61.

Kreisman, H., Mitchell, C., and Bouhuys, A., 1977, Inhibition of histamine-induced airway constriction. Negative results with oxtriphyline and ascorbic acid, *Lung* **154**:223.

Levene, C.I., and Bates, C.J., 1976, The effect of hypoxia on collagen synthesis in cultured 3T6 fibroblasts and its relationship to the mode of action of ascorbate, *Biochim. Biophys. Acta.,* **444**:446.

Loh, H.S., and Wilson, C.W.M., 1970, The origin of ascorbic acid stored in leucocytes, *Br. J. Pharmacol. Chemother.* **40**:169.

Manku, M.S., Oka, M., and Horrobin, D.F., 1979, Differential formation of prostaglandins and related substances from arachidonic acid. II. Effects of vitamin C, *Prostaglandins Med.* **3**:129.

Manzella, J.P., and Roberts, N.J., 1979, Human macrophage and lymphocyte responses to mitogen stimulation after exposure to influenza virus, ascorbic acid and hyperthermia, *J. Immunol.* **123**:1940.

Mathé, A.A., and Hedqvist, P., 1975, Effect of prostaglandins $F_{2\alpha}$ and E_2 on airway con-ductance in healthy subjects and asthmatic patients, *Am. Rev. Resp. Dis.* **111**:313.

McCall, C.E., De Chatelet, L.R., Cooper, M.R., and Ashburn, P., 1971, The effects of ascorbic acid on bacteriocidal mechanisms of neutrophils, *J. Infec. Dis.* **124**:194.

McCord, J.M., 1974, Free radicals and inflammation: Protection of synovial fluid by super-oxide dismutase, *Science* **185**:529.

Miller, T.E., 1969, Killing and lysis of Gram-negative bacteria through the synergistic effect of hydrogen peroxide, ascorbic acid and lysozyme, *J. Bact.* **98**:949.

Munster, A.M., Loadholdt, C.B., Leary, A.G., and Barnes, M.A., 1977, The effect of anti-biotics on cell-mediated immunity, *Surgery,* **81**:692.

Ogilvy, C.S., Douglas, J.S., Tabatabai, M., and Du Bois, A. B., 1978, Ascorbic acid reverses bronchoconstriction caused by metacholine aerosol in man: Indomethacin prevents this reversal, *Physiologist,* **21**:86.

Oliver, J.M., Spielberg, S.P., Pearson, C.B., and Schulman, J.D., 1978, Microtubule assembly and function in normal and glutathione synthetase deficient polymorphonuclear leukocytes, *J. Immunol.* **120**:1181.

Orehek, J., Douglas, J.S., Lewis, A.J., and Bouhuys, A., 1973, Prostaglandin regulation of airway smooth muscle tone, *Nature New Biol.* **245**:84.

Orehek, J., Douglas, J.S., and Bouhuys, A., 1975, Contractile responses of the guinea-pig trachea *in vitro:* Modification by prostaglandin synthetase-inhibiting drugs, *J. Pharmacol. Exp. Ther.* **194**:554.

Orr, C.W.M., 1967, Studies on ascorbic acid. I. Factors influencing the ascorbate-mediated inhibition of catalase, *Biochemistry,* **6**:2995.

Panush, R.S., and Delafuente, J.C., 1979, Modulation of certain immunologic responses by vitamin C, *Internat. J. Vit. Nutr. Res.,* Suppl. **19**:179.

Panush, R.S., Delafuente, J.C., Katz, P., and Johnson, J., 1981, Modulation of certain immunologic responses by vitamin C. III. Potentiation of *in vitro* and *in vivo* lymphocyte responses, *Int. J. Vit. Nutr. Res.* Suppl. **20**:35.

Pugh, D.M., Sharma, S.C., and Wilson, C.W.M., 1975, Inhibitory effect of L-ascorbic acid on the yield of prostaglandin F from the guinea-pig uterine homogenates, *Br. J. Pharmacol.* **53**:469.

Rebora, A., Crovato, F., Dallegri, F., and Patrone, F., 1980, Repeated staphylococcal pyoderma in two siblings with defective neutrophil bacterial killing, *Dermatologica* **160**:106.

Rebora, A., Dallegri, F., and Patrone, F., 1980, Neutrophil dysfunction and repeated infections: influence of levamisole and ascorbic acid, *Brit. J. Dermatol.* **102**:49.

Sacks, T., Moldow, C.F., Craddock, P.R., Bowers, T.K., and Jacob, H.S., 1978, Oxygen radicals mediate endothelial cell damage by complement-stimulated granulocytes. An *in vitro* model of immune vascular damage, *J. Clin. Invest.* **61**:1161.

Samuelsson, B., Hammarstrøm, S., Murphy, R.C., and Borgeat, P., 1980, Leukotrienes and slow reacting substance of anaphylaxis, *Allergy,* **35**:375.

Sandler, J.A., Gallin, J.I., and Vaughan, M., 1975, The effects of serotonin, carbamylcholine and ascorbic acid on leukocyte cyclic GMP and chemotaxis, *J. Cell Biol.* **67**:480.

Schwartz, E.R., 1979, Vitamin C: Effect on arylsulfatase activities and sulfated proteoglycan metabolism in cultures derived from human articular cartilage, *Int. J. Vit. Nutr. Res.* Suppl. **19**:113.

Schwartz, E.R, Ogle, R.C., and Thompson, R.C., 1974, Aryl sulfatase activities in normal and pathologic human articular cartilage, *Arthritis Rheum.* **17**:455.

Siegel, B.V., and Morton, J.I., 1977, Vitamin C and the immune response, *Experientia* **33**:393.

Simon, L.S., and Mills, J.A., 1980, Non-steroidal anti-inflammatory drugs, *N. Engl. J. Med.* **302**:1179.

Skosey, J.L., and Chow, D.C., 1981, A mechanism for the therapeutic actions of gold salts and D-penicillamine in rheumatoid arthritis, *Clin. Res.* **29**:559A (abstract).

Smorgorzewska, E.M., Layward, L., and Soothill, J.F., 1981, T-lymphocyte mobility: defects and effects of ascorbic acid, histamine and complexed IgG, *Clin. Exp. Immunol.* **43**:174.

Subramanian, N., Nandi, B.K., Majumder, A.K., and Chatterjee, I.B., 1973, Role of L-ascorbic acid on detoxification of histamine, *Biochem. Pharmacol.* **22**:1671.

Theron, A., Anderson, R., Grabow, G., and Meiring, J.L., 1981, *In vitro* and *in vivo* stimulation of neutrophil migration and lymphocyte transformation by thiamine related to inhibition of the peroxidase/H_2O_2/halide system, *Clin. Exp. Immunol.* **44**:295.

Thomas, E.L., and Aune, T.M., 1978a, Cofactor role of iodide in peroxidase antimicrobial action against *Escherichia coli, Antimicrob. Agents Chemother.* **13**:1000.

Thomas, E.L., and Aune, T.M., 1978b, Oxidation of *Escherichia coli* sulfhydryl components by the peroxidase-hydrogen peroxide-iodide antimicrobial system, *Antimicrob. Agents Chemother.,* **13**:1006.

Weening, R.S., Schoorel, E.P., Roos, D., Van Schaik, M.L.J., Voetman, A.A., Bot, A.A.M., Batensburg-Plenter, A.M., Willems, C., Zeijlemaker, W.P., and Astaldi, A., 1981, Effect of ascorbate on abnormal neutrophil, platelet, and lymphocyte function in a patient with the Chediak-Higashi syndrome, *Blood,* **57**:85c.

Wilkinson, P.C., 1976, Recognition and response in mononuclear and granular phagocytes, *Clin. Exp. Immunol.* **25**:355.

Yamamoto, I., and Ohmori, H., 1981, Degradation of histamine in the presence of ascorbic acid and Cu^{2+} ion: Involvement of hydrogen peroxide, *J. Pharm. Dyn.* **4**:15.

Yonemoto, R.H., 1979, Vitamin C and the immune response in normal controls and cancer patients, *Int. J. Vit. Nutr. Res.* Suppl. **19**:143.

Zurier, R.B., Weissmann, G.M., Hoffstein, S., Kammerman, S., and Tai, H.H., 1974, Mechanisms of lysozomal enzyme release from human leukocytes. II. Effects of cAMP and cGMP autonomic agonists and agents which affect microtubule function, *J. Clin. Invest.* **53**:297.

Zuskin, E., Lewis, A.J., and Bouhuys, A., 1973, Inhibition of histamine induced airway constriction by ascorbic acid, *J. Allergy Clin. Immunol.* **51**:218.

Zuskin, E., Valic, F., and Bouhuys, A., 1976, Byssinosis and airway responses due to exposure to textile dust, 1976, *Lung,* **154**:17.

Chapter 3

Epidemiologic Studies on Vitamin A and Cancer

Curtis Mettlin

1. Introduction

Many observations have led epidemiologists to suggest that environmental factors, in the sense of those elements exogenous to the individual, are responsible for as much as 90% of cancer incidence. Recently, Doll and Peto (1981) attempted to define what proportion of cancer was "avoidable" because it is caused by exposures which might be controlled. The many causes of cancer they identified included tobacco, alcohol, food additives, reproductive and sexual behavior, occupation, pollution, medicines and medical procedures, and viral infections. They also specified a number of currently promising hypotheses regarding the ways that diet might affect the incidence of cancer: diet might contain carcinogens or their precursors; aspects of the diet might affect the formation, transport, activation or deactivation of carcinogens in the body; the diet might contain substances which promote or inhibit the carcinogenic processes. Among this host of possible sources of environmental cancer risk and possible mechanisms whereby dietary factors may influence carcinogenesis, one aspect of diet, the intake of vitamin A, has been of considerable interest to laboratory scientists, epidemiologists, and nutritionists. We will review some of the bases for this interest and, in particular, examine the epidemiologic evidence which might link vitamin A to cancer in a protective role. Finally, we will discuss the implications of these findings and

---

Curtis Mettlin • Chief of Epidemiologic Research, Roswell Park Memorial Institute, Buffalo, New York 14263.

research questions for the future.

This topic is of interest because it may shed light on the nature of the carcinogenic process. Greater interest, perhaps, is generated by the possibilities for application of some preventive measures, by dietary or chemopreventive means, to reduce the mortality and morbidity from cancer, the second leading cause of death in the United States. Such practical application will require development of a body of knowledge derived from events occurring first in the research laboratory followed by corroborating observations in human population. To the present, there have been many laboratory studies and a number of epidemiologic studies on this subject. Finally, there is discussion and planning of prophylactic trials which might provide evidence of the efficacy of various means of cancer prevention (Magnus and Miller, 1980). In this review we will emphasize the current state of knowledge with respect to the role of vitamin A in human carcinogenesis.

Vitamin A is used here as a generic term referring to β-carotene, retinol, and the synthesized variations of the basic molecular form. These are distinct biochemical entities and there are hypotheses that these different forms of vitamin A have different biological effects. We will examine this question to the extent that one can differentiate among epidemiologic studies which have been carried out to date with respect to exposure to various types of vitamin A.

2.   Animal Studies

Experimental studies on laboratory animals are not epidemiological in their approach but they are generally regarded by epidemiologists as important sources of hypotheses. Also, in spite of many instances in which observations from *in vivo* or *in vitro* studies have not been confirmed in human populations, laboratory investigations may be supportive of epidemiologic studies when the findings are in agreement. For this reason, the early laboratory studies on cancer inhibition by vitamin A or its analogues will be reviewed. For a more extensive discussion of experimental work on this topic the reader is referred to Newberne and Rogers (1981).

One of the earliest studies on the physiological importance of vitamin A is that by Wolbach and Howe (1925). They observed that deprivation of vitamin A in the diets of albino rats generated abnormalities in the epithelium of respiratory, alimentary and genitourinary systems. Since then, the role of vitamin A deficiency in promoting lesions and, conversely, the potential of vitamin A excess to maintain normal epithelium when tissues are challenged by tumor-producing substances have been examined in several experimental situations.

Saffiotti *et al.* (1967) exposed Syrian golden hamsters to intratracheal instillations of the common carcinogen benzopyrene (BP). An additional group received the same BP treatment followed by stomach tube feedings of vitamin A palmitate continued for life. Among the first (control) group of 53 animals at risk, 13 cases of squamous tumors and 13 instances of metaplasia were observed. In the experimental group of 16 hamsters, only one exhibited tumor activity and another showed evidence of metaplasia. The investigators suggested that vitamin A had a systemic inhibitory effect on the induction of squamous changes, benign and malignant, in the mucous epithelium of the respiratory tract.

In experiments by Cohen *et al.* (1976) groups of rats were fed diets containing normal, high and deficient levels of vitamin A with and without the carcinogen N-[4-(5-nitro-2-furyl)-2-thiazolyl] formamide (FANFT). Vitamin A deficiency accelerated the carcinogenic action of FANFT. Both levels of vitamin A prevented the appearance of squamous cell metaplasia and neoplasia, but even excess vitamin A did not stop the transitional hyperplasia and neoplasia seen in rats receiving normal levels of vitamin A in conjunction with FANFT.

Newberne and Rogers (1973) fed rats receiving normal and vitamin A deficient diets the carcinogen aflatoxin B. Vitamin A did not affect the occurrence of liver tumors that are normally produced by aflatoxins. However, significantly more colon tumors were observed among the vitamin A deficient groups.

Nettesheim and Williams (1976) studied the susceptibility of rats to pulmonary carcinogens in the absence of current vitamin A intake when considerable amounts of vitamin A were stored in the liver and none of the normal signs of vitamin A deficiency were evident. They observed that the occurrence of 3-methylcholanthrene-induced metaplastic lung nodules was significantly reduced by moderate amounts of retinoic acid. Vitamin A in excess provided no greater protection against the effects of the carcinogen.

Decreases in the carcinogenic effect of polycyclic hydrocarbons in rodents following administration of vitamin A have also been observed with respect to mouse tumors of the cervix and forestomach (Chu and Malmgren, 1965), skin (Bollag, 1972), and prostate (Lasnitzki and Goodman, 1974). Further, vitamin A deficiency has been shown to enhance salivary gland and bowel carcinogenesis in the rat (Hill and Shih, 1974).

Although several similar animal studies have been conducted, the exact mechanism of the effect of vitamin A and its synthetic analogs on carcinogenesis is not known. It is suspected that the enhancement of carcinogenesis occurs in conjunction with increased DNA synthesis and mitosis in epithelial cells. These cells normally have little such activity but during vitamin A deficiency activity increases greatly. This, it is hypothesized, creates greater opportunity for cell growth and differentiation to be influenced by exposure to carcinogens. It has been suggested that vitamin A utilization is part of the "intrinsic physiological controls" in epithelial tissue that prevent the development of malignancy (Sporn, 1977).

Data from many animal studies such as those reviewed suggest an influence of vitamin A on carcinogenesis. However, data from rodent populations subjected to high doses of pure forms of carcinogenic chemicals combined with gross manipulation of vitamin A intake are inadequate as a basis for intervention in human populations. More pertinent data are obtained from human populations experiencing more normal exposures to risk. Several studies on human populations are now available.

3. Human Blood Level Studies

One means of assessing the possible protective effect of vitamin A in human carcinogenesis is by comparison of levels of retinol in the blood of cancer patients with those in healthy persons or persons having diseases other than cancer. While serum levels of retinol are governed largely by factors other than dietary intake of vitamin A, findings of lower levels of vitamin A in the sera of cancer patients would at least be consistent with the hypothesis that vitamin A, however derived, may lend some protective effects. A major weakness of such studies is that cancer patients may have modified their dietary habits or that the disease itself may alter metabolism so that the retinol levels observed are the result of the disease process rather than of the intake of dietary vitamin A.

Ibrahim *et al.* (1977) studied blood levels of retinol and β-carotene in 203 patients with squamous cell carcinoma of the oral cavity and oral pharynx and in 112 controls matched for age and sex. They found the cancer patients to have significantly lower levels of both constituents. They judged that, according to WHO criteria, over half the patients manifested deficient levels of plasma vitamin A, whereas a much smaller proportion of the control patients had less than adequate levels.

A small study of patients with *in situ* cervical cancer was conducted to compare their blood levels of carotene and retinol to those of patients with no evidence of neoplastic disease (Lambert *et al.,* 1981). Carotene levels were found to be somewhat higher in the cancer patients and retinol levels somewhat lower. However, the differences were not statistically significant and the small sample used provides little evidence one way or the other with respect to the role of vitamin A in cervical cancer.

Basu *et al.* (1976) obtained blood samples from 28 histologically confirmed cases of bronchial carcinoma, 10 healthy subjects, and a patient with non-malignant bronchial disease. Patients had significantly lower plasma vitamin A levels than controls. Although the report is unclear as to the degree the cases and controls were matched for smoking status, they did differ in age.

In a similar study, Cohen *et al.* (1977) measured the serum vitamin A levels of 67 non-resectable lung cancer cases. For 43 of these patients, daily intake of vitamin A was estimated from a dietary history. No controls were used. No association was found between level of plasma vitamin A and histologic type of cancer and all 67 patients had serum vitamin levels within the normal range. Eighteen of the sub-sample of 43 patients reported a vitamin A intake of less than the recommended five thousand international units per day.

To overcome the problem of a possible effect of the disease on blood levels of vitamin A, two studies have been conducted in which blood was available from large numbers of persons among whom diagnoses of cancer occurred several years later. In England, serum samples were collected from a population of 16,000 men and stored (Wald *et al.*, 1980). After five years, 86 of these subjects were identified as having developed cancer. Retinol levels in the blood of these men and of 172 persons from the population who did not develop cancer were assayed. Because some sample vials were cracked in storage and handling, separate analyses of the damaged (series A) and intact (series B) vials were performed. Low retinol levels were associated with an increased risk of cancer. Table I shows the relative risks of cancer at any site for men with different retinol levels. Overall, a 2.2-fold increase in risk was observed among men in the highest quintile compared to those in the lowest.

Using a similar design, Kark *et al.* (1981) followed 3102 men from Evans County, Georgia, for 12 to 14 years. Blood samples had been obtained from these men as part of their participation in a prospective study of coronary heart disease risk factors. The subsequent documented finding of 129 cases of cancer in the sample made it possible to study the relationship between serum retinol levels and cancer risk in this population. Persons who eventually developed cancer were found to have had significantly lower mean serum retinol levels at least 12 months prior to cancer diagnosis. This effect was found in all age and sex groups investigated but was most prominent among males. The lower serum retinol levels in eventual cancer patients were consistent across all sites and cell types with the exceptions of leukemia and Hodgkin's disease.

4.   Epidemiologic Studies

The fact that serum retinol levels in cancer patients and populations studied retrospectively or prospectively appear to be associated with cancer risk does not necessarily indicate that the intake of vitamin A influences risk. Serum retinol levels are regulated by intrinsic controls and some observed variation in the populations studied is attributable to these controls rather than to different levels of vitamin A in the diet. The only way

Table 1.   Number of Subjects and Controls by Quintile of Serum Retinol Concentration
and Relative Risk of Cancer

Retinol Concentration		Series A		Series B		
Limits						Relative
Quintile	IU/dl	Subjects	Controls	Subjects	Controls	Risk[b]
1st	20-187	15	6	10	21	1.40
2nd	188-210	6	6	12	26	1.14
3rd	211-233	8	4	10	30	1.09
4th	234-266	7	6	7	33	0.72
5th	267-346	1	5	10	35	0.67
All	20-346	37	27	49	145	1.00

[a]From Wald et al. (1981) by permission. [b]Relative risk was standardized according
to whether the serum sample was in Series A or B. The reend was statistically
significant (P<0.025).

of determining whether vitamin A intake influences risk is by observing
the levels of retinol or carotene in the diets and vitamin A supplements
ingested by populations at different risks. This may be done retrospectively by comparing the diet histories of cancer patients to those of
comparable control patients, or prospectively by observing cancer incidence in large populations of known dietary habits. In either of these
situations, the data on exposure to vitamin A are derived from subjects'
reports of their dietary habits and uses of preparations containing vitamin
A. Although this data source may be less reliable than laboratory assays,
several alternative methods for gathering such data are available. The
strengths and weaknesses of these different approaches have been discussed elsewhere (Mettlin and Graham, 1979). While dietary intakes are
difficult to study epidemiologically, vitamin A is perhaps one of the most
easily studied by this approach because this nutrient tends to occur in
predictable amounts in a limited number of dietary items. It is neither so
ubiquitous that populations cannot be differentiated with respect to their
intake nor so scarce that no measure of intake can be obtained.

Bjelke (1975) reported the results of a five year prospective study of
8278 Norwegian men who reported their cigarette smoking and dietary
habits by mailed survey. Cigarette smoking and vitamin A intake were
negatively associated with age as well as with each other. Among these
men, 36 cases of bronchus carcinoma were identified by use of the Norway Cancer Registry. An index of vitamin A intake was obtained by

weighting the frequency that respondents reported consuming several food items, by estimating the vitamin A content of a standard portion of each, and summing the products. Bjelke found a significant negative association between the vitamin index and cancer mortality at all levels of cigarette smoking.

MacLennan *et al.* (1977) studied the high incidence of lung cancer among the Chinese population of Singapore. In addition to evaluating the relative risk of cigarette smoking, the investigators sought leads to other possible risk factors. Several questions were asked regarding the frequency of consumption of common dark green leafy vegetables such as Chinese mustard greens and kale. This research showed a 2.23-fold increase in risk associated with relatively low frequency of consumption of these vegetables. Since these foods are rich in vitamin A activity, the authors interpreted their findings as consistent with the report of Bjelke.

During the period from 1954 to 1965, all patients entering Roswell Park Memorial Institute in Buffalo, New York, were interviewed with respect to their usual dietary habits. Over 24,000 such records were obtained in an eleven-year period. Because Roswell Park historically has been a cancer hospital, the majority of patients ultimately were diagnosed with a malignant neoplastic disease; but many patients, after diagnostic workup, were diagnosed as having benign conditions or no pathology. Because data had been obtained in a similar manner from all individuals before the establishment of their diagnoses, it is possible to carry out case-control comparisons of these data. This is done by comparing individuals with cancer at given sites with persons with cancer diagnoses at other sites or with non-neoplastic diseases.

A measure of vitamin intake was obtained from questions regarding the patient's usual frequency of consumption of 21 different food items during the month one year prior to the onset of symptoms (Mettlin *et al.*, 1979). The vitamin A content of the diet was estimated by U.S. Department of Agriculture tables of food values for a standard portion of each item. The products of this weighting factor and the monthly frequency of consumption were summed for all food items. The dietary interview did not include all foods that might have been eaten by the respondents and the vitamin A index was an estimate of relative rather than absolute vitamin A intake, based on a substantial portion of food ingested.

In Table II the age-adjusted relative risk of lung cancer for persons who were heavy smokers and for persons who were less than heavy smokers, as well as the overall smoking-adjusted relative risks for various levels of vitamin A intake, are presented. The coefficients indicate an ascending lung cancer risk associated with descending levels of vitamin A intake. Milk and carrots, rich sources of dietary vitamin A, contributed significantly to the overall vitamin A index and these were the only food items that, independent of the index, were associated with lung cancer risk.

Table II. Relative Risks[a] of Lung Cancer by Level of Vitamin A Intake and Smoking Habit[b]

Vitamin Level[c]	One pack or less a day			More than one pack a day			
	No. of Cases	No. of Controls	Age-adjusted RR[d]	No. of Cases	No. of Controls	Age-adjusted RR[c]	Age and Smoking Adjusted RR[d]
≤74	97	326	1.4	60	55	2.4[e]	1.7[d]
75-124	54	176	1.4	28	40	1.6	1.5
≥125	33	158	1.0	20	46	1.0	1.0
Total	184	660		108	141		

[a]Relative risk. Food items included in the index and their vitamin A content (IU/100 g edible portion) were: cabbage (130), sauerkraut (50), coleslaw (130), brussels sprouts (520), kale (8,000), cauliflower (60), broccoli (2,500), red cabbage (40), kohlrabi (20), parsnips (30), carrots (11,000), rutabaga (550), cucumber (100), pickles (100), beets (20), tomatoes (900), lettuce (330), beef (50), chicken (200), fruit (517), and milk (340 IU/cup). [b]Adapted from Mettlin et al. (1979) [c]Estimate of monthly intake in thousands of International Units. [d]Mantel-Haenszel pooled RR. [e]Differs from RR of 1.0 (P≤0.05); Chi-square test.

Table III. Reported Frequency of Milk Drinking for Lung Cancer Patients and Controls by Smoking Status[a]

Cups of milk/day	One pack of cigarettes or less/day				More than one pack of cigarettes/day				
	Lung Cancer Patients		Controls		RR[b]	Lung Cancer Patients		Controls		RR[b]
	No.	%	No.	%		No.	%	No.	%	
<	99	53.8	290	43.9	1.5[b]	63	58.3	58	41.1	2.0[c]
≥	85	46.2	371	56.1	1.0	45	41.7	83	58.9	1.0
Total	184	100	661	100		108	100	141	100	

[a] From Mettlin et al. (1979)   [b] RR = relative risk   [c] Differs from 1.0 (P<0.05); Chi-square test.

Table III shows the elevated risks observed for persons who reported typically drinking less than one, as opposed to greater than one, cup of milk per day. Carrot consumption was associated with reduced lung cancer risk but only among the men who were heavy smokers. Among the lighter smokers or non-smokers, there was no difference in risk associated with the frequency of carrot consumption (Table IV). This study was the first to provide evidence of a dose-response relationship between cancer risk and vitamin A across a range of intakes.

Using the same sources of data, we examined the possible dietary risk factors associated with human bladder cancer (Mettlin and Graham, 1980). Our study of 569 bladder cancer patients and 1025 age-matched controls admitted to Roswell Park showed that sex-adjusted relative risks increased at lower levels of vitamin A intake measured in the same fashion as was done in the lung cancer study. A similar pattern of risk elevation was associated with infrequent milk and carrot intake. The levels of risk observed for each of these variables are reported in Table V.

Graham et al. (1981) and Marshall et al. (1982) carried out a similar analysis on the incidence of cancer of the larynx and of oral cancer, using the Roswell Park data. Table VI shows the relative risk associated with vitamin A intake when the data were adjusted for variations in cigarette smoking and alcohol consumption, other risk factors known to be associated with larynx cancer, and when both these factors were controlled simultaneously. These data indicate a three-fold increase in risk associated with the lowest levels of vitamin A intake. With respect to oral cancer, Marshall et al. (1982) observed elevations in risk associated with low vitamin A intake that could not be explained by a relationship of dietary

Table IV.   Reported Frequency of Carrot Consumption for Lung Cancer
Patients and Controls by Smoking Status[a]

Frequency of eating carrots	One pack of cigarettes or less/day					More than one pack of cigarettes/day				
	Lung Cancer Patients		Controls		RR[b]	Lung Cancer Patients		Controls		RR[b]
	No.	%	No.	%		No.	%	No.	%	
Less than once per week	62	33.7	239	36.3	0.9	46	43.4	41	29.1	1.9[c]
Once per week or more	122	66.3	419	63.7	1.0	60	56.6	100	70.9	1.0
Total	184	100	658	100		106	100	141	100	

[a]From Mettlin et al. (1979).

[b]RR = relative risk.

[c]Differs from 1.0 ($P \leq 0.05$); Chi-square test.

habits to other known risk factors for this cancer, such as smoking, alcohol consumption or dentition.

Other analyses of the Roswell Park data for an association between vitamin A and cancer risk at other sites have failed to show similar results. Graham et al. (1978), in a study of colon cancer, found no association between dietary vitamin A intake and risk of this disease. Mettlin et al. (1981) observed an apparent association between low vitamin A intake and esophageal cancer, but when other known risk factors such as smoking and alcohol consumption were accounted for, the risk associated with vitamin A consumption was insignificant. However, this research did implicate low levels of vitamin C intake as a risk factor in colon cancer. Other analyses of the Roswell Park data are ongoing in a survey of the dietary factors which may be associated with cancer of each major site.

The Roswell Park study on vitamin A and lung cancer has been repeated in England by Gregor et al. (1981). They obtained estimates of current dietary vitamin A intake from 100 lung cancer cases and 173 controls. Patients were interviewed regarding their current patterns of consumption of major dietary sources of vitamin A, including cheese, eggs, butter, margarine, milk, liver, carrots, green vegetables, as well as their use of vitamin pills. The authors found total current vitamin A intake to be lower in male lung cancer cases than in comparable control patients. Marked differences were observed between cancer cases and controls with respect to their consumption of liver and vitamin A supple-

Table V. Sex-adjusted Relative Risk (RR) of Bladder Canacer by Vitamin A, Milk and Carrot Consumption for 569 Cases and 1025 Age-matched Controls at Roswell Park Memorial Institute[a]

Vitamin A index level[b]	RR	Cups of milk per day	RR	Frequency of carrot consumption	RR
<25	2.07**	Never	1.71**	Never – once a month	1.62*
25–50	1.89**	<1	1.53**	Once every 3 weeks	
50–75	1.72**	1	1.35	Twice a week	1.41
75–100	1.23	2	1.09	>twice a week	1.0
100–150	1.39	≥3	1.00		
125–150	1.23				
>150	1.00				

[a]Adapted from Mettlin and Graham (1979). [b]Estimate of monthly intake in thousands of International Units. **$P < .01$ *$P \leq .05$

ments. Vitamin A intake from liver was significantly lower in male lung cancer patients than in controls, although this was not found to be true for females, among whom lung cancer patients reported a higher intake of vitamin A via liver consumption than did the age-matched controls. Among males (Table VII), irrespective of smoking habit, a 2.3-fold increase in risk was associated with low levels of vitamin A intake compared with the highest levels. Among male cigarette smokers, a 3.8-fold increase in risk was associated with low levels of vitamin A intake. There were no significant differences in vitamin A intake between female cancer patients and controls, either among current cigarette smokers or all subjects taken together. It should be noted that lung cancer is much less common in females than in males, and in this study the small numbers involved may have precluded the finding of a significant effect among females.

Smith and Jick (1978) carried out a retrospective investigation of the frequency of regular use of vitamin A preparations among 800 newly diagnosed cancer patients and 3433 patients with non-malignant conditions. The frequency of a reported history of use of vitamin A-containing preparations was lower among the male cancer patients than among the male control subjects. No significant difference between female cancer patients and controls with respect to use of vitamin A preparations was detected. In this study no information on possible dietary sources of vitamin A was collected. Lung cancer was the most common tumor in the male patients.

Table VI. Relative Risk of Larynx Cancer Associated With Diet[a]

	Cases	Controls	Mantel–Haenzsel relative risk adjusted for:		
			Cigarettes	Alcohol	Cigarettes & alcohol
Vitamin A (IU/month)					
<50,000	98	78	2.68*	2.67*	3.01*
100,000–150,000	134	145	1.77	1.69	1.91*
>150,000	67	71	1.76	1.87	2.09
Total	338	359	1.00	1.00	1.00
Mantel linear trend, P			<0.005	<0.005	<0.005

[a]From Graham et al. (1981) with permission.   *P<0.05.

Prompted by the results of these investigations, researchers involved in the Western Electric Study examined their data to determine the 19-year incidence of lung cancer among 1954 middle-aged men. Thirty-three men in this population sample developed lung cancer three to nineteen years after they had been interviewed with respect to their dietary habits. Vitamin A food sources were classified as sources of preformed vitamin A (retinol) or carotene. Foods classified as retinol sources were whole milk, cream, butter, margarine, cheese, ice cream, eggs, and liver. The carotene index was based on consumption of vegetables, soup, and fruit sources. Shekelle et al. (1981) reported that the mean intake of carotene was lower in men who subsequently developed lung cancer than in men who did not. Fourteen of the 488 men who reported the lowest levels of carotene intake subsequently developed cancer while only two of the 488 in the highest quartile of carotene intakes were diagnosed with this disease. It was found that the lung cancer patients and the remainder of the population were similar with respect to their intake of retinol and other nutrients. One hundred and seventy-five other cancer diagnoses were identified in their follow-up, but for no cancer site other than lung did the carotene index deviate significantly from that observed for the study population as a whole.

In Japan, green and yellow vegetable consumption accounts for an average of 44% of the typical dietary intake of vitamin A activity. Hirayama (1979), reporting a ten-year prospective study of 256,118 Japanese adults, found associations between the frequency of green and yellow vegetable consumption and rates of mortality from lung cancer. Among both males and females, persons who reported daily consumption of green and yellow vegetables had significantly lower rates of lung cancer mortality

Table VII. Estimated Age-standarized Relative Risk of Lung Cancer
(x100) by Level of Current Total Vitamin A Intake[a]

Retinol intake (μg/wk)	Males			Females		
	0	E	R	0	E	R
All subjects						
<7,500	27	25.83	108	5	5.66	86
7,501-15,000	41	35.32	131	6	9.11	57
>15,000	10	16.86	46	11	7.23	188
Current cigarettes smokers only						
<7,500	17	16.42	107	3	4.13	65
7,501-15,000	25	19.96	160	6	7.25	73
>15,000	4	9.61	28	8	5.61	228

[a]From Gregor et al.(1980) with permission.

[b]0 = observed; E = expected; R = estimated relative risk.

than those who reported occasional or less frequent consumption of these food items. This relationship was independent of cigarette smoking.

5. Discussion

Many of these findings on the association between vitamin A intake and cancer incidence or mortality have been derived from studies not originally designed to examine this question. In many instances, the measures of vitamin A intake were crude, omitting major potential sources of vitamin A. In the studies focusing on dietary habits, data were collected by retrospective questionnaires rather than by more direct observation of food intake, and such data are limited by the capacity of subjects to report accurately their dietary habits. Furthermore, it should be noted that because epidemiology deals with human populations, it is a non-experimental science, and observed effects may be confounded by many interacting variables which cannot be controlled in a natural setting. Given these limitations, none of the investigations just reviewed and summarized in Table VIII should be regarded as definitive. However, taken together they may provide a more complete picture of the nature and extent of the possible relationship of vitamin A to cancer induction in humans.

Table VIII. Summary of Controlled Studies of Human Populations Associating Vitamin A and Cancer Risk

Investigator	Type of Study	Population Studied	No. & Sites	Measures of Vitamin A	Results – Comments
Bjelke(1975)	Prospective	8,278 Norwegian men	53 lung	Retinol weighted frequency of food consumption	Cases significantly more likely to report less than the mean level of intake. No specific sources of retinol specified.
Basu et al. (1975)	Retrospective	U.K. patients	28 lung	Plasma retinol	Cases had significantly lower retinol levels than controls.
Ibrahim et al. (1977)	Retrospective	Indian patients	203 oral oro- pharynx	Plasma retinol and carotene	Mean values for retinol and carotene significantly lower in patients than controls.
Hirayma(1979)	Prospective	256,118 Japanese adults	807 lung	Frequency of green and yellow vegetable consumption	Lower lung cancer mortality for daily vs. less frequent consumption among males and females, smokers and non-smokers.
Smith and Jick (1978)	Retrospective	Boston patients	800 all sites	History of use of retinol containing preparations	Significantly less frequent use reported by male cases compared to controls, no differences for females; lung cancer most frequent site among males, breast cancer most frequent among females.
Mettlin and Graham(1979)	Retrospective	RPMI patients	569 bladder	Retinol weighted frequency of food consumption	2.07-fold sex-adjusted increase in risk for low frequency dose-response milk and low carrot intake frequency.
MacLennan et al. (1977)	Retrospective	Singapore Chinese	233	Frequency of green vegetable consump- tion	2.23-fold increase risk associated with infrequent consumption.
Mettlin et al. (1979)	Retrospective	RPMI patients*	292 lung	Retinol weighted frequency of food consumption	1.7-fold age and smoking-adjusted increase in risk, for low frequency; dose-response relationship, increased risk associated with low milk and low carrot intake frequency.

Reference	Study type	Population	Cases/site	Method	Findings
Gregor et al. (1980)	Retrospective	U. K. patients	100 lung	Retinol weighted frequency of food consumption and use of retinol preparations	Male patients reported significantly less frequent intake than controls, mainly due to lower consumption of liver and less use of vitamin pills; no significant differences for females.
Wald et al. (1980)	Prospective	16,000 U.K. men	86 all	Serum retinol	Risk of cancer at any site for lowest level of serum retinol was 2.2 times greater than for highest level; dose-response relationship.
Graham et al. (1981)	Retrospective	RPMI patients*	374 larynx	Retinol weighted frequency of food consumption	Three-fold increase in risk for low retinol intake adjusted for cigarette and alcohol consumption; dose-response relationship.
Kark et al. (1981)	Prospective	3,102 Georgia adults	85 all	Serum retinol	Significantly lower serum retinol levels among cases; similar tendency observed for all sex-race groups; similar tendency observed for all sites and cell types except leukemia and Hodgkin's disease.
Shekelle et al. (1981)	Prospective	1954 Illinois men	208	Carotene and retinol weighted frequencies of food consumption	Increased risk of lung cancer (33 cases) for low carotene intake; no effects for retinol intake or for other sites of cancer.
Marshall et al. (1982)	Retrospective	RPMI patients*	440 oral	Retinol weighted frequency of food consumption	Significant elevated risk for infrequent consumption; dose-response relationship.

*Roswell Park Memorial Institute patients

Hill (1965) has defined the criteria that one may apply to assess the significance of epidemiologic data on etiologic factors in chronic disease. These criteria include such factors as evidence on the strength of the association, the presence of a dose-response relationship, the consistency of findings, the biological plausibility of the association, corroboration from other sources, and the outcomes observed when interventions based on the putative association are conducted. In the case of the vitamin A and cancer relationship, several of these criteria have been fulfilled.

First, in regard to the strength of the association, several investigations have revealed increases in risk which exceed 100% for low levels of serum vitamin A or of low dietary vitamin A. The doubling or tripling of risk for a disease such as lung cancer is modest relative to the effect of cigarette smoking, but nevertheless is not likely due to chance. In some investigations, significant cancer risks for low vitamin A consumption or low serum retinol were not observed, but these were usually at sites or for populations that were observed in too few numbers to permit detection of a significant risk were one to have occurred.

In one respect, findings on the association between vitamin A and cancer are remarkably consistent. In studies focusing on lung cancer, or on populations of men in which lung cancer predominated, prospective and retrospective studies of serum retinol, vitamin A intake or intake of vitamin A-containing preparations all supported the hypothesized relationship. The inconsistencies in findings have been observed mainly for tumors other than lung cancer or for lung cancer among females. Furthermore, in some instances dose-response relationships have been observed with risk ascending as vitamin A levels declined.

The congruence of the epidemiologic findings and the effects predicted from animal studies provides evidence that a relationship between vitamin A and cancer may be corroborated by researchers using altogether different approaches and that the association may be biologically plausible. The instances are rare in which findings from experiments in which rodents or other animal models were exposed to high levels of some carcinogenic or protective factor have been borne out in general human populations. More often, laboratory studies on carcinogenesis are corroborated only in human populations in which high levels of exposure are involved. For example, chemical carcinogenesis predicted by animal experimentation is usually observed only in humans among workers who have chronic high level exposures not widely shared in the population. However, in the case of vitamin A and cancer, the human populations studied have not been unique, but have been populations exposed to the "normal" range of interacting environmental factors. The congruence of the laboratory findings with those of investigations on such unspecialized human populations strengthens the evidence for a relationship between vitamin A and cancer.

Although the hypothesis that adequate vitamin A can reduce cancer risk is supported by epidemiologic evidence of various sorts, it remains to be seen whether intervention based on this knowledge will reduce cancer incidence. Controlled trials of cancer prophylaxis at the community level are the subsequent steps which will be required to determine the answer to this question. In designing such interventions, researchers confront many uncertainties that are not easily resolved on a basis of available information. For example, Peto *et al.* (1981) contend, on theoretical grounds, that β-carotene rather than retinol is the preferable agent to use in such trials. However, the evidence from several studies is that non-carotene sources of vitamin A such as liver, milk, and vitamin A-containing preparations are associated with reduced risk of cancer at different sites. Our studies of lung and bladder cancer, for example, showed that milk, a major source of retinol vitamin A in the typical American diet, was associated with reduced risk, and in the Gregor (1980) study of lung cancer only the retinol sources were associated with risk reduction. These findings, based on study of several hundreds of cancer patients, are contradicted only by the Western Electric Study of 33 cases of lung cancer. In addition, the large majority of animal studies which suggest a protective effect of vitamin A involved the use of preformed vitamin A to induce the observed effects. β-carotene may have lower toxicity and thus be more suitable in a clinical trial involving large numbers of persons, but available evidence on human populations does not support the conclusion that β-carotene would have a greater effect. It may be necessary to conduct intervention trials using different agents to resolve this question, but additional laboratory studies and epidemiologic studies in natural settings that would differentiate between sources of vitamin A activity also would be useful.

It is also an open question as to what populations and cancer sites are most suitable for prophylactic intervention. The bulk of the epidemiologic evidence relates to male populations and specifically to lung cancer. Most studies have focused on this site and on male populations because large samples are available, lung cancer among men being far more common than any other cancer. The Roswell Park studies, as well as some others, suggest that vitamin A consumption is characteristically lower among patients with cancer at other sites, but some of the more common cancers, such as colon cancer, have not shown such effects. Additional epidemiologic studies focusing on sites other than lung are needed.

6.   Summary

Data derived from epidemiologic studies on human populations are consistent with the protection from cancer afforded by vitamin A seen in animal studies. The populations studied are diverse, including groups living in India, Singapore, Norway, the United Kingdom, and the United States. The methodologies brought to bear on the question have been equally varied. Although there are inconsistencies in findings, and instances in which an association has not been observed, the weight of evidence suggests that the intake of vitamin A from dietary or other sources may inhibit the onset of lung cancer and possibly other cancers. However, the evidence from human populations is not experimental and it is conceivable that the associations observed are not causal. Additional epidemiologic research is needed to determine what sites of cancer may be inhibited by vitamin A and whether cancer growth at any other site is enhanced by high vitamin A intakes. It is also important that controlled trials using vitamin A as a chemopreventive agent be considered as a means of determining whether the epidemiologic findings are of clinical significance.

References

Basu, T.K., Donaldson, D., Jenner, M., Williams, D.C., and Sakulu, A., 1976, Plasma vitamin A in patients with bronchial carcinoma, *Brit. J. Cancer* **33**:119.

Bjelke, E., 1975, Dietary vitamin A and human lung cancer, *Int. J. Cancer* **15**:561.

Bollag, W., 1972, Prophylaxis of chemically induced papillomas and carcinomas of mouse skin by vitamin A acid, *Experientia (Basel)* **28**:1219.

Chu, E.W., and Malmgren, R.A., 1965, An inhibitory effect of vitamin A on the induction of tumors of the forestomach and cervix in the Syrian hamster by carcinogenic polycyclic hydrocarbons, *Cancer Res.* **25**:884.

Cohen, S.M., Wittenbrg, J.F., and Bryan, G.T., 1976, Effect of avitaminosis A and hypervitaminosis A on urinary bladder carcinogenesis of FANFT, *Cancer Res.* **36**:2334.

Cohen, M.H., Primack, A., Broder, L.E., and Williams, L.R., 1977, Vitamin A serum levels and dietary vitamin A intake in lung cancer patients, *Cancer Letters* **4**:51.

Doll, R., and Peto, R., 1981, The Causes of Cancer, Oxford University Press, New York.

Graham, S., Dayal, H., Swanson, M., Mittelman and Wilkinson, G., 1978, Diet in the epidemiology of the colon and rectum, *J. Natl. Cancer Inst.* **61**:709.

Graham, S., Mettlin, C., Marshall, J., Priore, R., Rzepka, T., and Shedd, D., 1981, Dietary factors in the epidemiology of cancer of the larynx, *Am. J. Epidemiol.* **113**:675.

Gregor, A., Lee, P.N., Roe, F.J.C., Wilson, M.J., and Melton, A., 1981, Comparison of dietary histories in lung cancer cases and controls with special reference to vitamin A, *Nutr. Cancer* **2**:93.

Hill, A.B., 1965, The environment and disease: association or causation, *Proc. Roy. Soc. Med.* **58**:295.

Hill, D.L., and Shih, T., 1974, Vitamin A compounds and analogues as inhibitors of mixed-function oxidases that metabolize carcinogenic polycyclic hydrocarbons and other compounds, *Cancer Res.* **34**:564.

Hirayama, T., 1979, Diet and Cancer, *Nutr. Cancer* **1**:67.

Ibrahim, K., Jafrey, N.A., and Zuberi, S.J., 1977, Plasma vitamin A and carotene levels in squamous cell carcinoma of oral cavity and oro-pharynx, *Clin. Oncology* **3**:203.

Kark, J.D., Smith, A.H., Switzer, B.R., and Hanes, C.G., 1981, Serum vitamin A (retinol) and cancer incidence in Evans County, Georgia, *J. Natl. Cancer Inst.* **66**:7.

Lambert, B., Brisson, G., and Bielmann, P., 1981, Plasma vitamin A and precancerous lesions of cervix uteri: a preliminary report, *Gynecol. Oncology* **11**:136.

Lasnitzki, I., and Goodman, D.S., 1974, Inhibition of the effects of methylcholanthrene on mouse prostate in organ culture by vitamin A and its analogues, *Cancer Res.* **34**:1564.

MacLennan, R., Dalosta, J., Day, N.E., Law, C.H., Ng, Y.K., and Shanmugaratnam, 1977, Risk factors for lung cancer in Singapore Chinese, a population with high female incidence rates, *Int. J. Cancer* **20**:854.

Magnus, K., and Miller, A.B., 1980, Controlled prophylactic trials in cancer, *J. Natl. Cancer Inst.* **64**:693.

Marshall, J., Graham, S., Mettlin, C., Shedd, D., and Swanson, J., 1982, Diet in the epidemiology of oral cancer, *Nutr. Cancer* **3**:145.

Mettlin, C., and Graham, S., 1979, Methodologic issues in etiologic studies of diet and colon cancer, *Nutr. Cancer* **1**:46.

Mettlin, C., and Graham, S., 1980, Dietary risk factors in human bladder cancer, *Am. J. Epidemiol.* **110**:255.

Mettlin, C., Graham, S., and Swanson, M., 1979, Vitamin A and lung cancer, *J. Natl. Cancer Inst.* **62**:1435.

Mettlin, C., Graham, S., Priore, R., Marshall, J., and Swanson, M., 1980, Diet and cancer of the esophagus, *Nutr. Cancer* **2**:143.

Newberne, P.M., and Rogers, A.E., 1973, Rat colon carcinomas associated with aflatoxin and marginal vitamin A, *J. Natl. Cancer Inst.* **50**:439.

Newberne, P.M., and Rogers, A.E., 1981, Vitamin A, retinoids, and cancer, *in* Nutrition and Cancer (G.R. Newell and N.M. Ellison, eds.), pp. 217-232, Raven Press, New York.

Nettesheim, P., and Williams, M.L., 1976, The influence of vitamin A on the susceptibility of the rat lung to 3-methylcholanthrene, *Int. J. Cancer* **17**:351.

Peto, R., Doll, R., Buckley, J.D., and Sporn, M.B., 1981, Can dietary beta-carotene materially reduce human cancer rates? *Nature* **290**:201.

Saffiotti, U., Montesano, R., Sellakumar, A.R., and Burg, S.A., 1967, Experimental cancer of the lung, *Cancer* **20**:857.

Shekelle, R.B., Lepper, M., Liu, S., Maliza, C., Raynor, W.J. Jr., Rossof, A.H., Paul, O., Shryock, A.M., and Stamlen, J., 1981, Dietary vitamin A and risk of cancer in the Western Electric study, *Lancet* **2**:1185.

Smith, P.G., and Jick, H., 1978, Cancers among users of preparations containing vitamin A, *Cancer* **42**:808.

Sporn, M.B., 1977, Retinoids and carcinogenesis, *Nutr. Rev.* **35**:65.

Wald, N., Idle, M., Boreham, J., and Bailey, A., 1980, Low serum-vitamin-A and subsequent risk of cancer, *Lancet* **2**:813.

Wolbach, S.B., and Howe, P.R., 1925, Tissue changes following deprivation of fat-soluble A vitamin, *J. Exp. Med.* **42**:753.

Chapter 4

Metabolic Bone Disease Associated with Total Parenteral Nutrition

Gordon L. Klein and Jack W. Coburn

1. Introduction

Since the work of Wilmore and Dudrick in 1968, the administration of a concentrated nutrient solution by central venous catheter has been employed widely to prevent and treat malnutrition in patients who are unable to absorb sufficient nutrients orally from the gastrointestinal tract.

Innovations in techniques for the administration of these solutions have made it possible to continue such nutrient treatment at home for months or years (Jeejeebhoy et al., 1973; Broviac and Scribner, 1974; Jeejeebhoy et al., 1976; Byrne et al., 1979). As experience with total parenteral nutrition (TPN) has accumulated, a number of metabolic abnormalities have been encountered. These were generally due to the omission of necessary nutrients from the TPN solution: they include deficiencies of phosphate (Travis et al., 1971), copper (Karpel and Peden, 1972), zinc (Strobel et al., 1978), and essential fatty acids (Paulsrud et al., 1972). Other recently identified deficiencies include those of biotin (Mock et al., 1981) and possibly molybdenum (Abumrad et al., 1981). Hyperammonemia has been detected in premature infants receiving various amino acid solutions (Johnson et al., 1972), and hypouricemia has also been identified (Vedig et al., 1981; Koretz, 1981).

Gordon L. Klein • Department of Pediatrics, City of Hope National Medical Center, Duarte, California 91010.   Jack W. Coburn • Department of Pediatrics and Medicine, UCLA School of Medicine and the Medical and Research Services, VA Wadsworth Medical Center, Los Angeles, California 90073.

Total parenteral nutrition, by definition, bypasses both the intestine and the portal system; thus, the major phylogenetic apparatus for the processing and distribution of nutrients is not operative. Instead, arbitrary quantities of various nutrients aie administered directly into the systemic circulation, and there is little knowledge of the consequences of this route of administration to various homeostatic mechanisms. Patients receiving treatment with long-term total parenteral nutrition have, on the whole, done well (Jeejeebhoy *et al.*, 1973; Broviac and Scribner, 1974; Jeejeebhoy *et al.*, 1976; Byrne *et al.*, 1979). However, abnormalities of unknown pathogenesis are now being detected in some patients. These include cholestatic liver disease (Rodgers *et al.*, 1976; Postuma and Trevenen, 1979; Dahms and Halpin, 1981), cholelithiasis (Roslyn *et al.*, 1981), and the bone disease that is the topic of the present review.

We (Targoff *et al.*, 1979; Klein *et al.*, 1980a) and others (Shike *et al.*, 1980) have described a metabolic bone disease associated with long-term total parenteral nutrition. Although its cause remains unknown, we have undertaken a series of studies in an attempt to define the abnormalities involved.

2.   Description of Index Case

The insidious manner of presentation and the puzzling course of this condition are illustrated by our index case (Klein *et al.*, 1980a). This patient was a 68-year-old white female with a short bowel resulting from radiation enteritis, a complication of cancer therapy. She began treatment with total parenteral nutrition late in 1975 and did very well for three months, gaining 12 kg while receiving minimal oral intake. Three months following the initiation of TPN, however, she developed pain in the feet, ankles, and knees on weight-bearing. Over the next two years the pain gradually intensified to such a degree that she was confined to bed, unable to move without severe pain. She had no recurrence of malignancy nor had she received any medication known to affect bone adversely.

Physical examination revealed marked tenderness at the articular margin and periarticular areas of the ankles, knees, and spine. There was no joint swelling or erythema. Nerve conduction studies and electromyography were normal. She was felt not to have any arthritic or neurologic abnormality, and it was suggested that her symptoms might represent a metabolic bone disease. Because the pain was intractable and failed to respond to narcotic analgesics, she was hospitalized at the end of 1977. The total parenteral nutrition regimen was discontinued with substitution of an intravenous solution of dextrose and electrolytes; oral intake increased. Within two months her pain improved to the degree that she

Table I. Standard TPN Solutions Employed in Los Angeles

	Adults	Children
Volume Infused	2-3 L/day	0.6-1 L/day
Glucose	200-750 g/day	60-250 g/day
Protein[a]	60-90 g/day	18-30 g/day
Na	70-150 mEq/day	18-30 mEq/day
K	60-90 mEq/day	15-25 mEq/day
Ca	200-300 mg/day	120-200 mg/day
Mg	240-360 mg/day	72-120 mg/day
Cl	70-105 mEq/day	17-29 mEq/day
P	1000-1500 mg/day	180-300 mg/day
Acetate	78-117 mEq/day	7-12 mEq/day
Cu	2-3 mg/day	0.6-1 mg/day
Zn	2-3 mg/day	1-1.8 mg/day
Fat[b]	500 ml of 10% emulsion twice weekly	1-4 g/kg of 10% emulsion twice weekly
Multivitamin Infusion (MVI)[c]	5 ml/day	5 ml/day

[a]As either crystalline amino acids or equivalent amount of utilizable protein from 90-150 g casein hydrolysate

[b]Added in 1978

[c]Vitamins K, and B_{12} administered by injection

From Klein et al. (1980[a]) and Cannon et al. (1980).

could ambulate. This improvement occurred despite the recurrence of severe diarrhea, weight loss, and nutritional deterioration.

Toward the end of 1977, this patient underwent an iliac crest bone biopsy with two pulse doses of tetracycline given three weeks apart to label the bone and to permit measurement of bone formation rate. She was found to have mild osteomalacia, but there was no separation of the two tetracycline bands, indicating a very slow rate of bone apposition. In addition, very few osteoclasts were seen, indicating diminished bone resorption.

3.   Prospective Study of Biochemical Abnormalities Related to Bone in TPN Patients

Following identification of the index case, we evaluated 26 of the approximately 100 patients in Los Angeles and Seattle who had been treated with total parenteral nutrition for more than three months. Eleven of this group were symptomatic, 15 were not. Approximately 20% of the total group of patients in both Los Angeles and Seattle programs had experienced bone pain in the lower extremities, lower back, or ribs. Most patients consumed less than 500 kcal/day via the oral route. Prior to initiation of TPN, all had severe malnutrition due to extensive intestinal disease; most patients had undergone multiple abdominal operations. The TPN solution, administered over a period of 10 to 12 nighttime hours, had the composition given in Table I.

The purpose of the evaluation was to identify any biochemical parameters that might distinguish patients with symptomatic bone disease from those who lacked symptoms; this was done with the hope of finding clues to the pathogenesis of this unusual condition. We added data on six additional symptomatic patients who were identified after the initial survey.

Accordingly, all patients had monthly determinations of serum levels of sodium, potassium, chloride, total carbon dioxide, serum glutamic pyruvic transaminase, serum glutamic oxaloacetic transaminase, alkaline phosphatase, creatinine, calcium, phosphorus, magnesium, albumin, and total protein. Serum copper and zinc were measured on at least one occasion. To rule out essential fatty acid deficiency, the plasma ratio of eicosatrienoic acid to arachidonic acid was determined in one patient. Each patient underwent 24-hr urine collection for measurement of calcium, phosphorus, and creatinine. Seven symptomatic patients had anterior iliac crest bone biopsies with double tetracycline labeling (Sherrard *et al.*, 1974).

The clinical features of the 11 afflicted patients are listed in Table II. Their biochemical parameters as well as the mean values for the asymptomatic patients receiving TPN treatment are given in Table III. Serum levels of calcium and phosphorus were normal or mildly elevated, while levels of magnesium, total carbon dioxide, copper, zinc, and total protein and albumin were within the normal range for both groups. Serum alkaline phosphatase activity and transaminase levels were consistently elevated in both groups of patients. The ratio of eicosatrienoic acid to arachidonic acid in our index case was 0.36, with normal values being less than 0.40 (Paulsrud *et al.*, 1972), indicating no deficiency of essential fatty acids. In addition, serum levels of 25(OH)-vitamin D were normal in the six symptomatic individuals who were tested.

Table IV lists urinary excretion of calcium and phosphorus in symp-

Table II. Clinical Features of Initial Patients with TPN-Associated Metabolic Bone Disease

Case	Age/Sex	Underlying Intestinal Disorder	Time from TPN Onset to Bone Pain (Months)	Weight Gain (Kg)	Oral Energy Intake (kcal/day)	Clinical Features	Degree of Disability
1	67/F	Radiation Enteritis Ileal Resection	5	12	300-500	Pain: Knees, Ankles, Low Back, Ribs	Bed-ridden
2	41/F	Radiation Enteritis Ileal Resection	2	14	500	Pain: Heels, Knees, Back, Hands	Wheelchair
3	65/F	Crohns's Disease Multiple Entercrocutaneous Fistulae	4	7.5	<500	Pain: Knees, Ankles,	Wheelchair
4	20/M	Radiation Enteritis No Resection	2	14.8	Uncertain	Pain: Heels, Knees, Hips, Back	Ambulatory
5	39/F	Crohn's Disease Ileal Resection	22	13.6	500	Pain: Low Back, Feet, Knees	Wheelchair
6	53/F	Intestinal Vasculitis Jejunal, Ileal Resection	7	4.5	<500	Pain: Heels, Low Back, Ankles, Knees, Elbows, Ribs	Wheelchair
7	38/F	Crohn's Disease Jejunal, Ileal Resection	18	6	500-1000	Pain: Ribs, Heels, Low Back	Ambulatory
8	55/F	Radiation Enteritis	10	Unk.	<500	Pain: Knees, Ankles	Wheelchair
9	56/F	Chronic Idiopathic Intestinal Pseudoobstruction; Short Bowel	a	13	Uncertain	Pain: Elbows, Knees	Ambulatory
10	64/F	Crohn's Disease Jejunal, Ileal Resection	3	3	300-500	Pain: Left Hip, Lower Spine, Shoulders; Left Scapular & Rib fracture (Prednisone Treatment)	Wheelchair
11	43/F	Crohn's Disease Short Bowel	36	0	500-1000	Pain: Back Vertebral Compression Prednisone Treatment	Ambulatory

[a]Symptoms of skeletal disease appeared 6 years prior to initation of TPN. Modified from Klein et al. (1980).

Table III. Serum Biochemical Features at Onset of Musculoskeletal Symptoms in Patients on TPN

Symptomatic Patients	Ca (mg/dl)	P (mg/dl)	Mg	Alk. Phos. (u/L)	Albumin (g/dl)	Cu (ug/dl)	Zn (ug/dl)	HCO₃⁻ (mEq/L)
	Ca (mg/dl)	P (mg/dl)	Mg	Alk. Phos. (u/L)	Albumin (g/dl)	Cu (ug/dl)	Zn (ug/dl)	HCO_3^- (mEq/L)
1	10.3	4.7	2.0	201	4.6	167	155	25
2	10.1	5.0	1.6	275	3.8	229	115	28
3	9.0	4.7	4.5	145	3.5	181	146	28
4	10.3	5.8	2.6	673	4.0	170	151	20
5	10.4	2.4	1.8	426	3.8	144	124	23
6	10.4	3.8	1.4	267	4.1	65	119	25
7	10.2	4.8	2.1	251	4.4	102	116	32
8	11.0	5.5	2.6	175	4.4	97	96	24
9	9.3	4.2	1.5	137	4.3	101	79	18
10	9.1	4.5	1.7	280	3.8	132	67	30
11	10.0	3.9	1.2	190	3.6	71	65	24
Mean	10.0	4.5	1.8	274.5	4.03	132	112	24.9
±SE	0.16	0.06	0.14	106.0	0.20	10.8	18.1	1.3
Asymptomatic Patients (N=11-15)								
Mean	9.8	4.9	1.6	203.0	4.10	151	148	25.2
±SE	0.16	0.25	0.05	30.0	0.10	14.1	18.2	1.3
Normal	9.2	3.0-	1.5-	45-	3.6-	M70-140	50-	22-
Range	10.5	4.5	2.1	105	4.5	F85-156	150	29

Modified from Klein et al. (1980a)

Table IV. Urinary Mineral Excretion at Onset of Muscoloskeletal Symptoms in TPN Patients

Symptomatic Patients	Ca (mg/period)			P (mg/period)		
	On TPN 12 hr	Off TPN 12 hr	Total 24 hr	On TPN 12 hr	Off TPN 12 hr	Total 24 hr
1	368	20	388	1305	636	1941
2	572	8	580	1318	676	1994
3	708	20	728	760	465	1225
4	188	0	188	1184	0	1184
5	728	316	1044	ND[a]	ND	1070
6	420	184	604	831	527	1358
7	352	108	460	363	366	729
8	ND	ND	180	ND	ND	732
9	124[a]	28	152	617[a]	419	1036
10	ND	ND	276	ND	ND	601
11	248	5	253	50	155	205
Mean	456	84	444	800	462	1097
±SE	72	36	100	161	65	164
Asymptomatic Patients						
Mean			572			1128
±SE			172			115

[a] None detectable [b] Collected for 6 1/2 hours only, not included in mean. Modified from Klein et al. (1980c).

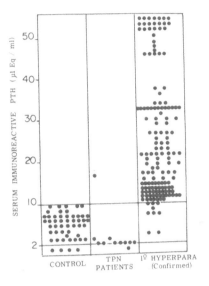

Fig. 1. Serum immunoreactive PTH (iPTH) levels in normal controls
(left), our TPN patients (center), and patients with confirmed
primary hyperparathyroidism (right). The TPN patients have serum
iPTH levels that are in the lower range of normal (Data modified
from Slatopolsky *et al.*, 1979).

tomatic and asymptomatic patients; there were no differences between the
two groups. However, in nearly all patients there was striking hypercalciuria, with a mean value of 440 mg/day; this marked hypercalciuria was
evident only during nighttime hours while the TPN solution was infused.

In a subsequent evaluation, serum immunoreactive parathyroid hormone (iPTH) levels were measured in 18 patients, both with and without
symptomatic bone disease (Fig. 1). Most of the values were either in the
low normal range or were undetectable, while only 8% of control values
were undetectable. There were no differences between the iPTH values in
those with and those without symptoms of bone disease. In addition,
serum levels of vitamin A were normal in all our patients (Klein *et al.*,
1981a), and bone biopsy findings were not compatible with those previously reported in vitamin A intoxication (Gerber *et al.*, 1954).

Thus, there was no parameter that distinguished patients who had
symptoms of bone disease from those who were asymptomatic. The results
of the six bone biopsies in the symptomatic patients were similar to that
described in the index patient: there was mild, patchy osteomalacia with
decreased bone turnover. Nine of the 11 symptomatic patients also had
roentgenographic evidence of diffuse osteopenia, and photon absorptiome-
try, obtained in our index case, revealed decreased bone mineral content

of the radius. This investigation failed to demonstrate any differentiating feature of patients with bone pain from other patients receiving long-term TPN; however, it did tell us that patients with TPN bone disease had hypercalciuria, occasional hyperphosphatemia and hypercalcemia, patchy osteomalacia and low bone turnover. In seven symptomatic patients, the TPN treatment was suspended for 4 to 8 weeks and a dextrose-electrolyte solution was substituted. In all cases the bone pain improved when TPN was discontinued.

4.   Observations Made by Others

At the same time that we were conducting our preliminary investigations, similar patients were studied by Shike *et al.* (1980) in Toronto. They prospectively studied 16 patients with severe gastrointestinal disease treated with total parenteral nutrition for 7 to 89 months. The composition of the TPN solution used by Shike and his colleagues was similar to that used in Los Angeles, with the exceptions noted in Table V.

In the Toronto patients, serum levels of calcium, phosphorus, magnesium, iPTH and 25(OH)-vitamin D were followed sequentially. In 11 of 16 patients, urinary calcium excretion exceeded 300 mg/day. Also, there was intermittent, mild hypercalcemia and hyperphosphatemia with normal levels of magnesium. Serum levels of 25(OH)-vitamin D were normal, and iPTH levels were low or normal. Two patients had severe back pain and roentgenographic findings of lumbar vertebral compression.

Bone biopsies were performed on two separate occasions: at 1 to 4 months and again between 6 and 73 months after initiation of treatment with total parenteral nutrition. Significant changes in bone histomorphometry from first to second bone biopsies were observed. A "hyperkinetic" state was reported in the initial bone biopsy in 9 of 12 patients; increased bone surfaces showed resorption and formation without increased osteoid. The second biopsy showed increased width of unmineralized osteoid, increased osteoid as a percentage of total bone surface, and a decreased percentage of osteoid taking up the single tetracycline label. These changes were interpreted as being consistent with osteomalacia. Double tetracycline labeling was not carried out.

There were certain similarities between the clinical and biochemical findings in the patients reported from Toronto and those seen in Southern California. Marked hypercalciuria and intermittent mild hypercalcemia and hyperphosphatemia were observed in both groups of patients. The observation of increased bone surfaces showing resorption and formation by the group in Toronto was not seen in the bone biopsies from patients in Southern California. The findings from Toronto may reflect changes in bone induced by abnormalities of calcium metabolism secondary to the

Table V.   Apparent Differences in TPN Solutions Employed in Los Angeles and Toronto

	Los Angeles	Toronto
Protein	Casein hydrolysate (1975–1981)	Casein hydrolysate (1970–1976)
	Crystalline amino acids (2g/kg/d)[a]	Crystalline amino acids (1976–1981) (1g/kg/d)
Ca	300 mg/day	400 mg/day
P	1500 mg/day	500–700 mg/day
Lipid	10% emulsion 30% of calories 2 times/week	10% emulsion 30% of total calories/day
Vitamin D_2	1000 IU/day	250 IU/day

[a]Occasional use

Data from Klein et al.(1980[a]),Jeejeebhoy et al.(1976) and Shike et al.(1980, 1981).

gastrointestinal disease and malabsorption. The fact that the biopsies performed in Southern California were carried out later during treatment with total parenteral nutrition may account for this difference. It is likely that pre-existing gastrointestinal disease leading to malabsorption and secondary hyperparathyroidism is present prior to initiation of treatment with TPN. Of particular note was the finding of abnormalities in bone in many asymptomatic patients in Toronto. Such observations suggest that all or most patients receiving long-term therapy with TPN may, in fact, have abnormal bone, while only a small number manifest symptoms.

A conclusion that clearly can be gathered from both of these clinical observations is that the mechanism responsible for the occurrence of metabolic bone disease with patchy osteomalacia occurring in patients with adequate intake of calcium, phosphorus, and vitamin D remains unknown.

5. Bone Disease in Infants Treated with TPN

During the same period of time, we have also noted evidence of metabolic bone disease in infants receiving long-term treatment with TPN. A syndrome that resembled rickets (Klein *et al.*, 1982b) occurred in three of ten infants treated with long-term TPN. Five of them had short bowel syndrome secondary to surgical resection, two had hypoplastic villus atrophy with chronic secretory diarrhea, two had chronic idiopathic intestinal pseudo-obstruction, and one had congenital short bowel. The infants began TPN treatment between one and five months of age. The composition of their nutrient solution is given in Table I.

The three afflicted infants had serum alkaline phosphatase levels greater than 900 IU/L and roentgenographic changes of rickets with fraying, splaying, and pathological fractures of the long bones. There were no uniform abnormalities in serum calcium or phosphorus levels. X-rays in a fourth infant revealed marked demineralization of bone (Klein *et al.*, 1980b), and the serum alkaline phosphatase level was greater than 900 IU/L.

At the time of detection of the skeletal abnormalities, the infants had been receiving TPN treatment for 5 to 15 months. The vitamin D_2 provided was at least 400 IU/day, the recommended oral daily allowance. Our first case was a 17-month-old girl, who had received total parenteral nutrition for 15 months as nutrition support for chronic idiopathic intestinal pseudo-obstruction. At the time of detection of roentgenographic rickets, she was given additional vitamin D_2 intravenously to raise the daily total dose to 1000 IU. After four months of treatment with the higher dose, there was no change in serum alkaline phosphatase (2000 IU/L) or improvement in the roentgenographic lesions; the serum levels of

25(OH)-vitamin D remained in the low-normal range (8-12 ng/ml). However, when a further increment in the vitamin D_2 dosage was instituted (25,000 IU given intramuscularly once a month) the patient's serum alkaline phosphatase levels declined and the 25(OH)-vitamin D levels increased twofold within 30 days. Within four months, the roentgenographic lesions disappeared.

The two infants who subsequently developed this syndrome had received vitamin D_2, 1000 IU/day intravenously, but did not demonstrate a fall in alkaline phosphatase or roentgenographic healing until the intramuscular supplement of vitamin D_2 was added. Spectrophotometric analysis of samples of the TPN solutions obtained from the bottle and from the plastic tubing and catheter utilized for the infusion revealed recovery of all of the vitamin D_2 added to the solution, indicating no significant adherence of vitamin D to either the glass bottle or to plastic tubing (Klein et al., 1982b). We concluded that a bone disease, which was roentgenographically indistinguishable from rickets, developed in infants receiving long-term TPN despite apparently adequate amounts of vitamin D_2, calcium, and phosphate. Since the condition improved with the administration of additional vitamin D_2, it seemed that this condition could represent a vitamin D-resistant state, possibly arising from altered metabolism of vitamin D.

6.   Controversy Regarding the Role of Vitamin D

Shike et al. (1980) reported that withdrawal of vitamin D from the total parenteral nutrition solution was associated with symptomatic improvement, a reduction of urinary calcium excretion, and a decrease in serum calcium levels in three patients. Subsequently, they reported a reduction in the quantity of unmineralized osteoid on bone biopsies carried out in 11 patients after the withdrawal of vitamin D for at least six months (Shike et al., 1981). They concluded that an increased sensitivity to vitamin D itself was in some way responsible for the production of this bone disease in patients on long-term TPN. Because of the finding of normal serum levels of 25(OH)-vitamin D in their patients, they concluded that this condition differed from the usual vitamin D toxicity.

In contrast, our clinical experience with infants with rickets during TPN treatment suggested that there was a requirement for more, rather than less, vitamin D. Thus, there was the implication in both reports that vitamin D metabolism was abnormal, but the exact nature of the abnormality was unknown.

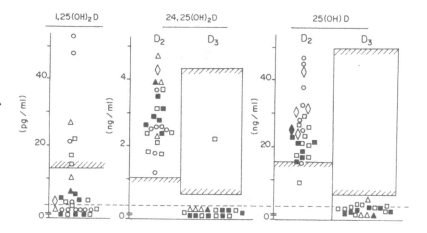

Fig. 2. Serum levels of vitamin D sterols in patients treated
with TPN. Each symbol represents a separate determination; open
symbols are samples obtained before infusion of the TPN solution;
solid symbols are those at the end of infusion;   and   are
patients with symptomatic bone disease;   and   are determinations
done at the National Animal Disease Center (NADC);   and   are
determinations done at the University of California, Riverside
(UCR). The lower limits of $1\alpha,25(OH)_2D$, the upper limits of
$24,25(OH)_2D_2$ and $25(OH)D_2$, and the normal ranges for $24,25(OH)_2D_3$
and $25(OH)D_3$ were determined at the NADC. The values for
$24,25(OH)_2D$ and $25(OH)D$, measured at UCR, are arbitrarily
designated as D_2 metabolites. The normal ranges are as follows:
$1\alpha,25(OH)_2D$, 21-53 pg/mL; $24,25(OH)_2D$, 1.4 to 5.4 ng/mL; and
$25(OH)D$, 17-57 ng/mL (mean $\pm$ SD). Reproduced from the Annals of
Internal Medicine (Klein *et al.*, 1981) with permission of the
publisher.

6.1.   Alterations in Vitamin D Metabolism and the Possible Role of Vitamin D in TPN Bone Disease

There are several observations which suggest that the metabolism of vitamin D is altered during total parenteral nutrition; however, it is not certain whether such changes relate to the pathogenesis of the bone disease
seen in TPN patients. Serum levels of 25(OH)-vitamin D_2 and
25(OH)-vitamin D_3, 24,25(OH)$_2$-vitamin D_2 and 24,25(OH)$_2$-vitamin D_3
and total 1,25(OH)$_2$-vitamin D have been measured in adults and children
under treatment with long-term TPN (Klein *et al.*, 1981a). Most samples

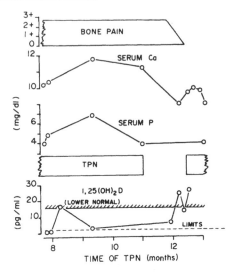

Fig. 3. The clinical course of a patient during long-term TPN
therapy which was interrupted because of an infection and replacement Broviac catheter. Serial levels of serum calcium, phosphorus,
and 1,25-dihydroxyvitamin D are indicated. Semiquantitative determination of bone pain is indicated by the degree of disability: 3+
= restricted to a wheelchair; 2+ = able to walk with a cane; and
0 = pain free. The broken line represents the lower detection
limits for 1α,25-dihydroxyvitamin D. Reproduced from the Annals
of Internal Medicine (Klein *et al.*, 1981) with permission of the
publisher.

were taken immediately or shortly before the usual nighttime infusion of
TPN solutions, although some were taken immediately after the infusion.
The results of these studies are shown in Fig. 2. The total levels of
25(OH)-vitamin D and 24,25(OH)$_2$-vitamin D are generally normal, but
there was a marked shift from sterols with the usual vitamin D$_3$ side
chain to a preponderance of those of the vitamin D$_2$ form. This shift
probably arose because all intravenous vitamin D is given in the form of
vitamin D$_2$; moreover, reduced exposure of these patients to sunlight may
have contributed to the low levels of the vitamin D$_3$ sterols. The only
patient with a normal serum level of 24,25(OH)$_2$-vitamin D$_3$ was a
woman who had been exposed to a sun lamp at regular intervals for several weeks. The biochemical differences between sterols of the natural
vitamin D$_3$ form and those of the vitamin D$_2$ form (which arises from a
plant precursor), exist only in the side chain and are minor; the two forms
are believed to act in the same way and to be equally potent in man
(Lawson, 1980). However, the actions of the vitamin D$_2$ and D$_3$ sterols,
when given intravenously to patients with minimal or no oral intake of
calcium or other nutrients, have not been defined.

Other striking findings are very low serum levels of 1,25(OH)$_2$-vitamin D in the TPN patients (Klein *et al.*, 1981a; Shike *et al.*, 1981). The serum levels of 1,25(OH)$_2$-D were low or undetectable in nearly 80% of our patients. Furthermore, there was a rise in serum 1,25(OH)$_2$-vitamin D from undetectable to normal in one patient within six weeks after stopping TPN treatment (Fig. 3). Four months after reinitiation of the TPN treatment, the serum levels of 1,25(OH)$_2$-vitamin D fell and have remained undetectable. The findings suggest that TPN treatment, in some way, accounts for low levels of serum 1,25(OH)$_2$-vitamin D.

It is quite certain that the reduction in serum levels of 1,25(OH)$_2$-vitamin D is more than a transient event produced by the TPN infusion *per se*. Thus, the serum levels of 1,25(OH)$_2$-vitamin D were essentially no different immediately before and after the 12-hr infusion of the TPN solution (Klein *et al.*, 1981a). Inasmuch as the plasma half-life of 1,25(OH)$_2$-vitamin D is approximately 3 hr (Mason *et al.*, 1980), these observations suggest that the abnormality produced by treatment with TPN is long-lasting.

6.2. Causes of Low Serum Levels of 1,25(OH)$_2$-vitamin D

Two possible explanations for the low serum levels of 1,25(OH)$_2$-vitamin D in these patients are either a decreased production or an increased degradation of 1,25(OH)$_2$-vitamin D. Most of our efforts have pursued the former possibility. The factors which may decrease the generation of 1,25(OH)$_2$-vitamin D include a lack of parathyroid hormone (Haussler, 1980), hyperphosphatemia (Tanaka and DeLuca, 1973; Hughes *et al.*, 1975), hypercalcemia or an increased calcium load (Henry *et al.*, 1974; Hughes *et al.*, 1975), and renal failure (Haussler *et al.*, 1976; Slatopolsky *et al.*, 1978). Reports from both Southern California (Klein *et al.*, 1980a, 1981b) and Toronto (Shike *et al.*, 1980, 1981) indicate that serum levels of iPTH are low or low-normal in most TPN patients and hypercalciuria and intermittent hyperphosphatemia are common; this constellation of findings is compatible with the induction of physiologic hypoparathyroidism. It is plausible that the reduced levels of serum iPTH are responsible for the subnormal serum levels of 1,25(OH)$_2$-vitamin D. The elimination of either calcium or phosphate from the TPN solution failed to produce changes in the serum levels of 1,25(OH)$_2$-vitamin D, suggesting that maneuvers which raise endogenous serum iPTH levels do not lead to increased serum levels of 1,25(OH)$_2$-vitamin D (Klein *et al.*, 1981b). However, studies with the exogenous administration of PTH, which can cause a sharp increase in serum levels of 1,25(OH)$_2$-vitamin D (Eisman *et al.*, 1979), have not been carried out.

The possibility that reduction in renal function may account for the diminished generation of 1,25(OH)$_2$-vitamin D is remote, as most patients treated with TPN have either normal or only minimally reduced renal function.

6.3. Do Low Serum Levels of 1,25(OH)$_2$-vitamin D Affect the Bone in TPN Patients?

Not only are reasons for the low serum levels of 1,25(OH)$_2$-vitamin D in the TPN patients uncertain, but it is also uncertain how such low levels of 1,25(OH)$_2$-vitamin D affect bone. *In vivo,* the administration of 1,25(OH)$_2$-vitamin D is reported to increase bone mineralization in the osteomalacia of vitamin D deficiency (Nagant de Deuxchaisnes *et al.,* 1979; Drezner *et al.,* 1981), but other investigators have reported conflicting results (Bordier *et al.,* 1978). When studied *in vitro,* the actions of 1,25(OH)$_2$-vitamin D are equally controversial. The major effect of this sterol has been to increase bone resorption (Frolik and DeLuca, 1972; Reynolds *et al.,* 1974). Other studies indicate an effect on bone formation; thus, Lieberherr *et al.* (1981) reported an increase in bone acid and alkaline phosphatases, and Manolagas *et al.* (1981) suggested that the osteoblastic production of alkaline phosphatase is increased as well. The injection of 1,25(OH)$_2$-vitamin D into eels with a high bone turnover promotes osteoblastic activity (Lopez *et al.,* 1980), and the synthesis of a vitamin K-dependent bone protein by rat osteosarcoma cells is enhanced by the addition of 1,25(OH)$_2$-vitamin D to the incubation medium (Price and Baukol, 1980).

The relationship between low serum levels of 1,25(OH)$_2$-vitamin D and the bone disease associated with TPN is unclear. The findings of Teitelbaum *et al.* (1980), which suggest that the absence of both PTH and vitamin D in patients with renal failure may predispose such patients to the development of osteomalacia, may be pertinent to the situation with TPN. A trial of parenteral 1,25(OH)$_2$-vitamin D$_3$ to evaluate its effects on the skeletal lesion may be indicated, since patients who receive treatment with TPN almost invariably exhibit little or no intestinal absorption of fat because of intestinal resection or severe chronic disease. Thus, oral administration of 1,25(OH)$_2$-vitamin D would not be expected to be followed by adequate absorption of the sterol.

6.4. Does Vitamin D$_2$ Itself Contribute to the Bone Disease of TPN Patients?

An equally puzzling question is whether the parenteral administration of vitamin D$_2$ itself contributes to the pathogenesis of the bone disease seen in patients treated with TPN. Shike *et al.* (1980) reported that the withdrawal of vitamin D$_2$ from the TPN solution was associated with symptomatic improvement, a reduction in urinary calcium excretion, and a decrease in serum calcium levels in three patients. In a more detailed report, Shike *et al.* (1981) reported the results of withdrawal of vitamin D_2 in 11 TPN patients, three of whom had bone pain and vertebral fractures and eight of whom had no skeletal symptoms. These patients had received treatment with TPN for 14 to 109 months. The biochemical features and bone biopsies were evaluated just before withdrawal of vitamin D_2 from the TPN solution and again six months later. There were no significant changes in the urinary excretion of calcium or phosphorus or in the serum levels of calcium or phosphorus in the symptomatic or asymptomatic patients. Plasma levels of 25(OH)-vitamin D fell somewhat, while serum levels of 1,25(OH)$_2$-vitamin D and iPTH levels, which were initially low, rose slightly but not significantly. Also, there were no significant increments in bone mineral content as measured by neutron activation. In the three patients with skeletal symptoms, urinary calcium and phosphorus did decrease, although the hypercalciuria persisted. Serum levels of 25(OH)-vitamin D were normal in these patients, and it was hypothesized that vitamin D_2 itself or some other metabolite somehow was associated with increased unmineralized bone. These investigators also contended that the low serum levels of 1,25(OH)$_2$-vitamin D played little role in the pathogenesis of the bone disease, inasmuch as there was a reduction in unmineralized osteoid during the time that vitamin D_2 was withheld and while serum levels of 1,25(OH)$_2$-vitamin D remained relatively constant. Moreover, the serum levels of 25(OH)-vitamin D, which were not above normal initially, fell slightly following the removal of parenteral vitamin D_2; this indicates that increased quantities of 25(OH)-D could not be responsible for the abnormal mineralization of bone. The levels of 24,25(OH)$_2$-vitamin D were not measured, and whether this sterol could have played a role in the pathogenesis of the bone disease is unknown.

The results of bone biopsies reported by these investigators are of interest in light of earlier observations on bone biopsies in their TPN patients. The bone biopsies reported prior to removal of vitamin D_2 generally showed an increase in osteoid tissue and decrease in the percentage of bone surface showing uptake of tetracycline. Subsequent biopsies, carried out in some of the same patients following withdrawal of vitamin D_2, exhibited an increase in the percentage of osteoid surface showing tetracycline uptake, a greater surface with bone resorption and a concomitant decrease in the osteoid index (i.e., the osteoid tissue as a percentage of trabecular volume). These data were interpreted as showing change with prolonged TPN treatment from a "hyperkinetic" state to osteomalacia and then a tendency for the bone to show increased "turnover" after removal of vitamin D_2. No double tetracycline labeling was done for the determination of such dynamic measurements and therefore this conclusion should be regarded as tentative.

7.   Trace Elements in the Pathogenesis of TPN Bone Disease

The development of metabolic bone disease following initiation of TPN treatment and improvement following withdrawal of such treatment and resumption of oral intake is consistent with either the administration of a toxic substance or the development of a deficiency state that affects bone metabolism. For this reason, the levels of various trace elements in serum and in the TPN solution were determined. Analysis of the TPN solutions utilizing x-ray fluorescence revealed quantities of cadmium, strontium, and lead which were less than 100 μ g/L (Klein et al.,1982a). Copper and zinc have been routinely added to the TPN solutions and serum levels of these trace substances were normal (Klein et al., 1980a; Shike et al., 1980). Decreased growth in zinc deficiency (Sandstead, 1981) and demineralization of bone in clinical copper deficiency (Aspin and Sass-Kortsak, 1981) have been reported. None of the patients exhibited typical signs of deficiencies of either copper or zinc. Manganese deficiency can impair skeletal development (Mena, 1981) and small amounts of manganese have been added to the TPN solutions used in Toronto (Shike et al., 1980).

7.1.   Aluminum Loading During Long-Term TPN as a Factor in Bone Abnormalities

Significant quantities of aluminum in the plasma, urine, and bone of patients treated with long-term total parenteral nutrition have been detected: the casein hydrolysate component of the TPN solution is the primary source for this aluminum (Klein et al., 1982a) (Table VI). Evidence for aluminum accumulation as a factor in metabolic bone disease has been developed primarily from experience with uremic patients undergoing long-term hemodialysis. The use of aluminum-contaminated water for the preparation of dialysate in conjunction with the oral intake of aluminum-containing gels for intestinal phosphate binding have been associated with a high incidence of osteomalacic bone disease in uremic patients (Alfrey, 1981). More direct evidence implicating aluminum in the pathogenesis of bone disease is the demonstration that injections of aluminum chloride into rats is associated with impaired mineralization of the growth cartilage (Ellis et al., 1979).

In uremic patients exposed to aluminum, the aluminum is localized along the calcification front, as determined by histochemical methods (Maloney et al., in press; Ihle et al., 1981) or by electron probe (Cournot-Witmer et al., 1981). Also, it has been found that the areas showing aluminum accumulation along the calcification front generally display no uptake of tetracycline (Maloney et al., in press), an observation which suggests that aluminum may interfere with mineralization of new bone. A

Table VI. Aluminum Content of Plasma, Urine and Bone in TPN Patients
          Receiving Casein Hydrolysate

	Plasma (µg/L)	Urine (µg/day)	Bone (mg/kg dry weight)
TPN Patients:	134±14[a]	1264±196	109±36
Normal Values	<10	<27	<10

[a] Mean ± SEM

Adapted from Klein et al.(1982[a])

similar aluminum accumulation in the bone of patients treated with total
parenteral nutrition suggests that aluminum may contribute to the diminished mineralization. Since substantial quantities of aluminum are found
in casein hydrolysates, the use of this source of protein has been discontinued. Whether aluminum had a role in skeletal disease in TPN patients
receiving casein hydrolysate is unknown. Also, it is unclear whether aluminum contributed to the bone disease reported in Toronto. The use of
casein hydrolysate was discontinued there in 1976 and it is not known how
long increased quantities of aluminum remain in bone.

    The relationship between aluminum accumulation and alterations in
parathyroid hormone and vitamin D levels is uncertain. Aluminum has
been found to accumulate preferentially in the parathyroid glands (Cann
et al., 1979) and it has been shown that very large concentrations of aluminum can diminish the secretion of PTH by parathyroid cells in vitro
(Morrissey et al., 1983). Further, it has been shown that aluminum can
decrease the generation of both alkaline and acid phosphatases by tissue
cultures of bone (Lieberherr et al., 1981). It is possible that aluminum
may suppress $1,25(OH)_2$-vitamin D levels, as has been suggested for lead
(Rosen et al., 1980); however, there are no data to support this postulate.

8.   Hypercalciuria in Patients Treated with Total Parenteral Nutrition

Significant hypercalciuria has been a consistent observation in patients on
long-term treatment with total parenteral nutrition (Klein et al., 1981c;
Shike et al., 1980). In many patients, urinary calcium exceeded calcium
intake, resulting in negative calcium balance. In other cases, urinary calcium was slightly lower than intake, but increased quickly with any increment in the quantity of calcium administered.

Several factors known to augment urinary calcium are present during TPN treatment. The renal tubular reabsorption of calcium decreases with consequent increases in urinary calcium as a result of: (1) physiologic hypoparathyroidism; (2) the administration of large quantities of glucose (Lindeman *et al.*, 1967; Lemann *et al.*, 1970; DeFronzo *et al.*, 1976); (3) saline infusions (Lindeman *et al.*, 1967); and (4) the administration of amino acids (Lutz and Linkswiler, 1981). Preliminary results from the administration of TPN solutions varying in glucose concentration from 100 to 600 mg/day showed no relationship between glucose load and urinary calcium (Klein *et al.*, 1980c). Other preliminary data indicated that a significant reduction in the quantity of protein or the removal of calcium from the TPN solution led to a significant decrease in urinary calcium. The reintroduction of either calcium or protein into the TPN solution caused an immediate increase in urinary calcium to the pre-existing level. Thus, calcium and protein, independently of each other, appear capable of increasing urinary calcium excretion. Similar findings on the effect of varying the calcium level in the TPN solution were reported in a preliminary communication by Sloan and Brennan (1980). Hence it appears that a number of factors which reduce the tubular reabsorption of calcium exist during infusion of TPN solutions, but which of these factors are paramount remains uncertain.

Moreover, it is unclear whether the large losses of calcium in the urine contribute to the pathogenesis of the skeletal disease seen in TPN patients. Sizeable quantities of calcium could appear in the urine as a consequence of decreased formation and turnover of bone with greater amounts of calcium filtered at the glomerulus; under such circumstances, an increase in urinary calcium could arise as a consequence of, rather than a cause of, decreased bone mineralization. It is also possible that a factor which reduces renal tubular calcium reabsorption, producing a renal tubular "leak" of calcium, might lead to deprivation of body calcium and thereby potentiate bone disease. In these circumstances, one would expect low serum calcium and increased serum levels of iPTH, findings that have not been seen in patients treated with total parenteral nutrition. Persistent hypercalciuria might increase the incidence of nephrolithiasis in patients under long-term TPN therapy (Adelman *et al.*, 1977).

9. Is There a Circulating Inhibitor of Bone Mineralization?

Another possible mechanism for the production of bone disease could be the presence of a circulating inhibitor of mineralization. Inorganic pyrophosphate is a naturally occurring compound that, in high concentrations, can inhibit both the formation and dissolution of bone crystals and thereby reduce bone formation and resorption. Plasma levels of pyrophosphate are elevated in patients with vitamin D deficient osteomalacia (Russell *et al.*, 1971), and ethane-1-hydroxy-1,1-diphosphonate (EHDP), an analog of pyrophosphate which is not degraded by pyrophosphatases, can impair bone mineralization. Plasma pyrophosphate levels in our patients were not significantly different from those of control subjects (McGuire *et al.*, 1982), suggesting that circulating pyrophosphate levels probably do not play a role in the pathogenesis of the bone disease, although the local accumulation of pyrophosphate in various tissues, including bone, cannot be ruled out.

10. Summary

Patients receiving long-term treatment with total parenteral nutrition often develop bony abnormalities characterized by patchy osteomalacia and low bone turnover. The patients present evidence of physiologic hypoparathyroidism, although low levels of iPTH cannot entirely explain the osteomalacia. Abnormally low serum levels of $1,25(OH)_2$-vitamin D have been demonstrated, but the significance of these reduced levels in the pathogenesis of the bone lesions is not defined. Aluminum has been detected in large quantities in the plasma, urine, and bone of some patients treated with TPN, and there is mounting evidence that aluminum may be associated with skeletal pathology, particularly osteomalacia. There is, however, no clear documentation that aluminum accumulation produces the skeletal lesions observed, although it could be a contributing factor.

There has been the unusual empiric observation that the removal of vitamin D_2 from the infusate is associated with a decrease in the quantity of unmineralized osteoid in TPN patients. A possible role of vitamin D_2 in producing osteomalacia is not easy to understand since normal serum levels of $25(OH)-D_2$, the circulating form of vitamin D_2, have been reported.

The long-term consequences of intravenous nutritional support for many aspects of metabolism remain unknown. Administration into the systemic circulation of predetermined quantities of calcium and phosphorus via a route that bypasses their passage across the intestinal mucosa, the portal system and the liver may have adverse consequences. It is possible that bypassing homeostatic mechanisms may affect bone formation and metabolism or lead to alterations in vitamin D sterols. Alternatively, a deficiency of an essential trace metal or the accumulation of a toxic trace substance could be responsible for the bony abnormalities.

Much remains to be clarified concerning calcium homeostasis and bone disease during total parenteral nutrition. Among various possible factors, it seems likely that the significance of the low levels of 1,25(OH)$_2$-vitamin D and of the accumulation of aluminum in this condition will soon be clarified.

11. Acknowledgments

The authors are grateful for the contributions made by the following individuals: Allen Alfrey, M.D., Marvin Ament, M.D., June Bishop, B.A., Merrie Burke, R.N., Thomas Hazlet, Pharm. D., Ronald Horst, Ph.D., Norma Maloney, Ph.D., Susan Ott, M.D., Anthony Norman, Ph.D., Donald Sherrard, M.D., and Eduardo Slatopolsky, M.D. Barbara Stabnow provided secretarial and editorial assistance. Supported, in part, by USPHS grants AM14750, AM29926, RR865 and by funds from the Veterans Administration.

References

Abumrad, N.N., Schneider, A.J., Steele, D., and Rogers, L.S., 1981, Amino acid intolerance during prolonged total parenteral nutrition (TPN) reversed by molybdenum, *Am. J. Clin. Nutr.* **23**:2551.

Adelman, R.D., Abern, S.B., Merten, D., and Halsted, C.R., 1977, Hypercalciuria with nephrolithiasis: a complication of total parenteral nutrition, *Pediatrics* **59**:473.

Alfrey, A.C., 1981, Aluminum and Tin, in: *Disorders of Mineral Metabolism* (F. Bronner and J.W. Coburn, eds.), Vol. I, pp 353-368, Academic Press, New York.

Aspin, N., and Sass-Kortsak, A., 1981, Copper, in: *Disorders of Mineral Metabolism* (F. Bronner and J.W. Coburn, eds.), Vol. I, pp. 60-86, Academic Press, New York.

Bordier, P., Zingraff, J., Gueris, J., Jungers, P., Marie, P., Pechet, M., and Rasmussen, H., 1978, The effect of 1α(OH)-D$_3$ and 1α,25(OH)$_2$-D$_3$ on the bone in patients with renal osteodystrophy, *Amer. J. Med.* **64**:101.

Broviac, J.W., and Scribner, B.H., 1974, Prolonged parenteral nutrition in the home, *Surg. Gynecol. Obstet.* **139**:24.

Byrne, W.J., Ament, M.E., Burke, M., and Fonkalsrud, E., 1979, Home parenteral nutrition, *Surg. Gynecol. Obstet.* **149**:593.

Cann, C.E., Prussin, S.G., and Gordan, G.S., 1979, Aluminum uptake by the parathyroid glands, *J. Clin. Endocrinol. Metab.* **49**:543.

Cannon, R.A., Byrne, W.J., Ament, M.E., Gates, B., O'Connor, M., and Fonkalsrud, E.W., 1980, Home parenteral nutrition in infants, *J. Pediatr.* **96**:1098.

Cournot-Witmer, G., Zingraff, J., Plachot, J.J., Escaig, F., Lefevre, R., Boumati, P., Bordeau, A., Garabedian, M., Galle, P., Bourdon, R., Drueke, T., and Balsan, S., 1981, Aluminum localization in bone from hemodialyzed patients: Relationship to matrix mineralization, *Kidney Int.* **20**:375.

Dahms, B.B., and Halpin, T.C., 1981, Serial liver biopsies in parenteral nutrition-associated cholestasis of early infancy, *Gastroenterology* **81**:136.

DeFronzo, R.A., Goldberg, M., and Agus, Z., 1976, The effects of glucose and insulin on renal electrolyte transport, *J. Clin. Invest.* **58**:83.

Drezner, M.K., Lyles, K.M., and Harrelson, J.M., 1981, Vitamin D resistant osteomalacias: evaluation of vitamin D metabolism and response to therapy, in: *Hormonal Control of*

Calcium Metabolism (D.V. Cohn, R.V. Talmadge, and J.L. Matthews, eds.), pp. 243-251, Excerpta Medica, Amsterdam.

Eisman, J.A., Prince, R.L., Wark, J.D., and Moreley, J.M., 1979, Modulation of plasma 1,25-dihydroxyvitamin D in man by stimulation and suppression tests, *Lancet* 2(8149):931.

Ellis, H.A., McCarthy, J.H., and Herrington, J., 1979, Bone aluminum in hemodialyzed patients and in rats injected with aluminum chloride: relationship to impaired bone mineralization, *J. Clin. Path.* 32:832.

Frolik, C.A., and DeLuca, H.F., 1972, Metabolism of 1,25-dihydroxycholecalciferol in the rat, *J. Clin. Invest.* 51:2900.

Gerber, A., Raab, A.P., and Sobel, A.E., 1954, Vitamin D poisoning in adults: with description of a case, *Am. J. Med.* 16:729.

Haussler, M.R., 1980, Clinical defects in vitamin D metabolism, in: *Vitamin D: Molecular Biology and Clinical Nutrition* (A.W. Norman, ed.), pp. 603-634, Marcel Dekker, New York.

Haussler, M.R., Baylink, D.J., Hughes, M.R., Brumbaugh, P.F., Wergedal, G.E., Shen, F.H., Nilsen, R.L., Counts, S.J., Barsac, K.M., and McCain, T.A., 1976, The assay of $1\alpha,25$-dihydroxyvitamin D_3: physiologic and pathologic modulation of circulating hormone levels, *Clin. Endocrinol.* (London) 5:151S.

Henry, H.L., Midgett, R.L., and Norman, A.W., 1974, Regulation of 25-hydroxyvitamin D_3-1 alpha hydroxylase in vivo, *J. Biol. Chem.* 249:785.

Hughes, M.R., Brumbaugh, P.F., Haussler, M.R., Wergedal, J.E., and Baylink, D.J., 1975, Regulation of serum $1\alpha,25$-dihydroxyvitamin D_3 by calcium and phosphate in the rat, *Science* 190:578.

Ihle, B., Buchanan, M., Plomley, R., Stevens, B., d'Apice, A., and Kincaid-Smith, P., 1981, Histology in dialysis osteomalacia(OM) secondary to aluminum toxicity, *Kidney Int.* 19:149 (abstract).

Jeejeebhoy, K.N., Zohrab, W.J., Langer, B., Phillips, M.J., Kuksis, A., and Anderson, G.H., 1973, Total parenteral nutrition at home for 23 months, without complication and with good rehabilitation, *Gastroenterology* 65:811.

Jeejeebhoy, K.N., Langer, B., Tsallas, G., Chu, R.C., Kuksis, A., and Anderson, G.H., 1976, Total parenteral nutrition at home: studies in patients surviving 4 months to 5 years, *Gastroenterology* 71:943.

Johnson, J.D., Albritton, W.L., and Sunshine, P., 1972, Hyperammonemia accompanying parenteral nutrition in newborn infants, *J. Pediatr.* 81:154.

Karpel, J.T., and Peden, V.H., 1972, Copper deficiency in long-term parenteral nutrition, *J. Pediatr.* 80:32.

Klein, G.L., Targoff, C.M., Ament, M.E., Sherrard, D.J., Bluestone, R., Young, J.H., Norman, A.W, and Coburn, J.W., 1980a, Bone disease associated with total parenteral nutrition, *Lancet* 2(8203):1041.

Klein, G.L., Cannon, R.A., Ament, M.E., Norman, A.W., and Coburn, J.W., 1980b, Rickets and osteopenia with normal 25-hydroxyvitamin D levels in infants on parenteral nutrition, *Clin. Res.* 28:596A (abstract).

Klein, G.L., Ament, M.E., Norman, A.W., and Coburn, J.W., 1980c, Urinary mineral excretion in patients on parenteral nutrition; effect of varying glucose concentration, *Clin. Res.* 28:397A (abstract).

Klein, G.L., Horst, R.L., Norman, A.W., Amend, M.E., Slatopolsky, E., and Coburn, J.W., 1981a, Reduced serum levels of 1-alpha,25-dihydroxyvitamin D during long-term total parenteral nutrition, *Ann. Intern. Med.* 94:638.

Klein, G.L., Horst, R.L., Slatopolsky, E., Ament, M.E., and Coburn, J.W., 1981b, Vitamin D bioactivation and parathyroid hormone levels during total parenteral nutrition: effect of calcium withdrawal, *Calcif. Tissue Int.* 33:337 (abstract).

Klein, G.L., Ament, M.E., and Coburn, J.W., 1981c, Hypercalciuria during total parenteral nutrition: roles of calcium and protein, *Clin. Res.* **29**:467A (abstract).

Klein, G.L., Alfrey, A.C., Miller, N.L., Sherrard, D.J., Hazlet, T.K., Ament, M.E., and Coburn, J.W., 1982a, Aluminum loading during total parenteral nutrition, *Am. J. Clin. Nutr.* **35**:1425.

Klein, G.L., Cannon, R.A., Diament, M., Kangarloo, H., Ament, M.E., Norman, A.W., and Coburn, J.W., 1982b, Infantile vitamin D-resistant rickets associated with total parenteral nutrition, *Am. J. Dis. Child.* **136**:74.

Koretz, R.L., 1981, Hypouricemia— a transient biochemical phenomenon of total parenteral nutrition, *Am. J. Clin. Nutr.* **34**:2493.

Lawson, D.E.R., 1980, Metabolism of Vitamin D, in: *Vitamin D: Molecular Biology and Clinical Nutrition* (A.W. Norman, ed.), pp. 117-118, Marcel Dekker, New York.

Lemann, J., Jr., Lennon, E.J., Piering, W.R., Prien, E.L., Jr., and Ricanati, E.S., 1970, Evidence that glucose ingestion inhibits net renal tubular reabsorption of calcium and magnesium in man, *J. Lab. Clin. Med.* **75**:578.

Lieberherr, M., Pezant, E., Garabedian, M., and Balsan, S., 1977, Phosphatase content of rat calvaria after in vivo administration of vitamin D₃ metabolites, *Calcif. Tissue Res.* **23**:235.

Lieberherr, M., Grosse, B., Cournot, G., and Balsan, S., 1981, *In vitro* action of aluminum chloride on bone phosphatase activities, in: *Hormonal Control of Calcium Metabolism* (D.V. Cohn, R.V. Talmadge, and J.L. Matthews, eds.), p. 408, Excerpta Medica, Amsterdam (abstract).

Lindeman, R.D., Adler, S., Yiengst., M.J., and Beard, E.S., 1967, Influence of various nutrients on urinary divalent cation excretion, *J. Lab. Clin. Med.* **70**:236.

Lopez, E., MacIntyre, I., Martelly, E., Lallier, F., and Vidal, B., 1980, Paradoxical effect of 1,25-dihydroxycholecalciferol on osteoblastic and osteoclastic activity in the skeleton of the eel Anguilla anguilla L. *Calcif. Tissue Int.* **32**:83.

Lutz, J., and Linkswiler, H.M., 1981, Calcium metabolism in postmenopausal and osteoporotic women consuming two levels of dietary protein, *Am. J. Clin. Nutr.* **34**:2178.

Maloney, N.A., Ott, S., Miller, N., Alfrey, A., Coburn, J.W., and Sherrard, D.J., Histological quantitation of aluminum in iliac bone from patients with renal failure. (In press).

Manolagas, S.C., Burton, D.W., and Deftos, L.J., 1981, 1,25-dihydroxyvitamin D₃ stimulates the alkaline phosphatase activity of osteoblast-like cells, *J. Biol. Chem.* **256**:7115.

Mason, R.S., Lissner, D., Posen, S., and Norman, A.W., 1980, Blood concentrations of dihydroxylated vitamin D metabolites after an oral dose, *Br. Med. J.* **1(6212)**:449.

McGuire, M.K.B., Klein, G.L., Russell, R.G.G., Ament, M.E., and Coburn, J.W., 1982, Plasma pyrophosphate levels in osteomalacia, *Kidney Int.* **21**:230 (abstract).

Mena, I., 1981, Manganese, in: *Disorders of Mineral Metabolism* (F. Bronner and J.W. Coburn, eds.), Vol. I pp. 234-264, Academic Press, New York.

Mock, D.M., DeLorimer, A.A., Liebman, W.M., Sweetman, L., and Baker, H., 1981, Biotin deficiency: an unusual complication of parenteral alimentation, *N. Engl. J. Med.* **304**:820.

Morrissey, J., Rothstein, M., Mayor, G., and Slatopolsky, E., 1983, Suppression of parathyroid hormone release by aluminum, *Kidney Int.* **23**:699.

Nagant de Deuxchaisnes, C., Rombouts-Lindemans, C., Huaux, J.P., Withofs, H., and Meerssman, F., 1979, Healing of vitamin D-deficient osteomalacia by the administration of 1,25(OH)₂D₃, in: *Molecular Endocrinology* (I. MacIntyre and M. Szelke, eds.), pp. 375-404, Elsevier, New York.

Paulsrud, J.R., Pensler, L., Whitten, C.F., Stewart, S., and Holman, R.T., 1972, Essential fatty acid deficiency in infants by fat-free intravenous feedings, *Am. J. Clin. Nutr.* **25**:897.

Postuma, R., and Trevenen, C.L., 1979, Liver disease in infants receiving total parenteral nutrition, *Pediatrics* **63**:110.

Price, P.A., and Baukol, S., 1980, 1,25-dihydroxyvitamin D_3 increases synthesis of vitamin K-dependent bone protein by osteosarcoma cells, *J. Biol. Chem.* **255**:11660.

Reynolds, J.J., Holick, M.F., and DeLuca, H.F., 1974, The role of vitamin D metabolites in bone resorption, *Calcif. Tissue Res.* **15**:333.

Rodgers, B.M., Hollenbeck, J.I., Donnelly, W.H., Talbert, J.L., 1976, Intrahepatic cholestasis with parenteral alimentation, *Am. J. Surg.* **131**:149.

Rosen, J.F., Chesney, R.W., Hamstra, A., DeLuca, H.F., and Mahaffey, K.R., 1980, Reduction in 1,25-dihydroxyvitamin D in children with increased lead absorption, *N. Engl. J. Med.* **302**:1128.

Roslyn, J.J., Pitt, H.A., Mann, L.L., Ament, M.E., and Denbesten, L., 1981, Long-term total parenteral nutrition induces gall bladder disease, *Gastroenterology* **81(2)**:1264 (abstract).

Russell, R.G.G., Bisaz, S., Donath, A., Morgan, D.B., and Fleisch, H., 1971, Inorganic pyrophosphate in plasma in normal persons and in patients with hypophosphatasia, osteogenesis imperfecta, and other disorders of the bone, *J. Clin. Invest.* **50**:961.

Sandstead, H.H., 1981, Zinc in human nutrition, in: *Disorders of Mineral Metabolism* (F. Bronner and J.W. Coburn, eds.), Vol. I, pp. 94-140, Academic Press, New York.

Sherrard, D.J., Baylink, D.J., Wergedal, J.E., and Maloney, N., 1974, Quantititative histological studies on the pathogenesis of uremic bone disease, *J. Clin. Endocrinol. Metab.* **39**:119.

Shike, M., Harrison, J.E., Sturtridge, W.T., Tam, C.S., Bobechko, P.E., Jones, G., Murray, T.M., and Jeejeebhoy, K.N., 1980, Metabolic bone disease in patients receiving long-term total parenteral nutrition, *Ann. Intern. Med.* **92**:343.

Shike, M., Sturtridge, W.C., Tam, C.S., Harrison, J.E., Jones, G., Murray, T.M., Husdan, H., Whitwell, J., Wilson, D.R., and Jeejeebhoy, K.N., 1981, A possible role of vitamin D in the genesis of parenteral nutrition induced metabolic bone disease, *Ann. Intern. Med.* **95**:560.

Slatopolsky, E., Gray, R., Adams, N.D., Lewis, J., Hruska, K., Martin, K., Klahr, S., DeLuca, H., and Lemann, J., 1978, Low serum levels of $1,25(OH)_2D_3$ are not responsible for the development of secondary hyperparathyroidism in early renal failure, *Kidney Int.* **14**:733 (abstract).

Slatopolsky, E., Hruska, K., Martin, K., and Freitag, J., 1979, Physiologic and metabolic effects of parathyroid hormone, in: *Hormonal Function and the Kidney* (B.M. Brenner and J.H. Stein, eds.), pp. 169-192, Churchill Livingstone, New York.

Sloan, G.M., and Brennan, M.F., 1980, Positive calcium balance in patients receiving total parenteral nutrition, *Clin. Res.* **28**:621A (abstract).

Strobel, C.T., Byrne, W.J., Abramovits, W., Newcomer, V.J., Bleich, R., 1978, Zinc deficiency and parenteral nutrition, *Int. J. Dermatol.* **17**:575.

Tanaka, Y., and DeLuca, H.F., 1973, The control of 25-hydroxyvitamin D metabolism by inorganic phosphorus, *Arch. Biochem. Biophys.* **154**:566.

Targoff, C.M., Coburn, J.W., Ament, M.E., Byrne, W.J., Miller, D.G., Sherrard, D.J., Brickman, A.S., and Bluestone, R., 1979, Bone disease associated with total parenteral nutrition, a new syndrome, in: *Vitamin D: Basic Research and its Clinical Application* (A.W. Norman, K. Schaeffer, D. von Herrath, H.G. Grigoleit, J.W. Coburn, H.F. DeLuca, E.B. Mawer, and T. Suda, eds.), pp. 1171-1172, DeGruyter, Berlin.

Teitelbaum, S.L., Bergfeld, M.A. Freitag, J., Hruska, K.A., and Slatopolsky, E., 1980, Do parathyroid hormone and 1,25-dihydroxyvitamin D modulate bone formation in uremia? *J. Clin. Endocrinol. Metab.* **51**:247.

Travis, S.F., Sugerman, H.J., Rubert, R.L., Dudrick, S.J., Delivoria-Papadopoulos, M., Miller, L.D., and Oski, F.A., 1971, Alterations of red cell glycolytic intermediates and oxygen transport as a consequence of hypophosphatemia in patients receiving intravenous hyperalimentation, *N. Engl. J. Med.* **285**:763.

Vedig, A.E., Phillips, P.J., Munday, L., Peisach, A.R., Hartley, T.F., and Nee, A.G., 1981, Effects of parenteral nutrition on urate metabolism, presented at the Australian Conference on Bone and Mineral Metabolism, March 1981.

Wilmore, D.W., and Dudrick, S.J., 1968, Growth and development of an infant receiving all nutritional requirements by vein, *J. Am. Med. Assoc.* **203**:860.

Chapter 5

Nutrition and Protein Turnover in Man

P.J. Reeds and P.J. Garlick

1. Introduction

The healthy young adult synthesizes and degrades at least 200 g of tissue protein each day (Waterlow *et al.*, 1978a). By comparison, the dietary intake of protein is typically only one-third of this value. Measurements in man and in a number of other species have shown that there is an approximate parallelism between the rate of protein turnover by the whole body and a number of other metabolic parameters, such as the requirement for dietary protein (Young *et al.*, 1975) and basal energy expenditure (Munro, 1964; Waterlow, 1968). Indeed, the expenditure of metabolic energy is required for the formation of peptide bonds; at a conservative estimate, the energy cost of protein synthesis accounts for 20% of oxygen consumption.

 Our main interest in protein turnover, however, lies with its relationship to the gain or loss of protein from body tissues during growth or in pathological states. This is illustrated in Fig. 1, which emphasizes the important fact that growth (or loss) of protein mass can occur by a variety of mechanisms. Of these, the simplest are either an increase in the rate of protein synthesis or a decrease in protein breakdown, but clearly combinations of these two extremes are equally likely. The significance of

P. J. Reeds • Protein Biochemistry Department, Rowett Research Institute, Bucksburn, Aberdeen, Scotland AB2 9SB. P. J. Garlick • London School of Hygiene and Tropical Medicine, Clinical Nutrition and Metabolism Unit, Hospital for Tropical Diseases, 4 St. Pancras Way, London NW1.

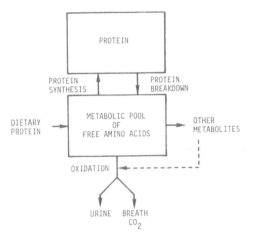

Fig. 1. An illustration of how different combinations of changes
in rates of protein synthesis and breakdown can result in the same
loss of protein from the body. Rates of synthesis and breakdown
are normally in balance (A). Negative protein balance can result
from an increase in breakdown (B), a decrease in synthesis (C) or
from combinations of changes in synthesis and breakdown (D,E,F).

the two extremes, however, is that protein synthesis and protein breakdown are discrete processes occurring in different parts of the cell, and
even though each has the capacity to alter the protein content, they may
be regulated in different ways.

A third process, which is also important to protein balance, is the
oxidation of amino acids. At present, though, we do not know if the rate
of oxidation is controlled independently of protein synthesis and degradation or whether it is merely a system for removing an excess of amino
acids after synthesis and breakdown have already determined the uptake
or loss from body protein. The interrelationships between synthesis,
breakdown, oxidation and dietary intake are shown in simplified form in
Fig. 2. It is generally assumed that the fifth process shown in the figure
(conversion to other metabolites such as nucleic acids, hemoglobin and
hormones), can be neglected because its rate is small compared with the
rates of the other processes. However, the validity of this assumption has
considerable bearing on the choice of labeled amino acids for quantitating
protein metabolism.

These general relationships were first recognized by Shoenheimer and
his colleagues when they undertook experiments with stable isotopes in
man in the 1940's, but only in the last 15 years has there been an acceleration of interest. This has apparently been related tó the increased
awareness of the role of nutrition in maintaining body functions, both in

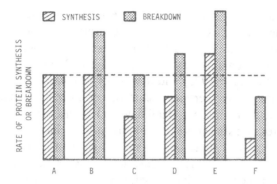

Fig. 2. Simple model to represent protein metabolism in the whole body. In the analysis of isotope labelling experiments the following letters will be used to denote the rates of protein synthesis (Z), breakdown (B), intake from diet (I) and oxidation (or N excretion) (E).

health and disease (e.g., Waterlow, 1968). Our purpose in this review is to describe what is known of protein turnover in man and some of the factors that influence it. We will concentrate particularly on the influence of nutrition. We feel that this is of primary importance, since dietary intake, or lack of it, is a variable which is almost certain to influence the outcome and interpretation of studies of any other factors (e.g., age, trauma).

We have, however, started by describing in some detail the methods that are most commonly used to measure protein turnover in man. These have been reviewed previously (e.g., Waterlow *et al.*, 1978a; Garlick, 1980a, 1980b), but there have been further developments. Continual regard for the accuracy of these techniques is necessary, not only because of the difficulties of finding adequate models and probes, but also because of the limitations of sampling. When, even in extreme instances, there are only two possible alternative results (a change in synthesis or in breakdown), any method which is not capable of detecting this distinction is not worth pursuing.

2.   Methods

2.1.   General Considerations

The search for methods for quantitating human protein metabolism has been governed very largely by what is practicable within the limitations of experiments on man. Sampling is a major difficulty, being restricted mostly to blood and urine. Hence, most of the techniques we will consider will be concerned with *whole-body protein turnover,* which can be determined from measurements on blood and/or urine only. Blood samples also can be used to measure plasma protein turnover, but there is already an extensive literature on this (e.g., Rothschild and Waldmann, 1970; Wolstenholm and O'Connor, 1973) and it will not be considered in detail here. It is also possible under certain circumstances to remove biopsy samples of tissue, particularly muscle (Halliday and McKeran, 1975), and this enables the rates of tissue protein synthesis to be estimated.

A second practical limitation in studies on man is the time required for measurements to be made. Apart from theoretical limitations (e.g., with regard to isotope recycling), methods must be compatible with the anticipated time scale of observed changes. For example, the method developed in the early 1950's by Olesen *et al.* (1954), which required at least 10 days to derive a valid answer, was used to measure changes in turnover during 10 days of immobilization (Shønheyder *et al.,* 1955). This clearly would be inappropriate for a study of the short-term responses to meal feeding. In fact, recent developments have concentrated upon methods which take only a few hours, both to enable short-term fluctuations in dietary intake to be accounted for and also to minimize the inconvenience for both the subject and the investigator. This latter point is crucially important if valid results are to be obtained outside the well controlled environment of the laboratory or metabolic suite (e.g., in a hospital ward).

The choice of isotope has been governed both on ethical grounds and also by the techniques available for measurement. Initial studies were with stable isotopes, e.g., ^{15}N, because they could be measured with mass spectrometers. Since it became readily available ^{14}C has been used, because radioactivity counting is relatively easy. Also, the techniques for separating and measuring $[^{14}C]$-labeled compounds are more sensitive, enabling measurements to be made more readily on metabolites in 5-10 ml of blood. The most recent development is the use of $[^{13}C]$-labeled amino acids, which avoid any potential radiation hazards to the host. Modern gas chromatograph/mass spectrometers are not only capable of measuring the isotope abundance, but do so on very small samples (e.g., $[^{13}C]$leucine in 0.1 ml plasma) (Mathews *et al.,* 1980). However, within these possibilities, the exact choice of isotope depends on the particular conditions of the individual experiment.

The models and approaches to analyzing the experimental data are very similar whichever isotope is given and whether it is given as a constant infusion or as a single dose. As discussed in detail by Waterlow *et al.* (1978a) multicompartmental analysis has been used by some investigators (e.g., Long *et al.,* 1977) but the majority of methods currently in use employ a simpler form of analysis. In this approach the turnover rates are calculated from the disappearance of isotope from the metabolic pool. This can be obtained either by direct sampling of this pool via the plasma or by inference from measurements on a metabolic end product in the urine. These two approaches, which have been termed the "precursor method" and the "end-product method" (Waterlow et al., 1978a), will be described in the following two sections.

2.2. Precursor Methods

This approach was first introduced by Waterlow (1967) who gave intravenous infusions of U-[^{14}C]lysine. It was shown that within 24 hr of infusion the specific radioactivity of free lysine in the plasma reached a constant (plateau) value, from which the rate of lysine turnover was calculated. Since then a number of modifications have been introduced, notably the change from uniformly labeled lysine to carboxyl-labeled leucine, and the collection of respiratory CO_2 to enable the rate of amino acid oxidation to be evaluated (James et al., 1974).

2.2.1. Analysis of the Model

The model shown in Fig. 2 is taken to represent the metabolism of a single amino acid in the whole-body. It is assumed that the free amino acid pool of the body is homogeneous and can be sampled by the plasma. It is also assumed that the only sources of amino acid entering this pool are from dietary protein (I) and from intracellular protein breakdown (B) (i.e., the amino acid is essential so there is no synthesis de novo). Similarly, the only pathways of removal of amino acid from the pool are by irreversible oxidation (E) and by synthesis into protein (Z). The conversion to other metabolic products is assumed to be zero. Over any period that the pool size of free amino acid stays constant, therefore, total entry must equal total exit. Hence:

$$B + I = E + Z = Q \tag{1}$$

Q has been termed the "flux," and is equal to the total turnover rate of the free amino acid pool. Its value is obtained from the plateau value for specific radioactivity of the free amino acid in the plasma (S_{pmax}) during a constant infusion of the tracer. At plateau the rates of entry and exit of tracer must be equal. Further, during the course of an experiment lasting of the order of one day or less there will be no significant reentry of label by breakdown of labeled protein. Hence:

$$i = Q.S_{pmax} \tag{2}$$

From Q, the rate of protein breakdown can be calculated by equation 1, since the value of I (the dietary intake of that amino acid) can be measured independently. Similarly, protein synthesis can be calculated from equation 1 if the rate of oxidation (E) is measured. When the amino acid is labeled with [^{13}C] or [^{14}C] this is done by collecting expired breath and assessing the rate of production of labeled CO_2, which also reaches a plateau value (e_{max}), since:

$$e_{max} = E.S_{pmax} \tag{3}$$

The rates of oxidation, protein synthesis, and protein breakdown thus calculated are all in terms of amount of the particular amino acid infused. To convert to amounts of protein, it is necessary to know the composition of that protein. For oxidation, the proportion of the amino acid in dietary protein can be used, provided the subject is reasonably close to nitrogen balance. For synthesis and breakdown, the composition of those body proteins that are turning over would be needed. Clearly, this is a potential source of error, since amino acid analysis can tell us only the average composition of body protein. However, the important point to recognize is that this is merely a constant factor which does not influence the accuracy of the basic data when expressed in terms of moles of a single amino acid.

2.2.2. Choice of Label and Amino Acid

The equation for flux rate applies to amino acid labeled with any isotope, provided that it does not undergo any exchange reactions. However, the majority of studies have used ^{14}C, or more recently ^{13}C, because this enables the oxidation pathway to be assessed from measurements on respiratory CO_2. With infusion of α-[^{15}N]lysine, for example (Halliday and McKeran, 1977; Conley et al., 1980), the label is not exchanged, but it is not possible to proportion the calculated rate of flux into its components: oxidation and protein synthesis. The criteria for choosing the most suitable amino acid are:

(1)   It is essential.
(2)   It undergoes no reactions other than oxidation to CO_2 and incorporation into protein.
(3)   The appearance of labeled CO_2 in the breath represents, quantitatively, the irreversible oxidation of the amino acid.
(4)   Plateau values for the free amino acid specific activity and for the production of labelled CO_2 should be reached as quickly as possible.

The studies by Waterlow (1967) with U-[^{14}C]lysine did not achieve criterion 3, which was not attempted. In addition the plateau was not reached until between 12 and 24 hr of infusion, resulting in inconveniently long periods of measurement. Later experiments with U-[^{14}C]tyrosine showed that plateau was reached very quickly in both plasma and CO_2 (in about 3 hr) and there was also a conveniently rapid chemical method for assaying the tyrosine specific activity (Garlick and Marshall, 1972).

However, this amino acid is not essential, and there are potentially many metabolic products other than protein and CO_2. Furthermore, during the oxidation process it is possible for labeled intermediates from the tricarboxylic acid cycle to be directed into fat, glycogen or other amino acids rather than to CO_2 (James *et al.*, 1976). These problems appear to have been largely overcome by the use of carboxyl-labeled leucine (1-[^{14}C]leucine or 1-[^{13}C]leucine) which appears to satisfy all the criteria (James *et al.*, 1974). In particular, plateau is reached in about 5 hr and release of labeled CO_2 represents leucine oxidation because decarboxylation is the first irreversible step in the oxidative pathway (James *et al.*, 1974).

2.2.3.   Method of Infusion

It has been usual to give an intravenous infusion of labeled amino acid dissolved in sterile saline over a period of 8-12 hr. However, Golden and Waterlow (1977) have shown that with 1-[^{14}C]leucine an intragastric infusion results in exactly the same values for turnover rates. With intravenous infusion of 1-[^{14}C]leucine we have typically observed a plateau in about 5 hr (Fig. 3). Infusion for a further 5 hr enables a number of samples of blood to be obtained so that a mean value can be calculated. The standard deviation about this plateau has generally been of the order of 5%, although this can be reduced if the diet is given, not as hourly meals, but as a constant intragastric infusion (e.g., Golden and Waterlow, 1977). In studies such as that illustrated in Fig. 3, in which the patient was infused for a period in excess of 24 hr to monitor diurnal variations, the 5-hr period before the plateau is attained is not a problem. However, in some recent studies the method has been adopted in which a priming dose of label is given immediately before the infusion, with the intention of reaching plateau immediately (Matthews *et al.*, 1980; Clague, 1980). With the method of Matthews *et al.* (1980) the plateau values for leucine in plasma and CO_2 in the breath were "anticipated," as were the total body pool sizes of free leucine and bicarbonate. The amounts of 1-[^{13}C]leucine and H[^{13}C]O$_3^-$ required to raise these pools to the anticipated plateau specific activities were then injected. By this means the plateau values seemed to have been achieved within 2 hr and repeated measurements were made during the next 2 hr. The variation about the plateau in the plasma was about 5% (SD) which is similar to that we have observed without the priming dose.

When it is important that an experiment is of short duration (e.g., when the patient will not tolerate a long infusion) the primed continuous infusion has clear advantages. However, for this to work the priming dose must be judged accurately or the time taken to attain the correct plateau

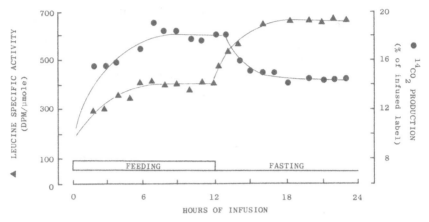

Fig. 3. The specific radioactivity of leucine in plasma and the rate of production of $^{14}CO_2$ in the breath during constant infusion of $1-^{14}C$-leucine for 24 hr into a normal adult male. The subject was fed hourly for 12 hr and then fasted. Unpublished data of G.A. Clugston and P.J. Garlick.

will not be significantly reduced by priming. It is also important that measurements are continued for long enough to be sure that plateau has indeed been reached. This could be significant if a systematic error in the priming dose caused the plateau to be approached from above with one group and from below with another. A systematic difference then might be apparent between groups which could converge to the same value with longer periods of infusion. This is less likely to occur with an unprimed infusion, since the plateau is always approached from below.

2.2.4. Assessment of Labeled CO_2 Production

As can be seen from Fig. 3, the production of labeled CO_2 in the breath reached a plateau shortly after that of leucine in the plasma. Usually this has been measured by taking frequent samples of expired breath for determination of specific activity of CO_2 by the method of Kaihara and Wagner (1968). The specific activity is then multiplied by the total CO_2 output, which is also measured at frequent intervals (James *et al.*, 1976). In our initial studies we estimated the total CO_2 output by measuring the total volume and CO_2 concentration of air collected in a Douglas bag (James *et al.*, 1976).

However, this method is prone to error, as the resistance of the valve and tubing frequently causes marked overbreathing and high rates of CO_2 output, typically 35% in excess of the steady state value (G.A. Clugston and P.J. Garlick, unpublished). Because of the large size of the body

Table I. The Recovery of Label During a Constant Infusion of $H[^{14}C]O_3^-$

Infusion time	State	Number of observations	Percent of dose recovered ± SD
0-12 hr	Fed	10	88.2 ± 4.9
12-24 hr	Fasted	10	82.6 ± 6.1
24-36 hr	Refed	6	91.3 ± 2.9

The specific activity of expired CO_2 was measured by the method of Kaihara & Wagner (1968) and total CO_2 production was measured by samples taken from a ventilated tent. Unpublished data of Clugston & Garlick.

bicarbonate pool, these errors can persist over quite long periods of adaptation to this system. We have found it preferable to erect a large plastic tent (oxygen tent) of about 1000 liters volume around the patient. Air is drawn through at about 70 liters/min and the change in CO_2 concentration is measured with an infrared CO_2 analyser. With this system, movements are restricted to only a very limited extent and the patients readily accept long periods (e.g., 24 hr) of measurement (Clugston, 1980).

Another source of uncertainty is the observation by a number of groups that when labeled bicarbonate is infused, the recovery of label in the breath reaches a plateau that is lower than the rate of infusion. For example, Issekutz et al. (1968) in sheep and James et al. (1976) in man reported recoveries of only about 80%. Clague et al. (1979) found a recovery of 87%, but reported in addition an apparent 12% loss of activity which they suggested resulted from the inefficiency of counting of CO_2 in certain scintillators (Nuclear Enterprises NE 233). Our own measurements in obese patients have shown the recovery to be nearer 90%. This value was little altered during a 36-hr period of infusion, during which subjects were fed for 12 hr, then fasted for 12 hr and then refed for a further 12 hr (Table I). Furthermore, there appeared to be no difference in recovery between those who received food containing normal protein and energy and those who received a protein-free, low-energy diet (Clugston and Garlick, unpublished). The low recovery from the body can be attributed to the presence of a slowly turning over compartment of the bicarbonate pool (Tolbert et al., 1959). However, its main significance for these studies is that it must be corrected for by assuming that labeled CO_2 released by amino acid decarboxylation will behave in the same way as infused bicarbonate. Thus, in most studies an 80% or 90% correction factor has been applied to apparent oxidation rates. An alternative approach has been suggested by Clague (1980). A primed infusion of $[^{14}C]$bicarbonate was given for 1.5 hr before commencing a primed infusion of $1-[^{14}C]$leucine. At plateau during each infusion a plasma sample was taken and the total $H[^{14}C]O_3^-/ml$ was assessed. Thus, on the assumption that the $H[^{14}C]O_3^-/ml$ was proportional to the rate of entry of $[^{14}C]O_2$ into the plasma, the infusion of a known amount of $H[^{14}C]O_3^-$ could be used to calibrate the rate of production of $[^{14}C]O_2$ from leucine by making measurements on plasma rather than on expired air. This technique has the advantage that a calibration is performed on each separate subject; this can be seen to be of value from the appreciable variation among the patients shown in Table I. However, the technique works only on the assumption that total CO_2 production and plasma bicarbonate concentration are identical at the two times of measurement; in many circumstances this might be very difficult to ensure.

2.2.5. The Precursor Pool Problem

One of the original assumptions of the method is that the free amino acid pool is homogeneous and can be sampled via the blood. Experiments on animals show that this is not true in general and that intracellular specific radioactivities may vary anywhere between 10% and 100% of those in plasma, depending on the amino acid and the tissue (See Waterlow *et al.*, 1978a). For example, Simon *et al.* (1978) found that in pigs the specific activity of free leucine ranged from 22% (liver) to 100% (heart) of that of leucine in plasma, compared with 11% to 56% for free lysine. Potentially this variation could create serious error, since we do not know for certain what is the source of amino acids for protein synthesis or for amino acid oxidation. Some studies have suggested that the specific activity of the free amino acid at the site of protein synthesis is the same as that in plasma; others have indicated that it is the same as that in the tissue; and yet others that it is somewhere in between. (For review see Waterlow *et al.*, 1978a).

The problem is less serious for amino acid oxidation, since N excretion in the urine provides an independent assessment. Table II shows a comparison between leucine oxidation measured directly by infusion of $1-[^{14}C]$leucine, compared with that deduced from the excretion of N in the urine (assuming that the ratio between leucine oxidation and N excretion is the same as leucine/total N in the diet). The directly measured value was close to 70% of the value derived from N excretion. Similarly, Golden and Waterlow (1977) reported a value of 80%. This ratio does not appear to alter appreciably when subjects are fasted for short periods of time (Garlick and Clugston, 1981). In similar experiments in pigs given diets containing different amounts of protein and energy, the regression of leucine oxidation (measured directly) against N excretion passed through the origin, indicating that this relationship was constant with different diets (Fig. 4).

The data in Fig. 4 imply that the specific activity of free leucine at

Table II. Leucine Oxidation, Measured by Constant Infusion of [^{14}C]leucine for
24 hr Compared with that Calculated from Urinary N Excretion in 9
Obese Subjects

	Mean ± SD
Measured leucine oxidation (mmol/d)	40.4 ± 4.8
Urinary N excretion (g N/d)	13.1 ± 1.1
Leucine content of the diet (mmol/g N)	4.54
Calculated leucine oxidation (mmol/d)	59.7 ± 4.1
Measured/calculated (%)	67.7 ± 6.7

For details of calculation see Section 2.2.5. Data of Clugston & Garlick
(1982).

the site of leucine oxidation was some 85% of that in plasma, but we do
not know the equivalent value for leucine at the site of protein synthesis.
However, if we assume that the specific activities at the sites of oxidation
and protein synthesis are equal, we can recalculate the rates of flux and
protein synthesis. These estimates are higher than those obtained from the
specific activity in plasma. We do not know which of these two estimates
is nearer the truth, but the potential error would invalidate measurements

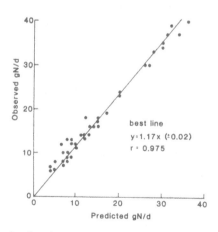

Fig. 4. Young female pigs were infused via the aorta for 6 hr with
1-^{14}C-leucine. The rate of leucine oxidation was calculated from
the product of the leucine flux and the proportion of the dose of
^{14}C excreted as CO_2. Thsi value was converted to protein on the
assumption that body protein contains (by weight) 6.7% leucine and
thence to nitrogen by dividing by 6.25. Nitrogen excretion in the
urine was measured over 3 days, the animals being offered feed
every hour. Data of Reeds and Fuller (unpublished).

of changes in synthesis or oxidation only if the ratio of the two specific activities varied with different treatments. As was shown above, with short periods of starvation or with diets of different composition this does not appear to occur. The problem occurs when breakdown is measured, since this is calculated from the flux, which is susceptible to this error, minus dietary intake, which is not. For example, this consideration altered the interpretation of an experiment contrasting the effects of feeding and fasting on whole-body protein synthesis and breakdown (Clugston and Garlick, 1982).

When plasma was used to calculate the flux, the rate of synthesis during feeding was 36% higher than during fasting and the rate of breakdown was 27% lower. When the specific activity of the metabolic pool was assumed to be 67% of that in plasma (from Table II) the change in synthesis was proportionally the same as before, but the change in breakdown was entirely eliminated. The conclusion we make is that, with this technique, measurements of protein synthesis or leucine oxidation are unlikely to be invalidated by uncertainty regarding the heterogeneity of the metabolic pool, but that rates of breakdown are more susceptible to error.

2.2.6.   The Steady State

Another important assumption is the maintenance of a steady state (i.e., that the metabolic pool size and the rates of inflow and outflow remain constant during the period the plateau is measured). Without this assumption the measurement is likely to be invalidated (Waterlow et al., 1978a). In particular this requires that food intake be maintained at a steady rate during the infusion, e.g., by giving regular meals at hourly intervals (James et al., 1976) or that the subject is totally fasted. As Fig. 3 shows, the rate of food intake can have a profound and acute effect on protein metabolism, particularly the rate of oxidation, and failure to regulate the food intake is likely to result in poor or nonexistent plateaus.

2.3.   End Product Methods

With few exceptions (e.g., Golden and Waterlow, 1977) these methods use [^{15}N]-labeled amino acids, and the behavior of the isotope in the metabolic pool is deduced from the pattern of excretion of isotope in an end product of N metabolism in the urine. The assumptions, described in detail by Waterlow et al. (1978a, 1978b), are: first, that the metabolic pool is homogeneous, so that the specific activity (isotope enrichment) at the site of protein synthesis is the same as that at the site of synthesis of the end product; second, that in general ^{15}N from an amino acid will to some extent redistribute itself among other amino acids, so that the label can be taken as a tracer for total amino acid N. The calculations are therefore very similar to those described previously for precursor methods, with the exception that rates are expressed not as amounts of an amino acid, but as amounts of N.

2.3.1.   Analysis of the Model

The model is that shown in Fig. 2, where the two pools are taken to represent the total free amino N of the body (metabolic pool) and the nitrogen in protein. In the method of Picou and Taylor-Roberts (1969) [^{15}N]glycine is infused intragastrically or intravenously. Alternatively, small doses can be given at regular intervals of 3 or 4 hr (Steffee et al., 1976). The chosen end product is urinary urea and measurements are made for 1-3 days, during which time a plateau of enrichment with [^{15}N] is reached. This plateau value and equation 2 are then used to calculate the flux. Rates of synthesis and breakdown of protein then can be calculated from equation 1 if the rates of oxidation and dietary intake of N are measured. The rate of oxidation is taken to be the rate of excretion of total N in the urine. When required, rates of N metabolism are converted to equivalent rates for protein by multiplying by 6.25.

Constant infusions, or even regular oral doses, are inconvenient over long periods, and a single dose procedure has been suggested instead (Waterlow et al., 1978b). After a single dose of [^{15}N]glycine the enrichment in urinary end products of N metabolism does not reach a plateau but rises to a peak followed by a return to baseline values. However, if the cumulative excretion of label in an end product is measured, this does reach a plateau, corresponding to the time when all the isotope has been cleared from the metabolic pool and either excreted or synthesized into protein. In fact, the enrichment does not quite reach zero because of reentry of label from protein breakdown, and possibly also because of components of the metabolic pool which turn over relatively slowly. However, for a particular end product a time can be detected after which there is relatively little further excretion of isotope, and this plateau in cumulative excretion (e_x) can be used to calculate the flux using the following equation:

$$Q = E_x.d/e_x \tag{4}$$

where E_x is the rate of excretion of unlabeled N in that end product and d is the dose of ^{15}N given. Rates of synthesis and breakdown are then calculated from Q as for constant infusion.

Single dose and constant infusion methods are mathematically equivalent and should therefore give identical results. Proof of this has been obtained in children (Waterlow et al., 1978b) and in obese adults (Garlick et al., 1980a).

2.3.2.   Urinary End Products

With the method of Picou and Taylor-Roberts (1969) described above, plateau labeling is not achieved until between one and three days in adults, making the method very inconvenient. The reason for this is that urea was chosen as urinary end product, and this compound has a large body pool which delays the excretion of isotope. However, if urinary ammonia is used as the end product the value of flux rate obtained is similar to that for urea, but plateau is reached in only 12 hr (Golden and Waterlow, 1977; Waterlow et al., 1978b). A similar state of affairs exists with the single dose method; with ammonia as end product the excretion of isotope is almost complete in 9-12 hr and a single collection for this period can be used to determine flux (Waterlow et al., 1978b). With urea as end product the urea pool would delay the excretion of isotope, again resulting in inconveniently long measurements. However, the time of measurement for urea can be reduced to the same as that for ammonia (Fern et al., 1981). The theory of the single dose method requires only that the isotope has cleared the metabolic pool: any isotope subsequently delayed in the urea pool can be measured and added to that already excreted in the urine. This can be done by measuring the concentration and enrichment of urea in the plasma at the end of the collection period. On the assumption that the body urea pool is homogeneous, and from a knowledge of the total water content, the amount of ^{15}N retained in the body as urea can be calculated readily (Fern et al., 1981). The single dose approach therefore is not only simple, in that a single collection of urine over a short (9-12 hr) period is sufficient to determine turnover rates, but is also adaptable for either ammonia or urea as end product without increasing the time required for measurement. How adequately these techniques work in practice will be considered in the general discussion of how well they satisfy the primary assumption of a homogeneous metabolic pool.

2.3.3.   The Assumption of a Homogeneous Metabolic Pool

This concept was first introduced by Sprinson and Rittenberg (1949) and implies that distribution of isotope from the metabolic pool into protein synthesis and oxidation (excretion) occurs in the same ratio as that of unlabeled N. This further implies that measured rates of turnover will not depend on which labeled amino acid is given, the end product chosen or the route of administration of isotope.

The above condition could be achieved fully if the label were immediately distributed uniformly throughout the free amino acid pools of the body. This is most likely to be achieved by giving dietary protein uniformly labeled with ^{15}N, but relatively few experiments appear to have been done with this (Picou and Taylor-Roberts, 1969). For reasons of cost and availability most work has been done with [¹⁵N]glycine. It is clear that this label is not uniformly distributed in the total free amino acid pool (Aqvist, 1951; Jackson and Golden, 1980; Matthews *et al.*, 1981). After infusion of [¹⁵N]glycine for 40-60 hr no other free amino acid except serine achieved an enrichment higher than 15% of that of glycine (Matthews *et al.*, 1981).

It is still possible, however, for the proportioning of isotope between excretion and protein synthesis to represent that of total N metabolism, providing the composition of the amino acid mixture entering these two pathways is the same. This is likely to be true as long as the diet is not unusually high in glycine (Golden and Jackson, 1981) or if there is not a high requirement for glycine, as is apparently the case with premature infants (Jackson *et al.*, 1981).

Comparison of the metabolism of [¹⁵N]glycine with other [¹⁵N]amino acids has shown that with [¹⁵N]aspartate a greater proportion of the dose is excreted as urea than with glycine, while with leucine the proportion is smaller (Taruvinga *et al.*, 1980). In general, the rate of protein turnover determined with [¹⁵N]glycine is in the middle of the range of those calculated from the metabolism of other [¹⁵N]amino acids. (For a fuller discussion see Waterlow *et al.*, 1978a.)

One practical result of the peculiarities of [¹⁵N]glycine metabolism is that it is possible to obtain different estimates of body protein synthesis from measurements based on different end products of glycine catabolism. Thus in premature infants the rate of protein synthesis calculated from the labeling of urinary urea was much higher than that calculated from urinary ammonia (Jackson *et al.*, 1981) and a similar, but smaller, difference has been observed in malnourished children (Waterlow *et al.*, 1978b), in normal adults (Table III) and in the elderly (Golden and Waterlow, 1977). When urinary metabolites are measured there is, in addition, a physical heterogeneity superimposed upon the metabolic heterogeneity. This arises because urinary urea is derived from *hepatic* aspartate, glutamate, and ammonia (Lund, 1981) whereas urinary ammonia is derived from glutamine in the kidney. This glutamine is synthesized in its turn by the peripheral tissues. Thus the route of administration of the [¹⁵N]glycine can influence the relationship between the labeling of urea and ammonia. In the experiments summarized in Table III, in postabsorptive subjects the rates of protein synthesis calculated from either urea or ammonia were similar when labeled glycine was given orally, but higher with urea when the dose was given intravenously. In the latter case the label is presented initially to the peripheral tissues and preferentially labels the glutamine pool. The difference with both routes of administration is even greater when the subjects ingested food during the period of

Table III.   Rates of Protein Synthesis Calculated from the Excretion of
^{15}N in Urinary Urea and Ammonia and from the End-product
Average

	Fed		Fasted	
	oral	i.v.	oral	i.v.
Ammonia	3.2 ± 0.7	2.3 ± 0.6	2.5 ± 0.5	2.1 ± 0.3
Urea	4.6 ± 1.0	6.2 ± 1.2	2.6 ± 0.7	3.4 ± 0.8
End-product average	3.9 ± 0.5	4.2 ± 0.5	2.5 ± 0.2	2.7 ± 0.4

Four normal adults were given a single dose of 200 mg [^{15}N]glycine. Rates of
protein synthesis (g protein/kg per d) were calculated from the exretion of
^{15}N in both urea and ammonia during the following 9 hr. Data of Fern et al.
(1981).

labeling, for under this circumstance dietary amino acids dilute the label
to a greater extent in the liver than in the peripheral tissues and hence
have a disproportionate effect upon the labeling of urea. In fact, there
appeared to be an overall inverse relationship between the results with the
two end products, and Fern et al. (1981) concluded that the metabolic
pool is better regarded as being divided into two functional parts, one
acting as a precursor for urea, the other for urinary ammonia. They went
on to suggest that the inconsistencies with route of administration could
be minimized if the two rates of protein synthesis given by urinary urea
and ammonia were averaged, and they suggested that this is the best basis
for calculation.

The condition that the metabolic pool is homogeneous is therefore not
satisfied, and both metabolic and physical compartmentation exists. However, in spite of this, results obtained with [^{15}N]glycine agree remarkably
well with those obtained with precursor methods such as the constant
infusion of 1-[^{14}C]leucine. This shows that in practice these apparent difficulties can be overcome and that end product methods with
[^{15}N]glycine are a useful addition to the techniques for measuring protein
turnover.

2.3.4.   The Assumption of a Steady State

Apart from the above theoretical considerations, it is still important to
comply with the assumption of a steady state. This requires that during
any period of measurement food is given either continuously or at regular
intervals. With long infusions, lasting several days, this will clearly present
problems. Even when the isotope is given as a single dose the steady state
still must be maintained by giving regular meals during the period over
which urine is collected. A suitable protocol for this type of study has
been set out by Fern et al. (1981).

Table IV. A Comparison of Values for Whole Body Protein Synthesis in Adult Man Measured Either with [^{15}N]glycine or with carboxyl-labelled leucine

Label	Mode of administration		Measurements on	Time (hr)	Protein intake (g/kg per d)	Age (yr)	Number of observations	Protein synthesis (g/kg per d)	Ref.
[^{15}N]glycine	S.D.	oral	End-product average	9	1.1	31-34	4	3.9 ± 0.5	a
[^{15}N]glycine	S.D.	i.v.	End-product average	9	1.1	31-34	4	4.2 ± 0.5	a
[^{15}N]glycine	R.D.	oral	Urinary urea	60	0.75	20-25	6	3.0 ± 0.3	b
[^{15}N]glycine	C.I.	i.g.	Urinary urea	24	0.8	66-91	6	3.3 ± 0.6	c
[^{15}N]glycine	C.I.	i.g.	Urinary NH$_3$	24	0.8	66-91	6	2.3 ± 0.6	c
[^{14}C]leucine	C.I.	i.v. or i.g.	Plasma leucine	24	0.8	66-01	6	2.7 ± 0.7	c
[^{14}C]leucine	C.I.	i.v.	Plasma leucine	12	1.0	43-67	5	3.4 ± 0.3	d
[^{14}C]leucine	P.I.	i.v.	Plasma leucine	25	0.84	45-77	5	3.2 ± 0.6	e
[^{13}C]leucine	P.I.	i.v.	Plasma leucine	4	0.6	19-21	4	4.0 ± 0.6	f

S.D.:Single dose R.D.:repeated dose C.I.:constant infusion P.I.:primed infusion i.v.:intravenous i.g.:intragastric.

[a]Fern et al.(1981) [b]Steffee et al.(1976) [c]Golden & Waterlow(1977) [d]Clugston & Garlick(1982) [e]Clague(1980) [f]Motil et al.(1981).

2.4. Comparison of Methods

Table IV shows the results of a number of recent measurements of whole-body protein synthesis by a variety of methods. The subjects are all adult, but are not standardized for sex or age. Where possible, studies with similar rates of dietary protein intake (1 g/kg/12 hr) have been selected. Only results obtained with [^{15}N]glycine or with 1-[^{14}C] or 1-[^{13}C]leucine have been shown; for a fuller discussion of the other labeled amino acids see Waterlow *et al.* (1978a) and Garlick (1980a). It is immediately apparent that most methods give values for whole-body synthesis between 3.0 and 4.0 g/kg/d. The three values that were obtained simultaneously (Golden and Waterlow, 1977) in the same subjects indicate that [^{15}N]glycine and [^{14}C]leucine give quite similar values when compared directly. Similarly, the two values for the end-product average obtained in the study with [^{15}N]glycine by Fern *et al.* (1981) (Table III) were obtained in the same group of subjects. These results demonstrate that even though the values obtained from urinary urea and ammonia may differ quite appreciably under certain circumstances, the end-product average is very close to the rate given by labeled leucine in young subjects. It should be pointed out, however, that for this comparison, the rates obtained by different methods have to be expressed in the same units. In particular, this requires the assumption that leucine comprises 8% by weight of body protein (James *et al.*, 1974). If the suggestion by Reeds and Harris (1981) that this value should be 6.9% is correct, then the rates derived from labeled leucine should be 16% higher.

For a method to be useful in practice, however, it must not only give the "correct" rate of synthesis in a normal individual, but must also show the appropriate change when the conditions are altered. Waterlow *et al.* (1978b) showed that the increase in synthesis brought about when malnourished children were refed could be detected with either constant infusion or a single dose of [^{15}N]glycine, and when either urea or ammonia was used as end product. Similarly, Garlick *et al.* (1980a) showed that the effects of protein and energy restriction in obese subjects could be demonstrated equally well by repeated or single dose of [^{15}N]glycine with measurements on urinary ammonia, or by constant infusion of 1-[^{14}C]leucine. However, this agreement between [^{15}N]glycine with 1-[^{14}C]leucine methods does not always hold (e.g., with the acute response to feeding) and Fern *et al.* (1981) have suggested that the end-product average may give more reliable values than either urea or ammonia taken separately.

The primary uses, advantages and disadvantages of the main methods are shown in Table V. The major advantage of precursor methods, such as infusion of 1-[^{14}C]leucine, is that they are more precise. Although we cannot be sure that the specific activity of leucine in the plasma is the

Table V. Practicality of Various Methods for Measuring Body Protein Synthesis in Man

Feature	Precursor Methods (e.g., [14C]glycine)		End-Product Methods (e.g., [15N]glycine)		
	Constant Infusion	Primed Infusion	Constant Infusion		Single Dose
			End Product		
			Urea	Amonia	
Accuracy (absolute)	predictable underestimate		variable	variable	variable
Precision	good	good	moderate	moderate	moderate
Quality of plateau	good	good	moderate	moderate	not measured
Time required , hr	5-12	5-12	48+	9-12	9-12
Samples	blood; CO_2	blood; CO_2	urine	urine	urine; blood
Frequency of food intake	hourly	hourly	every 4 hr	hourly	every 1 or 2 hr
Cooperation of subject	considerable	considerable	considerable	moderate	moderate
Behaviour of subject	abnormal	abnormal	fairly normal	fairly normal	almost normal
Example of use	continuous monitoring of acute changes	intermittent measurement of acute and chronic changes	chronic changes	intermittent measurement of chronic changes	field studies
Reference to example of use	Garlick et al. (1980b)	Motil et al. (1981a)	Winterer et al. (1976; 1980)	Waterlow et al. (1978b)	Albertse et al. (1979)

correct value for calculating flux rates, the specific activity at any other site must be lower. The flux from plasma is therefore a minimum possible rate. In addition it is possible from the rates of N excretion and leucine oxidation to assess the potential extent of any error incurred (see 2.2.5). By contrast, with [^{15}N]glycine the heterogeneity of the metabolic pool could result in values which are either under- or over-estimates of true rates. The value of the ^{15}N methods is their simplicity. After a single oral dose of isotope it is relatively easy to collect urine and to take one or two blood samples, without altering the subjects' normal pattern of behaviour or activity, even in the confines of a busy hospital ward. More specialized equipment for infusion, regular blood sampling and respiratory gas collection makes the constant infusion of 1-[^{14}C]leucine more appropriate to studies in a well controlled metabolic laboratory, and when some restriction of normal behaviour can be tolerated.

3.   Protein Synthesis and Turnover

3.1.   Protein Turnover in Man Related to That in Other Mammals

In the later sections of this review we discuss the relationship between nutrition and the dynamics of protein metabolism, both in relation to the short term effects of food intake and in relation to the long term, adaptive changes which accompany alterations in the nature of the diet. In so doing we will be drawing, at times, upon relevant information obtained in animals. It is of some importance, therefore, to attempt to establish how closely the rate of protein turnover in man is related to that in other mammalian species. The framework for this discussion will be the general relationships between protein turnover, development and the energy exchanges of the body.

It has been pointed out a number of times (for example, by Waterlow *et al.,* 1978a; Reeds and Lobley, 1980; Reeds and Harris, 1981) that it is difficult to find information on whole-body protein synthesis which allows a strict comparison to be made between human beings and animals. This is partly because data on normal individuals are comparatively sparse and also because the methodological approaches that have been adopted have been different, usually involving, in animals, precursor methods with [^{14}C]labeled amino acids. Quantitative comparisons of the two methods require that the values are in the same units (g protein or g nitrogen) but this is complicated by the differences of opinion which exist regarding the factors that should be used to convert to whole-body protein information obtained with a single amino acid (Section 2.4) (Reeds and Harris, 1981). The one direct comparison in the same individuals of the two approaches

Table VI. Body Protein Synthesis Estimated with Leucine or Tyrosine in Adult Mammals and in Immature Individuals Restricted to a State of Energy Equilibrium

Species	Body wt (kg)	Protein Synthesis[a]			Ref.
		g/d	b/kg/d	$g/kg^{.75}/d$	
Rat	0.2	5.6	28.1	18.8	c
	0.35	7.7	22.0	16.9	c
	0.82	11.1[b]	13.5	12.8	d
Rabbit	3.6	33	9.2	12.6	e
Pig	32	268	8.1	18.9	f
Sheep	63	351	5.6	15.7	g
Man	71	310	4.4	13.4	h
Cattle	575	1740	3.0	14.8	i

[a]Results calculated on an assumption that body protein contains (by weight) 6.9% leucine and 2.8% tyrosine
[b]Measured with tyrosine

[c]Reeds & Lobley (unpublished)   [d]Millward (quoted by Waterlow, 1980c)
[e]Lobley, Milne & Reeds (unpublished)   [f]Reeds et al. (1980a)
[g]Reeds & Chalmers (unpublished)   [h]James et al(1976); Motil et al.(1981a,b)
[i]Reeds et al.(1981b)

to the measurement of protein synthesis has shown close agreement (Golden and Waterlow, 1977). When we consider the literature as a whole, it can be demonstrated that similar values are obtained between studies. It is possible to make some comparisons between species in which both the method of measurement and the experimental conditions, at least with respect to nutrient intake, are similar. Such a comparison is shown in Table VI, and it demonstrates that there is no marked interspecies difference in basal whole-body protein synthesis, i.e., protein synthesis measured in individuals close to energy equilibrium. Furthermore, bearing in mind the different tracers used, there seems to be relatively little difference in the fractional rate of synthesis of skeletal muscle protein in adults (Table VII).

The information that is summarized in Table VI is noteworthy for the fact that differences in the rate of protein synthesis are minimized by expressing the values relative to body weight $(kg)^{0.75}$. This is the exponent of body weight which minimizes the differences in fasting heat pro-

Table VII. Fractional Rates of Protein Synthesis in the Skeletal Muscle
of Adult Mammals (K_s/d^{-1})

Species	Body wt (kg)	k_s	Ref.
Rat	0.55	0.049	a
Rabbit	3.6	0.019	b
Sheep	50	0.018	c
Man	70	0.014	d
Ox	450	0.009	e

[a] Millward et al.(1975), measured with [^{14}C]tyrosine

[b] Nicholas et al.(1977), measured with [^{3}H]tyrosine

[c] Buttery et al.(1977), measured with [^{3}H]lysine

[d] Halliday & McKeran (1975), myofibrillar protein synthesis, measured with [^{15}N]lysine. Based on the isotope abundance of free lysine in plasma.

[e] Lobley et al.(1980), measured with [^{3}H]tyrosine.

duction between adults of different species. It is to be expected then that the basal metabolic rate and the rate of whole-body protein synthesis will be correlated (Waterlow, 1968; Kien et al., 1978a). Indeed, Waterlow (1968), in commenting upon this general parallelism, made the important point that whole-body protein synthesis calculated from the irreversible loss of an amino acid is, conceptually, very similar to the metabolic rate. Both represent the result of the activities of many different cells, each with its own characteristic rate of metabolism. We can, however, extend the predictable relationship between the basal metabolic rate and the basal protein synthesis rate to include a remarkably close relationship between changes in oxygen consumption and changes in whole-body protein synthesis, both between individuals and between species.

Fig. 5 summarizes the results of three experiments in which measurements of both oxygen consumption and protein synthesis were made in the same subjects. Despite the fact that these studies involved immature pigs (Reeds et al., 1980), prepubertal and adolescent human beings (Kien et al., 1978a) and adults (Clugston and Garlick, 1982), the similarity is striking. Somewhat different slopes were observed in each case (ranging from 4.6 to 6.9 kcals/g protein synthesized) but these were not significantly different. Overall the relationship was:

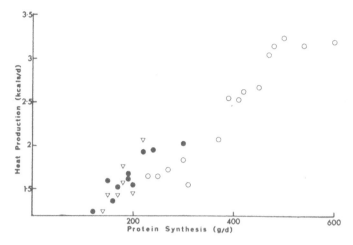

Fig. 5. Data of Reeds *et al.* (1980) in young pigs (o), of Clugston
and Garlick (1982) in adult human beings (•) and of Kien *et al.*
(1978a) in young human beings after reconstructive surgery of the
skin (▽). Heat production measured by open-circuit indirect
calorimetry in all cases. Protein synthesis calculated from the
flux of leucine (corrected for leucine oxidation; o,•) or by
the method of Picou and Taylor-Roberts (▽).

$$\text{Heat production (kcals)} = 4.16\,(\pm 0.33)\ \text{Protein synthesis (g)}$$
$$+\ 746\,(\pm 103)$$

$$r = 0.907 \quad N = 37$$

It should be stressed that this equation describes a statistical relationship
and does not mean that the energy cost of protein synthesis is 4.2 kcal/g.
Furthermore, this equation defines the relationship in normal individuals
receiving a diet of a constant composition. Under abnormal circumstances,
such as in obesity (Clugston and Garlick, 1982) and in severe trauma
(Kien *et al.*, 1978a), the relationship is no longer statistically significant.
Similarly, in the study described by Reeds *et al.* (1981a) alterations in the
ratio of protein:non-protein energy in the diet led either to markedly
increased energy expenditure with a small change in body protein synthesis (low protein:high energy) or to markedly increased protein synthesis
with a small increase in heat production (high protein diet). What seems
most likely is that oxygen consumption and protein synthesis are both
markers of the level of metabolic activity. Protein synthesis is a significant
contributor to the energy expenditure of the animal but its contribution
seems not to exceed 26% of the total (Reeds *et al.*, 1982).

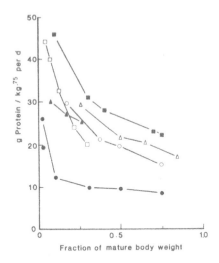

Fig. 6. References: ■ Rats - Millward (quoted by Waterlow, 1980); ○ Rabbits - Lobley, Milne, and Reeds (unpublished); □ Sheep - Soltesz *et al.* (1973), Davis *et al.* (1981); ▲ Pigs - Reeds *et al.* (1980); △ Cattle - Lobley *et al.* (1980, 1982); ● Man - Pencharz *et al.* (1977, 1981), Golden *et al.* (1977a), Kien *et al.* (1978a, 1978b).

3.2. Protein Synthesis During Development

The information discussed above was related largely to measurements in comparatively mature individuals who, with the exception of some of the individuals represented in Fig. 5, were neither depositing protein nor at an age at which body protein was being lost. As the influence of nutrition may be of most importance at the extremes of the developmental spectrum, it is pertinent to discuss changes in protein synthesis which occur during development and whether these also are similar in man and other mammals.

As individuals mature, the fractional rate of growth and the rate of energy expenditure per unit of body weight become smaller. The information summarized in Fig. 6 shows that the fractional rate of protein synthesis in the whole body (as well as in skeletal muscle) shows a reduction with increasing age, but it is noteworthy that immature human beings synthesize less protein per unit of metabolic body weight than do immature animals (Fig. 6). Presumably this fact is related primarily to the much longer time scale of development and hence much slower growth rate of the human, but this explanation leaves unanswered the question

Table VIII.   The Relationship Between Amino N Flux and Protein Intake

Species	Wt or age	N intake / Amino N flux	Ref.
Rat	55 g	.38	a
	299 g	.30	b
	350 g	.33	b
Pig	30 kg	.36	c
	60 kg	.36	c
	90 kg	.38	c
Sheep	15–22 kg	.42	d
Man	Premature	.27	e
	1–2 yr	.44	f
	4–6 yr	.41	g
	10–15 yr	.38	g

[a] Reeds, Haggarty & Wahle (unpublished). Protein turnover (Farlick et al. 1980).

[b] Reeds & Lobley (unpublished). Protein turnover from flux of leucine.

[c] Reeds et al. (1980). Protein turnover from flux of leucine.

[d] Davis et al. (1981). Protein turnover from flux of leucine; calculated intake.

[e] All human data based on method of Picou & Taylor-Roberts (1969). Pencharz et al. (1981).

[f] Recovered malnourished children (Golden et al. (1977).

[g] Kien et al. (1978a,b).

whether the relationship between the rate of accretion and the rate of synthesis of body protein is any different in animals than in man.

The question cannot be answered unequivocally for the simple reason that unless young animals are severely restricted in feed intake their rate of nitrogen retention is always much higher than is that of children. (See Section 4.3 and Reeds and Harris (1981) for one comparison.) However, the slow rate of growth of children also entails a much lower *ad libitum* intake of food and we can question whether the relationship between the synthesis of body protein and the intake of dietary protein is similar in different species. Young *et al.* (1975) concluded from a survey of measurements of protein synthesis in humans of different ages that there is a relatively fixed relationship between protein requirements (which cover

Table IX.   Calculations of the Contribution of Muscle Protein Synthesis to Total Body Protein Synthesis

Species	Wt (kg)	Muscle Protein Synthesis / Body Protein Synthesis	Ref.
Rat	0.055	.23	a
	0.100	.18	b
	0.500	.20	c
Rabbit	3	.19	d
Sheep	16	.23	e
	21	.26	e
Cattle	350	.17	f
Man	10	.25	g
	38	.30	h
	70	.26	i

[a] Reeds, Haggarty & Wahle (unpublished). Synthesis by methods of Garlick et al. (1980c). Muscle mass calculated from the creatine pool (28% of body weight).

[b] Muscle synthesis data Garlick et al. (1980c); whole body data McNurlan et al.(1979); muscle mass by creatine pool size (Reeds unpublished; 33% of body weight).

[c] Muscle protein synthesis Millward et al. (1975). Whole body synthesis Milward (quoted by Waterlow, 1980); muscle mass taken as 40% of body weight.

[d] Lobley & Reeds (1980). Muscle mass by dissection.

[e] Muscle and whole body data Davis et al. (1981). Muscle mass taken as 35 and 38% of body weight at 15 and 21 kg respectively.

[f] Lobley & Reeds (1980); Lobley et al. (1980).

[g] All human data based on 3-methyl histidine excretion and body protein synthesis measured with [^{15}N]glycine. Recovered malnourished children; methyl histidine excretion Nagabushan & Narasinga-Rao (1978); protein turnover Golden et al. (1977a); muscle mass Reeds et al. (1978). See Reeds (1980).

[h] Normal children. Methyl histidine excretion Tomas & Ballard (1979); protein turnover Kien et al. (1978b).

[i] Uauy et al. (1978).

both maintenance needs and the needs for growth) and protein synthesis. Further information, summarized in Table VIII, suggests that when the intake is sufficient to sustain an adequate rate of growth there seems to be a close interspecies relationship between protein intake and protein synthesis. Since the protein:energy ratio in the diets was similar, there is also a close relationship between energy intake and protein synthesis. Significantly, man seems not to be exceptional.

Before concluding this section we must make one final point. Thus far the discussion has been confined to measurements of whole-body protein synthesis and we must not lose sight of the fact that this is the sum of the protein synthetic activities of different tissues, each with a characteristic rate of protein synthesis (Waterlow *et al.*, 1978a). There exists the possibility that the division of whole-body protein synthesis between different tissues may be different in different species. As individual tissues show similar qualitative but different quantitative changes in protein synthetic activity following alterations in nutrition (McNurlan and Garlick, 1980; Preedy and Waterlow, 1981) (Section 4.2.3), differences in the division of total protein synthesis between, say, visceral tissues, with their characteristically high rates of protein synthesis, and skeletal muscle may influence the magnitude of whole-body protein synthetic responses of individuals of different species to nutritional changes. Unfortunately, we know of no extensive direct information in man. Recent data in rats (Millward *et al.*, 1980) have suggested that in this species the excretion of 3-methyl histidine may not be as satisfactory an index of the breakdown of the proteins of skeletal muscle as hitherto assumed. However, this has not been extended to other species and has been challenged by Harris (1981). The information in Table IX suggests that there is no obvious interspecies variation in the contribution of skeletal muscle to whole-body protein turnover.

Thus, if we accept the limitations imposed by the heterogeneous nature of the available information, it seems that man is unusual neither in his rate of protein synthesis nor in the relationship of this process to energy expenditure and development, and we can, with due caution, use information gained in animals to aid our understanding of the human being.

4.   Nutrition and Protein Turnover

Statistically, if not physiologically, body protein turnover can be regarded as consisting of two components: (1) that related to maintenance, i.e., intracellular degradation and resynthesis, together with the synthesis of gastrointestinal and other exocrine secretions and integumental losses; (2) that related to alterations in growth.

Because of differences in the relative magnitudes of these two components, the magnitude of the response of body protein turnover to nutritional change may well be different at different developmental stages. There is, for example, no compelling reason to suppose that the rate of protein turnover in normal adults will respond greatly to changes in intake above that required for the maintenance of nitrogen and energy equilibrium (Young *et al.*, 1975; Reeds *et al.*, 1981b), but it may alter when

Table X.   Nitrogen Excretion and Apparent Balance of Nitrogen in
Adults during Feeding and Following a Short Fast[1]

State	Intake	Excretion	Apparent Balance
Fed	11.2	5.43	+5.77
Fast	0	3.22	-3.22

[1]Results of Fern et al.(1981). All values as g N per 9 hr. Note that
these values do not represent complete 24 hr collections and are
uncorrected for changes in the urea pool.

intake is reduced. Conversely, protein turnover in immature individuals
may well respond to changes in intake both above and below maintenance.
In other words, changes in protein synthesis and degradation may occur
only when the rate of nitrogen retention itself is altered. The importance
of this will become evident when we discuss the effects of dietary nonprotein energy on protein metabolism.

4.1.   The Response to Food Intake

It has been pointed out (Garlick et al., 1980a; Young and Bier, 1981)
that because of the intermittent nature of food intake, the adult, who over
24 hours remains in nitrogen equilibrium, passes through successive periods in which the rate of nitrogen excretion either exceeds or is exceeded
by the rate of nitrogen intake (Table X). What mechanisms underlie these
cyclical changes in nitrogen retention?

Most nutritionists naturally would assume that these changes are
related to food intake, but the possibility has been considered seriously
that they are related to sleep (Adam, 1980). This seems to us unlikely for
the following reasons. First, Golden and Waterlow (1977) measured body
protein turnover continuously over a 24-hr period during which their elderly subjects received nutrients solely via the stomach as a continuous
infusion. Irrespective of whether the subjects were alert or asleep the rates
of body protein synthesis and leucine oxidation were the same. Second, in
separate published studies, alterations in the balance of leucine associated
with a short fast were the same whether the periods of feeding or fasting
corresponded to day or night (Garlick et al., 1980a; Clugston and Garlick,
1982) or whether both periods occurred during the day (Motil et al.,
1981a, 1981b).

Given that changes in nitrogen retention are related to food intake,
do they represent true changes in protein balance? Cessation of food
intake in the short term does lead to a contraction of the pool of free

amino acids in the body, and the period of positive nitrogen retention during feeding could be due to an expansion of the free amino acid pool. However, when we consider single amino acids (leucine, for example) the changes in the free pool are too small to account for the changes in the balance of that amino acid which accompany feeding and fasting. It appears, then, that the changes in nitrogen retention shown in Table X are due to changes in *protein* balance. As such, they must reflect changes in the relationship between protein synthesis, breakdown and amino acid catabolism. Two series of measurements have been made to investigate the changes which occur in these three processess; the results of these experiments are summarized in Table XI.

The first point to be made is that under the conditions of these experiments (small, frequent, highly digestible meals) the metabolic changes which follow the cessation of food intake are rapid (Fig. 3). Second, the results of both experiments suggest that a *short* fast is associated with an increase in body protein breakdown. The studies do, however, differ with respect to the magnitude of the changes in protein synthesis and leucine oxidation. In the studies of Clugston and Garlick (1982) a significant reduction in both protein synthesis and leucine oxidation was a consistent finding, but the same was not true for those of Motil *et al.* (1981a, 1981b).

Unfortunately, there are technical differences between these studies which may have led to apparent differences in response:

(1) In the study of Motil *et al.* (1981a, 1981b) the "fed" and "fasted" groups were different individuals while in the experiment of Clugston and Garlick (1982) the same individuals were studied in each state. Consideration of the results obtained in different subjects receiving the same intake (lines 3 and 4 in Table XI) shows the degree to which individual differences obscure comparatively small changes in metabolism.

(2) There were differences in the absolute intakes of both protein and energy and in the ratio of protein:energy in the diets. It seems reasonable to suppose that the composition of the diet, in influencing the rate of protein synthesis in the fed state, will also influence the pattern of metabolic change upon fasting. Furthermore, in the papers published by Motil *et al.* (1981a, 1981b) the subjects had been adapted to the experimental diets for seven days, whereas the subjects described by Clugston and Garlick (1982) had had no period of adaptation. Indeed, Motil *et al.* discuss the influence of preceding diet at some length.

Table XI. Changes in Protein Metabolism Associated with Short Fasts

Dietary protein (g/kg/d)	Protein/Energy (mg/kcal)	Protein synthesis (µmoles leucine/kg/hr)		Protein breakdown (µmoles leucine/kg/hr)		Leucine oxidation (µmoles leucine/kg/hr)		Ref.
		Fed	Fast	Fed	Fast	Fed	Fast	
0.1	2.25	64	76	66	84	12	13	a
0.6	11.1	92	81	65	92	18	16	b
0.6	13.5	88	81	66	92	12	16	b
0.6	13.5	102	89	82	107	22	22	a
1.5	33	113	113	67	125	46	18	a
1.0	37	85	66	57	76	33	14	c
1.18	45	99	74	65	89	39	16	c,d

[a] Motil et al. (1981a). Different subjects for fed and fast. 1-[^{13}C]leucine.
[b] Motil et al. (1981b). Different subjects for fed and fast. 1-[^{13}C]leucine.
[c] Clugston & Garlick (1982). Same subjects fed and fast. 1-[^{14}C]leucine.
[d] Clugston and Garlick (1982).

(3)   The methods of collecting the expired gases were different and it has been shown that the method of gas collection can influence the estimate of leucine oxidation (see Section 2).

(4)   The frequency of food intake also differed between the two experiments.

At present we cannot ascribe the differences in results to any one of these factors and therefore we have no unequivocal answer as to whether protein synthesis changes during a short fast. However, on the basis of supporting evidence from measurements of protein synthesis in rats (Garlick *et al.*, 1973; Waterlow *et al.*, 1978a), we believe that when the preceding diet is not deficient in protein (Winterer *et al.*, 1980) changes in protein balance during *short* fasts involve alterations in both protein synthesis and protein degradation.

4.2. Diet Composition

The foregoing discussion is related to the short term effects of the ingestion of meals. This section now examines the effects of the composition of the diet (with regard to energy and protein) on protein metabolism in subjects receiving food.

Both dietary protein and non-protein energy (NPE) influence nitrogen retention (Munro, 1964; Fuller and Crofts, 1977). In man, a large proportion of the work relating the intake of these dietary constituents to changes in body protein turnover has been concerned with adults. At the outset, we must point out that adults do not normally deposit protein to any great extent. Recent studies have, by and large, concentrated upon the changes in protein metabolism which accompany alterations in intake *below* that required for nitrogen equilibrium. Even with extreme nutritional changes the alterations in nitrogen balance were not great. Indeed, as pointed out in Section 4.1 and by Young *et al.* (1981) the changes in protein balance are very small when compared with the amounts of protein synthesized by the adult, in whom the turnover of body protein is related largely to the maintenance requirement. It is possible that the precision of the methods currently available for the measurement of protein synthesis and nitrogen balance limits our ability to draw unequivocal conclusions regarding the precise nature of the responses of the adult to changes in diet composition.

4.2.1. Non-Protein Energy

Recent papers on the effects of dietary NPE in adults have dealt with somewhat different circumstances. Motil *et al.* (1981b) adapted young

Table XII. Changes in Body Protein Turnover and Leucine Balance in
Young Adults Associated with the Ingestion of Additional
Non-protein Energy

Energy Intake (Kcal/kg/d)	Protein Synthesis	Protein Breakdown	Leucine Oxidation	Leucine Balance
43	88	66	18.0	+22
54	92	64	12.4	+28

From Motil et al.(1981b) using 1-[^{13}C]leucine. All values are in μmoles
leucine/kg body wt per hr.

men for seven days to two diets which supplied protein at a rate of 0.6
g/kg body wt/d and NPE at either 43 or 54 kcals/kg body wt per d. At
the lower level of energy intake, the subjects were in a state of negative
nitrogen balance. An increase in their intake of NPE induced a state of
nitrogen equilibrium and a small increase in the retention of leucine.
Table XII shows a summary of the results that were obtained in the fed
state. As is inevitable when nitrogen retention is increased at a constant
intake of protein, the rate of leucine oxidation was reduced during the
"high" energy period, and it is to be noted that the change in the balance
of leucine was greater than the change in nitrogen balance. The increase
in protein synthesis was, however, small, particularly in relation to the
total amount of protein that was synthesized by each individual, and was
no greater than the change in nitrogen retention. There seemed to be no
change in the rate of protein breakdown.

In the study by Motil et al. (1981b) only a small change in the
intake of NPE was effected and it is possible that more extreme changes
in energy intake might produce a more marked response. In three studies,
much greater changes in energy intake have been imposed on the subjects. The results of two of these studies, carried out in obese subjects
(Garlick et al., 1980b; Winterer et al., 1980), are summarized in Table
XIII. It should be noted that different methods were applied to the
measurement of protein turnover in each of these experiments. Winterer
et al. (1980) used the method of Picou and Taylor-Roberts (1969) while
Garlick et al. (1980b) used either [^{14}C]leucine, administered intravenously, or frequent oral doses of [^{15}N]glycine with urinary ammonia as
end product. Waterlow et al., (1978b) discuss this method. Remarkably
(perhaps because the subjects were obese) there was little change in
nitrogen balance associated with substantial reductions in energy intake
from 2000 to 500 kcals per d (Garlick et al., 1980b) or from 2400 to 580
kcals per d (Winterer et al., 1980). The changes in protein synthesis were
equally small. When measured with [^{14}C]leucine the rate of protein

Table XIII. The Effect of a Prolonged Period of Low Energy Intake on Protein
Turnover and Nitrogen Balance in Obese Subjects

Dietary Intake (Kcals/day)	[^{14}C]leucine Protein synthesis (mmoles Leucine per hour)	Protein breakdown	Overall nitrogen balance (g N/d)	Ref.
2000	7.9	6.6	-0.70	a
500	6.9	6.0	-0.80	a

Dietary Intake (Kcals/day)	[^{15}N]glycine Protein synthesis (g N/d)	Protein breakdown (g N/d)	Overall nitrogen balance (g N/d)	Ref.
2000	56	53	-0.4	a
500	53	58	-1.6	a
2400	25	24	-0.4	b
580	23	23	-0.4	b,c

[a] Garlick et al. (1980b) and Clugston & Garlick (unpublished).

[b] Winterer et al. (1980).

[c] Diet consisted only of protein.

synthesis was reduced by 14% and with [^{15}N]glycine by between 5% and 6%. It appeared, then, that a change in the intake of NPE in itself had little effect upon body protein turnover.

In the three studies discussed above, the changes in nitrogen retention were very small and we would expect equally small changes in body protein turnover. There is, however, one report in which a marked change in nitrogen retention was brought about by an alteration in the amount of NPE supplied to adult subjects. Sim et al. (1979) (Table XIV) maintained young adults with an intravenous infusion of amino acids which supplied 10.7 g N/d (i.e., a total energy intake of 380 kcals/d). Body protein turnover was measured by the method of Picou and Taylor-Roberts (1969). A second measurement was made in the same subjects when the intravenous infusate was supplemented with an additional 2000 kcals/d as glucose. The provision of additional energy was associated with a substantial increase in nitrogen retention (+5.7 g N/d) and also with an increase in protein synthesis (+7.8 g N/d). The rate of protein breakdown was unaltered, a consistent observation in all these reports.

Table XIV. The Effect of Additional Energy Supplied as Glucose on
Protein Turnover and Nitrogen Balance in Young Adults
Maintained Solely by Intravenously Infused Nutrients

Intake	Protein synthesis g N/d	Protein Breakdown g N/d	Balance g N/d
Amino acids	10.8	19.4	-7
Amino acids + glucose	17.4	19.7	-1

Taken from Sim et al.(1979). Method of Picou & Taylor-Roberts (1969).

The studies which have been carried out in adult human beings are characterized, by and large, by a small effect of NPE on the retention of nitrogen and the turnover of body protein. In the one exception (Sim *et al.*, 1979), changes in nitrogen retention seemed to involve changes only in protein synthesis. This observation is consistent with findings in growing animals (Reeds *et al.*, 1981a).

4.2.2. Dietary Protein

There have been a number of studies on the effect of variation in dietary protein intake on protein turnover in normal adults and in the obese. With one exception the studies have given the same general result. Motil *et al.* (1981a) (Table XV) offered young adults protein at rates of 0.1, 0.6, and 1.5 g/kg/d at a constant and adequate intake of NPE and measured body protein synthesis with intravenous 1-[^{13}C]leucine as tracer. The subjects were studied both in the fed and postabsorptive states. Unfortunately, separate measurements of N balance were not made in each state. In the fed state each increment in protein intake led to an improvement in leu-

Table XV. The Effect of the Level of Protein Intake on Protein
Turnover and Leucine Balance in Young Adults

Intake	Protein synthesis	Protein breakdown	Leucine Oxidation	Leucine Balance
8	64.4	65.7	11.8	- 6.1
37.4	102.2	82.3	21.6	+13.5
90.1	113.3	67.2	46.3	+41.5

Taken from Motil et al.(1980a). All values are from subjects in the
fed state and are expressed as μmoles leucine/kg body wt/d.

Table XVI. The Effect of Protein Intake at Low Energy Intake on Protein Turnover and Nitrogen Balance in Obese Adults

Intake	Protein synthesis	Protein breakdown	Overall N balance	Ref.
		$[^{15}N]glycine$		
7.8	53.6	58	+1.6	a
0	39.4	35	−4.5	a
16.3	23.2	23.6	−0.4	b
0	17.9	22.8	−5.8	b,c
		$[^{14}C]leucine$		
2.95	5.83	4.27	+1.46	d
0	3.78	4.14	−0.36	d

[a]Garlick et al. (1980b) and Glugston & Garlick (unpublished). All values as g N/d. Subjects in the fed state. [b]Winterer et al. (1980). [c]Subjects fasting. [d]Garlick et al. (1980b) and Clugston & Garlick (unpublished). All values as mmoles leucine per hour.

cine balance and an overall increase in N retention. There was also an increase in body protein synthesis and leucine catabolism. However, the magnitude of the change in protein synthesis was different for each of the two increments in protein intake (g synthesized/g increase in dietary protein), being 1.28 between intakes of 0.1 to 0.6 g protein/kg, and 0.21 between intakes of 0.6 and 1.6 g protein/kg. The reverse was true for leucine catabolism, the increase being less for the lower of the two increments in dietary protein intake. The authors made the point that in the adult an increase in protein intake above maintenance affects primarily the rate of amino acid catabolism and has relatively little effect upon body protein synthesis. Similar results have been obtained in adult cattle (Reeds et al., 1981b).

In studies with obese adults (Garlick et al., 1980b; Winterer et al., 1980) a somewhat different approach was adopted. In these experiments the subjects received either a low energy diet (Garlick et al., 1980b; Garlick and Clugston, unpublished) or no non-protein energy (Winterer et al., 1980). The results of these two experiments (Table XVI) were similar, inasmuch as they demonstrated that there was a marked reduction in body protein synthesis associated with the reduced rate of protein intake and that this reduction was associated with a reduced rate of nitrogen retention.

The study of Steffee et al. (1976) is an exception to the consensus of the three investigations described above. These workers studied young adults who received diets supplying either 0.38 or 1.5 g protein/kg/d. The

Table XVII. A Summary of Changes in Body Protein Synthesis Following Alterations in Either the Rate of Protein Intake or the Rate of Non-protein Energy Intake

Effect of protein intake			
Change in protein energy intake	Change in protein synthesis		Ref.
(Kcal)	g/d	g/additional Kcal/d	
207	134	0.65	a,i
280	82	0.29	b,i
280	145	0.51	c
380	39	0.10	d,i
440	23	0.06	e
550	33	0.10	f

Effect of non-protein energy intake			
Change in energy intake	Change in protein synthesis		Ref.
(Kcal)	g/d	g/additional kcal/d	
810	23	0.029	g,i
1410	52	0.037	b
1410	24	0.017	c,i
2010	41	0.020	h

[a]Motil et al.(1981a); "low" protein to "adequate" protein; [^{13}C]leucine.
[b]Garlick et al.(1980b); Clugston & Garlick (unpublished); [^{14}C]leucine.
[c]As in b but based on [^{15}N]glycine.
[d]Motil et al.(1981a); "adequate" protein to "surfeit" protein.
[e]Steffee et al.(1976); [^{15}N]glycine; results recalculated by present authors.
[f]Winterer et al.(1980); zero protein to "adequate" protein; [^{15}N]glycine.
[g]Motil et al.(1980b); combined values for all supplements; [^{13}C]leucine.
[h]Sim et al.(1978); [^{15}N]glycine.
[i]All the results that are based on leucine have been calculated on the basis that body protein contains 6.9% leucine by weight.

reduction in protein intake was associated with an increase in the rate of body protein synthesis. However, the measurements of protein synthesis in this study were made by the method of Picou and Taylor-Roberts (1969) in which the calculation is based upon the plateau abundance of isotope in urinary urea during a constant infusion of [^{15}N]glycine. As pointed out in

Section 2.3, a problem with this method is that the rate of attainment of plateau *in urinary urea* is a function of the fractional rate of turnover of the body urea pool. Steffee *et al* (1976) showed that when dietary protein intake was reduced this rate was reduced by a factor of two and that this is sufficient to delay the plateau significantly. It seems likely that a plateau enrichment of urinary urea had not been attained in their low protein study. However, changes in urea turnover were reported and it is possible to assess to what degree the plateau had been underestimated and hence by how much the N flux had been overestimated. We calculate that, on average, in the low protein study the rate of protein synthesis was overestimated by 28%. This suggests, in its turn, that the reduction in protein intake from 1.5 to 0.38 g/kg/d reduced body protein synthesis by 1.14 g protein/kg/d.

By and large, the studies described above suggest that dietary protein intake (below maintenance) affects both body protein synthesis and breakdown in the adult. However, a problem in assessing this effect of dietary protein arises from the fact that the intake of protein also makes a contribution to the energy available to the organism. This is particularly true in the adult, in which nearly all dietary protein is catabolised. The question then remains whether a change in dietary protein has a greater or lesser effect per Kcal of additional ingested energy than does a change in NPE. The relevant information for protein and NPE supplements is summarized in Table XVII. There is a marked variability in the results, which perhaps is to be expected from such heterogeneous studies, but it would appear that protein intake does indeed have a greater effect than energy intake upon protein synthesis.

4.2.3. Studies in Animals of the Effects of Food and Protein Intake

While the experiments in man have been confined to measurements of

Table XVIII. The Effect of an Increase in the Intake of Either Protein or Non-protein Energy (N.P.E.) Upon Whole-body Protein Synthesis and Nitrogen Retention in Growing Pigs

Increase in energy intake (Kcal/d)	Nature of supplement	Change in protein synthesis (g/d)	Change in protein synthesis (g/kcal/d)	Change in N retention (g protein/d)
360	Protein	170	0.47	21
1400	N.P.E.	28	0.02	35

Taken from Reeds et al. (1981a); [^{14}C]leucine converted to g protein per day on the assumption that body protein contains 6.9% leucine by weight.

Table XIX.   The Effect of Either a Prolonged Fast or Severe Protein
Depletion on Protein Synthesis in Tissues of the Rat

Tissue	Treatment		
	None	Fast	Protein free diet
Jejunal Mucosa	123	92	95
Bone	90	62	ND
Liver	86	72	69
Skin	64	47	ND
Heart	20	12	10
Gastrocnemius muscle	17	6	4

Data of McNurlan, Preedy, Fern & Garlick (unpublished). Values are
fractional rates of protein synthesis (% of pool synthesis/d) using
the technique of Garlick et al. (1980), with [^{3}N]phenylalanine as tracer.

whole-body protein synthesis, with the exception of some experiments in pigs (Reeds et al., 1980, 1981a) and in cattle (Reeds et al., 1981b; Lobley et al., 1982), most of the measurements relating protein turnover and dietary intake in animals have concentrated upon changes in individual tissues (Waterlow et al., 1978a). Nevertheless, these studies allow us, partially at least, to clarify the somewhat confused nature of the observations that we have described above.

Studies on whole-body protein synthesis in growing pigs (Table XVIII) lead us to conclude that, per unit of additional energy ingested by the animals and per unit increase in nitrogen retention, dietary protein has a more marked influence upon body protein synthesis than does dietary nonprotein energy. In rats, as in man, experiments have been concentrated upon the metabolic responses to reductions in protein and energy intake either by offering young animals grossly protein-deficient diets or by imposing a complete fast. With flooding dose techniques (McNurlan and Garlick, 1980; Garlick et al., 1980c) responses in a wide variety of tissues have been defined (Table XIX) (McNurlan et al., 1979; Preedy and Waterlow, 1981). These results demonstrate that fasting and specific protein deprivation are both associated with significant reductions in the fractional rate of protein synthesis in many tissues. It is to be noted that different tissues give different quantitative responses, skeletal muscle being particularly sensitive to nutritional deprivation. These studies in rats also suggest much the same conclusion as we have drawn with respect to man: by and large, fasting leads to no greater reduction in tissue protein synthesis than does the prolonged ingestion of protein-free diets and the response of protein synthesis to dietary non-protein energy is less marked than is its response to dietary protein. It seems likely to us that the inevitable ethical constraints that are placed on the nature of the experiments that can be carried out with human beings limit the degree of dietary insufficiency that can be applied and thereby reduce the magnitude of the responses that can be measured.

4.3. Protein Turnover in Severe Undernutrition and During Recovery

Although it is ethically unjustifiable to expose normal individuals to prolonged periods of severe nutritional deprivation, such a circumstance occurs with only too great a frequency in economically disadvantaged populations. One end product is severe protein-energy malnutrition (PEM) in children. Such children are retarded with respect to body length and weight and have severe deficits in body and skeletal muscle protein (Halliday, 1967; Reeds et al., 1978).

Children presenting with PEM appear to have undergone some degree of adaptation, as they retain protein at an intake which, in the recovered child, is associated with negative N balance (Picou and Taylor-Roberts, 1969; Golden et al., 1977a). It is of interest to note that obese adults also appear to show a reduced maintenance protein requirement when subjected to long periods of low protein and/or energy intake (Motil et al., 1981a, 1981b; Garlick et al., 1980b; Winterer et al., 1980). Moreover, the studies of Golden et al. (1977a) demonstrate that children with PEM show a reduction in protein synthesis (Table XX) and, on the basis of 3-methyl histidine excretion, a reduced rate of skeletal muscle contractile protein breakdown (Nagbushan and Narasinga-Rao, 1978). Golden et al. (1977a) pointed out that the reduction in protein turnover also serves to minimize the energy requirement of the individual.

In addition to demonstrating that children are able to adapt to a prolonged reduction in food intake, studies performed during the period of rehabilitation allow us the rare opportunity to investigate the relationship

Table XX. Body Protein Synthesis and N Balance in Children with Severe Protein-energy Malnutrition (MN) and after Recovery (REC)

Nutritional state	Intake	Protein synthesis	Nitrogen balance
MN	0.6	4.0	+0.2
REC	0.6	7.0	−0.1

Taken from Golden et al. (1977a). Values are g protein (N x 6.25)/kg body wt/d. Measured with [^{15}N]glycine.

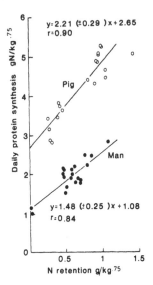

Fig. 7. Data of Reeds *et al.* (1980) in pigs (o) and of Golden *et al.* (1977b) and Golden (personal communication) in children (•).

between protein turnover and growth in children. We emphasized in Section 3.2 that normal children grow at a rate that is so slow that the changes in N retention and protein turnover associated with growth may lie within the limits of accuracy of the methods that can be used. Thus attempts to investigate the interaction between protein synthesis and protein retention in premature infants (Pencharz *et al.*, 1977, 1981) and in children recovering from burn injury (Kien *et al.*, 1978a, 1978b) have not revealed a clear quantitative relationship. However, rehabilitated malnourished children grow at a rate some 20 times in excess of age standards and ingest energy and protein at about twice the amount required for maintenance (Ashworth, 1969).

The results published by Golden *et al.* (1977a, 1977b) and by M.H.N. Golden (personal communication) demonstrate that, when the intake of the whole diet is the principal influence upon the rate of growth, a linear relationship exists between the rate of protein synthesis and the rate of nitrogen retention (Fig. 7). This result is similar to that obtained with growing pigs (Reeds *et al.*, 1980) (Fig. 7), although the slopes of the lines are significantly different. An important point to note is that the slope of both lines is significantly greater than unity and hence accelerated growth is also associated with an increase in the rate of protein breakdown in the whole-body. Further, if we accept the excretion of 3-methyl histidine as a reliable indicator of the rate of contractile protein breakdown in skeletal muscle, the calculations shown in Table XXI sug-

Table XXI.   Estimates of Whole Body and Skeletal Muscle Protein
             Breakdown in Children Before (MN), during (RWG) and
             After (REC) Recovery from Severe Protein-energy
             Malnutrition

Nutritional state	Body protein catabolism[a] (g/kg body wt/d)	Muscle protein catabolism[b] (g/kg body wt/d)
MN	3.7	0.6
RWG	9	2.9
REC	7.1	1.6

[a]Taken from Picou & Taylor-Roberts (1969) and Golden et al. (1977a,b).

[b]Based on the excretion of 3-methyl histidine (Nagabushan & Narasinga-Rao, 1978) and estimates of muscle mass (Reeds et al. (1978).

gest that catch-up growth is also associated with an acceleration in this process. This result is consistent with observations in the rat (Millward *et al.*, 1976). It appears, then, that the relationship between growth and protein turnover is qualitatively similar in at least three mammalian species. The factors which underlie the disproportionate acceleration of body protein turnover in relation to growth are unknown.

5.   Conclusion

The development of our understanding of the nutritional control of body protein turnover has been dominated by the search for methods of assessment which are not only valid but are also practicable. In adult man, protein turnover and nitrogen retention respond rapidly to the cessation of food intake. It seems reasonable to propose that a diurnal ebb and flow of protein synthesis and nitrogen retention occurs which follows the normal intermittent pattern of food intake. Strict control of this pattern of intake is therefore critically important for the interpretation of the results of nutritional and metabolic measurements.

Information regarding the influence of the composition of the diet upon the dynamics of protein metabolism in adults is equivocal. We propose that, in man, dietary protein has a greater influence on protein synthesis than does dietary non-protein energy. Experiments with animals support this contention. There is no readily comparable information in immature human beings but information gained in rats and pigs suggests that the same general conclusion applies to growth.

In this review we have not discussed the effects of trauma on protein metabolism although we recognize that this is an area of great medical and scientific importance. However, a more complete understanding of the effects of nutritional status on one hand and of food intake on the other is essential both for the design and for the interpretation of such studies. It is possible that nutritional considerations may be at least as great, if not greater in magnitude, than the other changes that occur.

Our ability to carry out detailed and perhaps more subtle measurements has increased of late and we look forward to an increase in our understanding of the basis of the nutritional influences upon protein metabolism.

6.   Acknowledgment

We are very grateful to Mrs. H. Burnett for her skilled secretarial assistance. The donation of unpublished results by Drs. G. Lobley and M. Golden is gratefully acknowledged.

References

Adam, K., 1980, Sleep as a restorative process and a theory to explain why, *Progr. Brain. Res.* **53**:289.

Albertse, E.C., Garlick, P.J., Pain, V.M., Reeds, P.J., Watkins, P.J., and Waterlow, J.C., 1979, Effect of insulin treatment on protein turnover in adult diabetics, *Proc. Nutr. Soc.* **38**:90A.

Aqvist, S.E.G., 1951, Metabolic interrelationships among amino acids studied with isotopic nitrogen, *Acta Chem. Scand.* **5**:1046.

Ashworth, A., 1969, Growth rates in children recovering from protein-calorie malnutrition, *Br. J. Nutr.* **23**:835.

Blaxter, K.L., 1978, Comparative aspects of nutrition, in: *Diet of Man: Needs and Wants* (J. Yudkin, ed.), pp. 145-158, Applied Science Publishers, London.

Buttery, P.J., Beckerton, A., and Lubbock, H., 1977, Rates of protein metabolism in sheep, in: *Protein Metabolism and Nutrition* (S. Taminga, ed.), pp. 32-34, European Association for Animal Production Publ. No. 22, Centre for Agricultural Publishing and Documentation, Wageningen, Holland.

Clague, M.B., 1980, *Protein Turnover and the Metabolic Response to Trauma in Man,* M.D. Thesis, University of Newcastle.

Clague, M.B., Kier, M.J., and Clayton, C.B., 1979, Apparent and actual ^{14}C retention in the slower turnover bicarbonate pool in man using liquid scintillation counting, *Int. J. Appl. Radiat. and Isot.,* **30**:647.

Clugston, G.A., 1980, *Whole Body Protein Turnover and Energy Metabolism in Man,* Ph.D. Thesis, University of London.

Clugston, G.A. and Garlick, P.J., 1982, The response of protein and energy metabolism to food intake in lean and obese man. *Human Nutrition:* Clinical Nutrition, **36C**:57.

Conley, S.B., Rose, G.M., Robson, A.M., and Bier, D.M., 1980, Effects of dietary intake and hemodialysis on protein turnover in uremic children, *Kidney Int.,* **17**:837.

Crane, C.W., Picou, D., Smith, R., and Waterlow, J.C., 1977, Protein turnover in patients before and after elective orthopaedic operations, *Br. J. Surg.* **64**:129.

Davis, S.R., Barry, T.N., and Hughson, G.A., 1981, Protein synthesis in tissues of growing lambs, *Br. J. Nutr.,* **46**:409.

Fern, E.B., Garlick, P.J., McNurlan, M.A., and Waterlow, J.C., 1981, The excretion of isotope in urea and ammonia for estimating protein turnover in man with [^{15}N]glycine, *Clin. Sci. Mol. Med.* **61**:217.

Fuller, M.F., and Crofts, R.M., 1977, The protein-sparing effect of carbohydrate. 1. Nitrogen retention of growing pigs in relation to diet, *Br. J. Nutr.* **38**:479.

Garlick, P.J., 1980a, Protein turnover in the whole animal and specific tissues, in: *Comprehensive Biochemistry* Vol. 19B, (M. Florkin and E.H. Stotz, eds.), pp. 77-152, Elsevier, Amsterdam.

Garlick, P.J., 1980b, Assessment of protein metabolism in the intact animal, in: *Protein Deposition in Animals* (P.J. Buttery and D.B. Lindsay, eds.), pp. 51-67, Butterworths, London.

Garlick, P.J., and Clugston, G.A., 1981, Measurement of whole-body protein turnover by constant infusion of carboxyl-labeled leucine, in: *Nitrogen Metabolism in Man* (J.C. Waterlow and J.M.L. Stephen, eds.), pp. 303-322, Applied Science Publishers, London.

Garlick, P.J., and Marshall, I., 1972, A technique for measuring brain protein synthesis, *J. Neurochem.* **19**:577.

Garlick, P.J., Millward, D.J., and James, W.P.T., 1973, The diurnal response of muscle and liver protein synthesis *vivo* in meal-fed rats, *Biochem. J.* **136**:935.

Garlick, P.J., Clugston, G.A., Swick, R.W., and Waterlow, J.C., 1980a, Diurnal pattern of protein and energy metabolism in man, *Am. J. Clin. Nutr.* **33**: 1983.

Garlick, P.J., Clugston, G.A., and Waterlow, J.C., 1980b, Influence of low-energy diets on whole-body protein turnover in obese subjects, *Am. J. Physiol.* **288**:E235.

Garlick, P.J., McNurlan, M.A., and Preedy, V.R., 1980c, A rapid and convenient technique for measuring the rate of protein synthesis in tissues by injection of [^{3}H]phenylalanine, *Biochem. J.* **192**:719.

Golden, M.H.N., and Jackson, A.A., 1981, Assumptions and errors in the use of [^{15}N]excretion data to estimate whole-body protein turnover, in: *Nitrogen Metabolism in Man* (J.C. Waterlow and J.M.L. Stephen, eds.), pp. 323-344, Applied Science Publishers, London.

Golden, M.H.N., and Waterlow, J.C., 1977, Total protein synthesis in elderly people: a comparison of results with [^{15}N]glycine and [^{14}C]leucine, *Clin. Sci. Mol. Med.* **53**:277.

Golden, M.H.N., Waterlow, J.C., and Picou, D., 1977a, Protein turnover, synthesis and breakdown before and after recovery from protein-energy malnutrition, *Clin. Sci. Mol. Med.* **53**:473.

Golden, M.H.N., Waterlow, J.C., and Picou, D., 1977b, The relationship between dietary intake, weight change, nitrogen balance and protein turnover in man, *Am. J. Clin. Nutr.* **30**:1345.

Halliday, D., 1967, Chemical composition of the whole-body and individual tissues of two Jamaican children whose death resulted primarily from malnutrition, *Clin. Sci.* **33**:365.

Halliday, D., and McKeran, R.O., 1975, Measurement of muscle protein synthetic rate from serial muscle biopsies and total body protein turnover in man by continuous intravenous infusion of L-α-[^{15}N]lysine, *Clin. Sci. Mol. Med.* **49**:581.

Harris, C.I., 1981, Reappraisal of the quantitative importance of non-skeletal-muscle source of N-methyl histidine in urine, *Biochem. J.* **195**:1011.

Issekutz, B., Paul, P., Miller, H.I., and Borz, W.M., 1968, Oxidation of plasma FFA in lean and obese humans, *Metabolism* **17**:62.

Jackson, A.A., and Golden, M.H.N., 1980, [^{15}N]Glycine metabolism in normal man: the metabolic α-amino-nitrogen pool, *Clin. Sci. Mol. Med.* **58**:517.

Jackson, A.A. Shaw, J.C.L., Barber, A., and Golden, M.H.N., 1981, Nitrogen metabolism in preterm infants fed human donor breast milk: the possible essentiality of glycine, *Pediatr. Res.* **15**:1454.

James, W.P.T., Sender, P.M., Garlick, P.J., and Waterlow, J.C., 1974, The choice of label and measurement technique in tracer studies of body protein metabolism in man, in: *Dynamic Studies with Radio Isotopes in Medicine,* Vol. 1, pp. 461-472, International Atomic Energy Agency, Vienna.

James, W.P.T., Garlick, P.J., Sender, P.M., and Waterlow, J.C., 1976, Studies of amino acid and protein metabolism in normal man with L-U-[14C]tyrosine, *Clin. Sci. Mol. Med.* 50:525.

Kaihara, S., and Wagner, H.N., 1968, Measurement of intestinal fat absorption with carbon-14 labeled tracers, *J. Lab. Clin. Med.* 71:400.

Kien, C.L., Rohrbaugh, D.K., Burke, J.F., and Young, V.R., 1978a, Whole body protein synthesis in relation to basal energy expenditure in healthy children and in children recovering from burn injury, *Pediatr. Res.* 12:211.

Kien, C.L., Young, V.R., Rohrbaugh, D.K., and Burke, J.F., 1978b, Whole body protein synthesis and breakdown rates in children before and after reconstructive surgery of the skin, *Metabolism* 27:27.

Lobley, G.E., and Reeds, P.J., 1980, Protein synthesis − are there species differences, in: *Proceedings of the 3rd EAAP Symposium on Protein Metabolism and Nutrition,* Vol. 1, (H.J. Oslage and K. Rohr, eds.), pp. 80-85, European Association for Animal Production Publ. No. 27, Braunschweig, Germany.

Lobley, G.E., Milne, J.M., Lovie, J.M., Reeds, P.J., and Pennie, K., 1980, Whole body and tissue protein synthesis in cattle, *Br. J. Nutr.* 43:491.

Lobley, G.E., Smith, J.S., Mollison, G., Connell, A., and Galbraith, H., 1982, The effect of an anabolic implant (trenbolone acetate + oestradiol-17β) on the metabolic rate and protein metabolism of beef steers, *Proc. Nutr. Soc.* 41:28A.

Long, C.L., Jeevanadam, M., Kim, B.M., and Kinney, J.M., 1977, Whole body protein synthesis and catabolism in septic man, *Am. J. Clin. Nutr.* 30:1340.

Matthews, D.E., Motil, K.J., Rohrbaugh, D.K., Burke, J.F., Young, V.R., and Bier, D.M., 1980, Measurements of leucine metabolism in man from a primed continuous infusion of L-1-[13C]leucine, *Am. J. Physiol.* 238:E473.

Matthews, D.E., Conway, J.M., Young, V.R., and Bier, D.M., 1981, Glycine nitrogen metabolism in man, *Metabolism* 30:886.

McNurlan, M.A., and Garlick, P.J., 1980, Contribution of rat liver and gastrointestinal tract to whole-body protein synthesis in the rat, *Biochem. J.* 186:383.

McNurlan, M.A., Tomkins, A.M., and Garlick, P.J., 1979, The effect of starvation on the rate of protein synthesis in rat liver and small intestine, *Biochem. J.* 178:373.

Millward, D.J., Garlick, P.J., Stewart, R.J.C., Nnanyelugo, D.O., and Waterlow, J.C., 1975, Skeletal muscle growth and protein turnover, *Biochem. J.* 150:235.

Millward, D.J., Garlick, P.J., Nnanyelugo, D.O., and Waterlow, J.C., 1976, The relative importance of protein synthesis and degradation in the regulation of muscle mass, *Biochem. J.* 156:186.

Millward, D.J., Bates, P.C., Grimble, G.K., and Brown, J.G., 1980, Quantitative importance of non-skeletal muscle sources of N-methyl histidine in urine, *Biochem. J.* 190:225.

Motil, K.J., Matthews, D., Rohrbough, D., Bier, D., Burke, J.F., and Young, V.R., 1979, Simultaneous estimates of whole-body leucine and lysine flux in young men: effect of reduced protein intake, *Fed. Proc.* 38:708, (abstract).

Motil, K.J., Matthews, D.E., Bier, D.M., Burke, J.F., Munro, H.N., and Young, V.R., 1981a, Whole-body leucine and lysine metabolism: response to dietary protein intake in young men, *Am. J. Physiol.* 240:E712.

Motil, K.J., Bier, D.M., Matthews, D.E., Burke, J.F., and Young, V.R., 1981b, Whole body leucine and lysine metabolism studied with 1-[13C]leucine and α-[15N]lysine; response in healthy young men given excess energy intake, *Metabolism* 30:783.

Munro, H.N., 1964, Regulation of protein metabolism, in: *Mammalian Protein Metabolism* Vol. 1, (H.N. Munro and J.B. Allison, eds.), pp. 381-481, Academic Press, New York.

Nagabushan, V.S., and Narasinga Rao, B.S., 1978, Studies on 3-methyl histidine metabolism in children with protein-energy malnutrition, *Am. J. Clin. Nutr.* **31**:1322.

Nicholas, G.A., Lobley, G.E., and Harris, C.I., 1977, Use of the constant infusion technique for measuring rates of protein synthesis in New Zealand White rabbits, *Br. J. Nutr.* **38**:1.

Nicholson, J.F., 1970, Rate of protein synthesis in premature infants, *Pediatr. Res.* **4**:389.

O'Keefe, S.J.D., Sender, P.M., and James, W.P.T., 1974, "Catabolic" loss of body nitrogen in response to surgery, *Lancet* **2**:103S.

Olesen, K., Heilskov, N.C.S., and Shønheyder, F., 1954, The excretion of ^{15}N in urine after administration of [^{15}N]glycine, *Biochim. Biophys. Acta* **15**:95.

Pencharz, P.B., Steffee, W., Cochran, W., Scrimshaw, N.S., Rand, W.H., and Young, V.R., 1977, Protein metabolism in human neonates, *Clin. Sci. Mol. Med.* **52**:485.

Pencharz, P.B., Masson, M. Desgrauges, F., and Papageorgiou, A., 1981, Total-body protein turnover in human premature neonates: effects of birth weight, intra-uterine nutritional status and diet, *Clin. Sci. Mol. Med.* **61**:207.

Picou, D., and Taylor-Roberts, T., 1969, The measurement of total body protein synthesis and catabolism and nitrogen turnover in infants in different nutritional states receiving different amounts of dietary protein, *Clin. Sci.* **36**:283.

Preedy, V.R., and Waterlow, J.C., 1981, Protein synthesis in the young rat — the contribution of skin and bones, *J. Physiol.* **291**:45P.

Reeds, P.J., 1979, Protein turnover in man, in: *Clinical and Scientific Aspects of the Regulation of Metabolism* (M. Ashwell, ed.), pp. 67-88, CRC Press, Boca Raton, U.S.A.

Reeds, P.J., and Harris, C.I., 1981, Protein synthesis in animals: man in his context, in: *Nitrogen Metabolism in Man* (J.C. Waterlow and J.M.L. Stephen, eds.), pp. 391-408, Applied Science Publishers, London.

Reeds, P.J., and Lobley, G.E., 1980, Protein synthesis: are there real species differences, *Proc. Nutr. Soc.* **39**:43.

Reeds, P.J., Jackson, A.A., Picou, D., and Poulter, N., 1978, Muscle mass and composition in malnourished infants and children and changes seen after recovery, *Pediatr. Res.* **12**:613.

Reeds, P.J., Cadenhead, A., Fuller, M.F., Lobley, G.E., and McDonald, J.D., 1980, Protein turnover in growing pigs: effects of age and food intake, *Br. J. Nutr.* **43**:445.

Reeds, P.J., Fuller, M.F., Cadenhead, A., Lobley, G.E., and McDonald, J.D., 1981a, Effects of changes in the intakes of protein and non-protein energy on whole-body protein turnover in growing pigs, *Br. J. Nutr.* **45**:539.

Reeds, P.J., Ørskov, E.R., and Macleod, N.A., 1981b, Whole body protein synthesis in cattle sustained by infusion of volatile fatty acids and casein, *Proc. Nutr. Soc.* **40**:50A.

Reeds, P.J., Wahle, K.W.J., and Haggarty, P., 1982, Energy costs of protein and fatty acid synthesis, *Proc. Nutr. Soc.* **41**:155.

Rothschild, M.A., and Waldmann, T., 1970, *Plasma Protein Metabolism*, Academic Press, New York.

Shønheyder, F., Heilskov, N.C.S., and Olesen, K., 1954, Isotopic studies on the mechanism of negative nitrogen balance produced by immobilization, *Scand. J. Clin. Lab. Invest.* **6**:178.

Sim, A.J.W., Wolfe, B.M., Young, V.R., Clarke, D., Young, V.R., and Moore, F.D., 1979, Glucose promotes whole body protein synthesis from infused amino acids in fasting man, *Lancet* **1**:68.

Simon, O., Munchmeyer, R., Bergner, H., Zebrowska, T.O., Buraczewska, L., 1978, Estimation of rate of protein synthesis by constant infusion of labeled amino acids in pigs, *Br. J. Nutr.* **40**:243.

Soltesz, G., Joyce, J., and Young, M., 1973, Protein synthesis in the newborn lamb, *Biol. Neonate* **23**:139.

Sprinson, D.B., and Rittenberg, D.E., 1949, The rate of utilization of ammonia for protein synthesis, *J. Biol. Chem.* **180**:707.

Steffee, W.R., Goldsmith, R.S., Pencharz, P.B. Scrimshaw, N.S., and Young, V.R., 1976, Dietary protein intake and dynamic aspects of whole-body nitrogen metabolism in adult humans, *Metabolism* **25**:281.

Taruvinga, M., Jackson, A.A., and Golden, M.H.N., 1979, Comparison of [^{15}N]labeled glycine, aspartate, valine and leucine for measurement of whole-body protein turnover, *Clin. Sci. Mol. Med.* **57**:281.

Tolbert, B.M., Kirk, M., and Upham, F., 1959, Carbon-14 respiration pattern analyser for clinical studies, *Review of Scientific Instruments* **30**:116.

Tomas, F.M., and Ballard, F.J., 1979, Muscle protein breakdown *in vivo* as assessed by 3-methyl histidine excretion, in: *Muscle, Nerve and Brain Degeneration* (A.D. Kidman and J.K. Tomkins, eds.), International Congress Series No. 473, Excerpta Medica, Amsterdam: 1979.

Uauy, R., Winterer, J.C., Bilmazes, C., Haverberg, L.N., Scrimshaw, N.S., Munro, H.N., and Young, V.R., 1978, *J. Gerontol.* **33**:663.

Waterlow, J.C., 1967, Lysine turnover in man measured by intravenous infusions of L-U-[^{14}C]lysine, *Clin. Sci.* **33**:507.

Waterlow, J.C., 1968, Observation on the mechanism of adaptation to low protein intakes, *Lancet* **1**:1091.

Waterlow, J.C., 1980, Protein turnover in the whole animal, *Invest. Cell. Pathol.* **3**:107.

Waterlow, J.C., Garlick, P.J., and Millward, D.J., 1978a, *Protein Turnover in Mammalian Tissues and in the Whole Body*, North Holland, Amsterdam.

Waterlow, J.C., Golden, M.H.N., and Garlick, P.J., 1978b, Protein turnover in man measured with ^{15}N: comparison of end products and dose regimes, *Am. J. Physiol.* **235**:E165.

Winterer, J.C., Steffee, W.P., Davy, W., Perera, A., Uauy, R., Scrimshaw, N.S., and Young, V.R., 1976, Whole body protein turnover in aging man, *Exp. Geront.* **11**:79.

Winterer, J., Bistrian, B.R., Bilmazes, C., Blackburn, G.L., and Young, V.R., 1980, Whole body protein turnover, studies with [^{15}N]glycine and muscle protein breakdown in mildly obese subjects during a protein-sparing diet and a brief total fast, *Metabolism* **29**:575.

Wolstenholm, G.E.W., and O'Connor, M., 1973, Protein Turnover, *CIBA Foundation Symposium No. 9*, Elsevier, Amsterdam.

Young, V.R., and Bier, D.M., 1981, Protein metabolism and nutritional state, *Proc. Nutr. Soc.* **40**:343.

Young, V.R., Steffee, W.P., Pencharz, P.B., Winterer, J.C., and Scrimshaw, N.S., 1975, Total human body protein synthesis in relation to protein requirements at various ages, *Nature (Lond.)* **253**:192.

Chapter 6

Zinc Binding Ligands and Complexes in Zinc Metabolism

Bo Lönnerdal, Carl L. Keen, and Lucille S. Hurley

1. Introduction

The essential role of the trace element zinc in both the prenatal and post-natal periods of mammalian development has been well documented (Underwood, 1977; Hurley, 1980; Prasad, 1976). Zinc is important in many ways: it is required for the activity of numerous enzymes (Parisi and Vallee, 1969), and is important for stability of biological membranes (Chvapil, 1973), for integrity and synthesis of nucleic acids (Dreosti et al., 1972; Dreosti and Hurley, 1975; Eckhert and Hurley, 1977; Eichhorn et al., 1973), and for synthesis of proteins and lipids (Underwood, 1977). Based on the large number of important reactions that involve zinc, it is reasonable to suppose that metabolism of this element is under some regulation. It has been suggested that the primary site of zinc homeostasis is the intestine (Cotzias et al., 1962; Cousins, 1979a, 1979b; Weigand and Kirchgessner, 1976a, 1976b, 1980). Currently, most investigators think that zinc homeostasis in the intestine, occurring through absorption or excretion of the element (or both), is mediated in part by low molecular weight (LMW) ligands. This area of nutrition research has been the subject of considerable recent interest, especially because of reports of dietary zinc deficiency in various age groups of several human populations (Prasad, 1976; Hambidge et al., 1972, 1976; Jameson, 1976, 1981). Zinc

Bo Lönnerdal, Carl L. Keen, and Lucille S. Hurley • Department of Nutrition, University of California, Davis, California 95616.

deficiency in humans has been correlated with a number of disorders, including congenital malformations and complications of pregnancy, poor growth and development during infancy, childhood, and adolescence, and impaired immunocompetence. Furthermore, genetic disorders of zinc metabolism have been described, such as acrodermatitis enteropathica in humans and Adema disease in cattle, in which the metabolic lesion may involve LMW zinc ligands.

In this chapter, we review the role of LMW zinc complexes and zinc ligands in zinc metabolism and homeostasis, with special reference to the neonatal period and to genetic disorders of zinc in man and animals.

2.   Zinc Binding Ligands in Infant Nutrition

2.1.   Identification of Zinc Binding Ligands in Milk

2.1.1.   Background

Acrodermatitis enteropathica (AE) is a rare inherited disorder transmitted by an autosomal recessive gene. The symptomatology of this disease was first described in detail by Brandt (1936), who suggested that it resulted from a toxic substance in cow's milk or from a deficiency of a nutrient found in breast milk, which children with this condition could not obtain from other food sources because of a gastrointestinal disorder. The disease was given the name acrodermatitis enteropathica by Danbolt and Closs (1943). The signs of the disease usually appear first in infancy and include acral and orificial vesiculobullous, pustular, and eczematoid skin lesions, alopecia, diarrhea, and immune dysfunction. If untreated, AE is usually lethal in early childhood (Roberts, 1970; Bergsma, 1973). The prognosis for AE patients improved dramatically when Dillaha *et al.* (1953) reported that diiodohydroxyquinoline (Didoquin) could control the disease. A nutritional factor was suggested by the observation that the development of symptoms was usually precipitated by weaning the infant from breast milk to cow's milk or other foods (Brandt, 1936). In addition, feeding breast milk to AE infants completely alleviated the symptoms (Entwisle, 1965).

Moynahan and Barnes (1973) showed that an infant with AE had severely reduced serum zinc levels and that treatment with small doses of zinc sulphate alleviated the signs of the disease. This effect of oral zinc supplementation on AE has been confirmed and is well documented (Moynahan, 1974; Michaëlson, 1974; Neldner and Hambidge, 1975; Portnoy and Molokhia, 1974; Chandra, 1980; Rodin and Goldman, 1969;

Sunderman, 1975; Entwisle, 1965; Freier *et al.*, 1973). Furthermore, the therapeutic benefits of Diodoquin could be explained by its contamination with zinc (Robertson and Sotos, 1975) or, probably more importantly, by its ability to form an absorbable zinc chelate (Moynahan, 1966) and thus to enhance zinc absorption (Jackson, 1977). However, there is some evidence that Diodoquin treatment may lead to optic atrophy and blindness (Berggren and Hansson, 1966).

The observation that AE could be treated with zinc, coupled with the knowledge that the onset of the disease usually occurs after weaning from breast milk to cow's milk (which generally contains more zinc than does breast milk), led Hurley *et al.* (1977a, 1977b) to the hypothesis that the zinc in human milk is bound differently from that in cow's milk. These investigators (Eckhert *et al.*, 1977) predicted that breast milk contained a specific zinc binding ligand not present in bovine milk, and postulated that this ligand-zinc complex produced higher bioavailability of zinc from human milk than from cow's milk. The difference in molecular localization of zinc in breast milk and in cow's milk has been the subject of considerable debate.

2.1.2.   Citrate

The search for the identity of the zinc binding ligand (ZBL) in human milk was initiated as a result of the observation that human milk, but not cow's milk, was useful in the treatment of AE, a genetic disorder of zinc metabolism. In the early studies by Hurley's group (Hurley *et al.*, 1977b; Eckhert *et al.*, 1977), fresh samples of milk from both humans and cows (Holstein) were examined by the method of gel filtration. Prior to gel filtration, milk samples were centrifuged at 10,000 g for 35 min at 4°C to separate the fat. When comparing the zinc content of skim milk from these two species, it was found that the average zinc concentration of cow's milk was 4.2 µg/ml, while the zinc concentration in mature human milk was 0.97 µg/ml. The skim milk from both species was subjected to gel filtration using Sepharose 2B, Sephadex G-200 and G-75.

Results from these studies showed that the molecular distribution of zinc was different in the two milks. In human milk, part of the zinc was associated with fractions of low molecular weight (LMW) and the remainder with high molecular weight (HMW) fractions. In cow's milk, virtually all zinc was bound to fractions of HMW. The molecular weight (MW) of the LMW complex was estimated to be 8,700 (Eckhert *et al.*, 1977). The early estimation of the MW of the LMW zinc complex was made with gels which had residual charges due to the presence of carboxylic groups on the gel matrix. The use of gels with residual charges can lead to poor recoveries, anomalous elution patterns, and ultimately to

Table I. Zinc and Citrate Concentrations in Mature Human and Bovine Milk

	N	Zinc Mean ± SEM		Citrate Mean ± SEM
		μg/ml	mM	mM
Human milk	8	1.07 ± 0.06	(0.016 ± 0.001)	
Human milk, ultracentrifuged[a]	8	0.51 ± 0.07	(0.008 ± 0.001)	
Human milk, ultrafiltered	8	0.20 ± 0.05	(0.003 ± 0.001)	0.28 ± 0.10
Bovine milk	8	3.47 ± 0.32	(0.053 ± 0.005)	
Bovine milk, ultracentrifuged	8	0.15 ± 0.02	(0.002 ± 0.001)	
Bovine milk, ultrafiltered	8	0.07 ± 0.01	(0.001 ± 0.000)	3.96 ± 0.47

[a]Defatted, decaseinated milk.

inappropriate conclusions regarding localization of trace metals. The same laboratory later conducted a second series of experiments concerning the identity of the ZBL in human milk; in these studies, the gels had been treated with sodium borohydride, a treatment that removes the carboxylic groups by reduction at alkaline pH and leads to recoveries of trace elements of 95-100% (Lönnerdal and Hoffman, 1981). Skim milk samples from both humans and cows were chromatographed on Sephadex G-50 and subsequently on Sephadex G-15; using this technique, a MW of 600-650 was obtained for the ZBL-zinc complex in both milks (Lönnerdal *et al.*, 1980a). To prepare sufficient amounts of this complex for further purification and identification, skim milk was subjected to ultrafiltration using a membrane with a MW cutoff of 1000. The concentration of zinc was 1.07 ± 0.06 μg/ml (mean ISEM) in human milk and 0.20 ± 0.05 μg/ml in human milk ultrafiltrate. The corresponding values for bovine milk were 3.47 ± 0.32 μg/ml zinc and 0.07 ±. 0.01 μg/ml zinc in the ultrafiltrate (Table I). These data are based on studies with mature human and bovine milk. A similar distribution of zinc is found for colostrum and transitional milk, although the absolute zinc concentrations are considerably higher during these periods (Lönnerdal *et al.*, 1981b).

Human and cow milk ultrafiltrates were applied to an anion-exchange column (DEAE-Sephadex A-25). The ZBL-zinc complexes eluted with a linear ionic strength gradient. Using IR and nmr spectroscopy, the organic moiety of the complex was identified as citrate. This identification was further confirmed by specific enzymatic assay which showed that citrate co-eluted with zinc in all chromatographic experiments. Spectroscopy data (IR, nmr) clearly demonstrated that no other ligand of significant quantity was present in the zinc fraction. The molecular weight of zinc citrate

$Zn_3(C_6H_5O_7)_2$ is 572 daltons, a figure which is in close agreement with the MW of 600 obtained from gel filtration chromatography. Synthetic zinc citrate applied to both gel filtration and ion-exchange columns eluted in exactly the same pattern as that shown by the human milk ZBL-zinc complex. The concentration of citrate was 0.28 ± 0.10 mM in human milk ultrafiltrates and 3.96 ± 0.47 mM in corresponding samples from bovine milk (Table I). Thus, although the citrate concentration was lower in human milk than in bovine milk, the amount of zinc complexed to citrate in human milk was about three times higher. Data have been presented indicating that at a lower pH (4.6), citrate is the LMW ligand binding zinc in human milk (Blakeborough et al., 1981). This implies that there also may be a physiological shift of zinc towards citrate during the digestive process.

The major HMW zinc binding compound of cow's milk is casein (Lönnerdal et al., 1981b), which is present in cow's milk at a concentration about 10 times higher than that of human milk (Hambraeus, 1977). Casein in raw cow's milk forms hard curds in the stomach of newborn infants, and a considerable portion, therefore, may pass through the gastrointestinal tract virtually undigested (Fomon, 1974). As yet, the impact of this phenomenon on infant nutrition and mineral bioavailability in general has not been well delineated. In particular, data have not been derived on developmental changes with regard to digestive capacity. Therefore, it is possible that zinc bound in this form is inaccessible to the young infant. Thus, even though the absolute amount of zinc is higher in cow's milk than in human milk, the amount available to the infant would be considerably less.

We have recently demonstrated that casein binds only a small fraction of the total zinc in human milk. The major zinc binding protein in human milk is serum albumin (Lönnerdal et al., 1981a, 1982). We have also found, as shown in Table I, that a significant portion of the zinc in human milk is bound to the fat fraction (Lönnerdal et al., 1981b). Alkaline phosphatase, bound to the fat globule membrane, appears to be the major zinc binding protein of milk fat (Lönnerdal and Fransson, 1981). The effect on zinc bioavailability of this difference in human zinc binding compounds between human milk and cow's milk has not yet been investigated, but it is possible that the zinc bound to serum albumin in human milk is more available than is the zinc bound to casein. These considerations may also relate to the therapeutic value of human milk in the treatment of AE.

The identification of the LMW ZBL in human milk as citrate is compatible with existing knowledge of this compound. Citrate is a well-known component of human milk, which was, in fact, the first material of animal origin in which this compound was detected about 100 years ago (Thunberg, 1953). Citrate has been shown to bind major cations in milk, such as calcium and magnesium (Jenness, 1974). It also plays an important role in certain important physiological processes such as bone formation (Sobel *et al.*, 1954). Citrate is a good chelating agent of iron, and this property has been exploited for iron supplementation using ferrous citrate (Brise and Hallberg, 1962). There are a number of reports showing that zinc citrate is effective in zinc absorption in turkeys (Vohra and Kratzer, 1966) and in rats (Giroux and Prakash, 1977, Wapnir *et al.*, 1981, Jackson *et al.*, 1981). We have analyzed the zinc and the citrate in milk from several species and have subjected skimmed milk samples to gel filtration chromatography. In the milk of one species, the rat, no measurable quantity of citrate has been found thus far. Further, rat's milk does not contain a LMW Zn complex (Lönnerdal *et al.*, 1981b).

2.1.3. Picolinic Acid

Another compound that has been implicated in the zinc absorption process is picolinic acid (pyridine-2-carboxylic acid). Evans and Johnson (1980b) reported the characterization and quantitation of picolinic acid in human milk and postulated that it is an important ZBL in human milk. These investigators reported the purification of picolinic acid from human milk and from cow's milk following ultrafiltration, ion-exchange chromatography, and gel filtration. The purified compound was identified as picolinic acid by the use of mass spectrometry, thin layer chromatography, and IR spectrophotometry. Following the identification by Evans and Johnson of picolinic acid as the ZBL in human milk, zinc picolinate was found to be efficacious when fed to children with disorders which responded to zinc therapy (Krieger, 1980). Furthermore, zinc picolinate has been reported to increase zinc absorption in rats (Evans and Johnson, 1980a). In a comparative study, when a diet that was marginally deficient in zinc was supplemented with zinc picolinate or zinc citrate, the apparent zinc retention in rats was similar. Similar results were obtained when zinc was given in the form of acetate, aspartate, or chloride (Hurley *et al.*, 1982). This indicates that when a marginally deficient diet is supplemented with zinc, the form of LMW zinc added may not be a primary determinant of zinc absorption. The availability to the rat of zinc from several ligands appears to be similar when measured using the vascular perfusion technique and by other methods (Menard and Cousins, 1982) and its availability to the human is similar using plasma zinc uptake (Casey *et al.*, 1981). It is evident that studies are needed on the relative ability of ligands to compete for dietary zinc which is bound to HMW zinc binding compounds (such as casein) from which zinc may be poorly absorbed.

The identification as picolinic acid of the LMW ligand binding zinc in milk has been questioned, partly because of the methodology employed in its isolation and subsequent quantitation (Hurley and Lönnerdal, 1980, 1981). One of the purification steps used was so-called "modified gel chromatography." This method is based on the addition of excess zinc ions to the elution buffer to saturate negative surface charges on the gel matrix (Yoza, 1977; Evans *et al.*, 1979b). This technique reveals all the ligands present in the sample, even those which do not actually chelate divalent cations *in vivo* such ligands may chelate and remove divalent cations originally present in the buffer from the gel, creating complexes not present *in vivo*. An additional problem is that ion-exchange chromatography, as well as gel chromatography, may occur on the columns. Thus, when a sample with higher ionic strength than the running buffer is applied, ligand exchange can occur. The combination of these problems can lead to the creation of "false peaks," yielding complexes which may have no relevance to the *in vivo* situation. Thus, while a zinc-picolinate complex may be present *in vivo*, the technique of modified gel chromatography should not be used as verification of its existence *in vivo*.

However, another laboratory, using the modified gel technique in a manner similar to that of Evans *et al.* (1979b), reported that the LMW zinc complex in human milk is zinc citrate (Martin *et al.*, 1981). This agrees with the findings of Lönnerdal *et al.* (1980a) using the borohydride treated columns. It has also been demonstrated that picolinic acid added to human milk ultrafiltrate does not elute at the same position as does the LMW zinc complex of human milk (Hurley and Lönnerdal, 1981). In a comparative study using both "modified gel chromatography" and conventional gel filtration on charge-free gels, in both cases zinc in human milk was found to elute in the same position as zinc citrate but not as zinc picolinate (Hurley and Lönnerdal, 1982). Picolinic acid is present in milk, but the amount estimated by Evans *et al.* (308 μM) seems to be too high by a large factor. Rebello *et al.* (1982), using high pressure liquid chromatography (HPLC), have reported the concentration of picolinic acid in human milk as <3.7 μM. This value is compatible with the concentration of other metabolites of tryptophan to which picolinic acid is related (Henderson and Swan, 1971). The low concentration of picolinic acid in human milk reported by Rebello *et al.* (1982) would appear to rule out any important role for this compound as a zinc complexing ligand *in vivo* in milk.

A negative influence of picolinic acid on the uptake of zinc by the small intestine of the rat has been reported by McMaster *et al.* (1981) using the vascular perfusion technique. Addition of picolinic acid to the vascular medium decreased the rate of zinc removal, while the same addition to the luminal medium resulted in secretion of zinc into the lumen. These investigators proposed that picolinic acid may cause damage to the mucosal cells when present at high concentrations. Jackson *et al.* (1981) and Wapnir *et al.* (1981) have performed comparative zinc absorption studies in rats fed zinc adequate diets; both picolinic acid and citrate were used as zinc chelators. A significant increase in zinc absorption was found for zinc citrate compared to control ($ZnCl_2$), while picolinic acid did not increase zinc absorption. These findings, together with other reports (Aggett *et al.*, 1982; Menard and Cousins, 1982), suggest that although picolinic acid is a chelator of zinc *in vitro*, zinc picolinate does not appear to have high bioavailability *in vivo* when administered to an animal fed a diet adequate in zinc. In contrast to the above, data recently have been presented showing that picolinic acid, when taken orally by adult males, may facilitate intestinal reabsorption of endogenously secreted zinc (Canfield *et al.*, 1982). Clearly, further studies are needed before the role of picolinic acid in zinc metabolism is elucidated.

2.1.4. Prostaglandins

Other investigators have identified prostaglandin $E_2(PGE_2)$ as the ZBL in human milk. This identification was based on gel filtration, ultrafiltration, thin layer chromatography, and infrared spectroscopy (Evans and Johnson, 1977; Song and Adham, 1978). The concept of PGE_2 as a ZBL is in itself interesting, as prostaglandins alter membrane permeability and have been reported to influence zinc absorption (Song and Adham, 1979). PGE_2 added to the mucosal medium *in vitro* increased the transport of zinc from the mucosal surface to the serosal surface by over 50%, while PGF_2 decreased it by 40%. In contrast, addition of PGE_2 to the serosal medium decreased zinc transport from serosa to mucosa by over 30%, while PGF_2 increased it by over 30%. *In vivo* studies have also been interpreted as suggesting a role for prostaglandins; oral administration of PGE_2 caused a twofold increase in zinc whereas PGF_2 resulted in decreased zinc concentrations of tissues (Song and Adham, 1978, 1979). This very indirect evidence may be interpreted in several other ways, however.

However attractive prostaglandin may be as an enhancer of zinc absorption, it is unlikely to be of importance in human milk, as its concentration is many times less than that of zinc (Zn/PG ratio $>10^3$) (Lönnerdal *et al.*, 1980a). In addition, it is known that zinc does not preferentially form coordination complexes with oxygen-rich molecules (Siegel and McCormick, 1970). One of the former proponents of the theory that zinc prostaglandin is the zinc complex in human milk suggested, in recent papers, that this proposal may have been based on improper chromatographic techniques (Evans, 1980) in a manner similar to that experienced by Eckhert *et al.* (1977). Despite the current consensus that zinc prostaglandin does not occur in milk, it must be pointed out that this does not rule out a role for prostaglandins in zinc metabolism at the mucosal site, perhaps by affecting membrane permeability (O'Dell, 1981).

2.2. Low Molecular Weight Zinc Complexes in Milk - A Role in Infant Nutrition

The search for the LMW zinc ligand in human milk was stimulated by reports on infants with AE, but information concerning the identity of this LMW zinc complex may be of importance for infant nutrition in general. Several clinical studies have demonstrated that the bioavailability of zinc from human milk is much higher than that from formulas based on cow's milk (Lönnerdal et al., 1981a). Most infants are in negative zinc balance during the first week of life (Cavell and Widdowson, 1964). Shortly after this time, full-term infants fed human milk may be in positive zinc balance (Dauncey et al., 1977). However, even with positive zinc balance, plasma zinc concentrations declined from adult levels at birth to levels about 30% lower at three months of age (Henkin et al., 1973). Walravens and Hambidge (1976) found that growth was better in male infants fed a zinc supplemented formula than in those given the unsupplemented formula. Even when formulas are supplemented with a level of zinc two to three times higher than that found in breast milk, plasma zinc values of formula-fed infants are lower (Walravens et al., 1978; Hambidge et al., 1979; MacDonald and Gibson, 1981). Findings of higher serum zinc levels in breast-fed infants than in infants fed a formula supplemented with zinc have also been reported in Japan and Canada (Ohtake, 1977; MacDonald and Gibson, 1981). A high bioavailability of zinc given in human milk, compared to cow's milk or formula, also has been reported in human adults using plasma zinc uptake studies (Casey et al., 1981). Thus, it appears that zinc in breast milk has a higher bioavailability than does zinc added to a formula in the form of a salt (usually zinc chloride). Zinc in breast milk also has been shown to have a higher bioavailability than does zinc in cow's milk or formula, using the rat as a model (Johnson and Evans, 1978b). The addition of zinc citrate to formulas may help to correct this difference.

3. Genetic Abnormalities of Zinc Metabolism

The potential role of LMW complexes in zinc absorption and zinc therapy is demonstrated by several genetic mutants that possess errors of zinc metabolism. Interactions occurring between zinc and genetic factors and affecting absorption can be classified into two groups (Hurley, 1976). The first involves strain differences which produce differential responses to a dietary deficiency of the element; the second type involves a single mutant gene, whose phenotypic expression may resemble the characteristic signs of a deficiency or toxicity of an element, and whose expression can be reduced or prevented by pre-, post-, or perinatal nutritional manipulation. In this review, we discuss four mutants which are representative of the second type of gene-nutrient interaction.

3.1.   Acrodermatitis Enteropathica in Human Beings

As discussed in 2.1.1., acrodermatitis enteropathica (AE) is an autosomal recessive disorder in man, characterized by lesions similar to those found in dietary zinc deficiency. These signs occur even though the infant (or adult) is consuming a diet which would normally be considered adequate in zinc content. Two apparently distinct types of genetic AE have been described in the literature.

3.1.1.   Genetic Acrodermatitis Enteropathica

The first type of AE is the "classical" one which is characterized by hypozincemia; plasma zinc levels may be lower than 50 µg/dl. The expression of the disease is usually precipitated by weaning the infant from breast milk to cow's milk, although there has been a report of one case of the disease occurring in an infant while still breast feeding (Ohlsson, 1981). The low plasma zinc levels can be increased to normal ($\geq$100 µg/dl) by dietary zinc supplementation. The average dosage used is approximately 150 µg of zinc sulfate per day. Serum alkaline phosphatase, a zinc metalloenzyme, which is subnormal in AE, also returns to normal following zinc treatment (Neldner and Hambidge, 1975). Concomitant with the increase in plasma zinc levels, the signs of the disease usually diminish. This type of AE is sometimes accompanied by severe diarrhea, which has been implicated in the etiology of the zinc deficiency (Ølholm-Larsen, 1979). However, it does not seem reasonable that the depression in plasma zinc levels should be ascribed to the diarrhea, as hypozincemia has also been found in some AE infants without diarrhea (Graves *et al.*, 1980; Ølholm-Larsen, 1978). It is possible that the diarrhea in AE infants is due to a cow's milk protein intolerance or to infections of the intestine exacerbated by the reduced immune competence of the infants.

The penetration of the AE gene is not well established. Some investigators (Hirsch *et al.*, 1976) have reported that heterozygous carriers of the gene have slightly reduced (25% lower than normal) serum zinc values. Other investigators have not been able to document the penetration of the gene (Ølholm-Larsen, 1979). Ohlsson (1981) recently reported a case of AE in which the mother had significantly lower than normal serum zinc, but the author suggested that this may have been due to a dietary cause rather than to gene penetration. The question of the penetration of the AE gene is an important one to resolve, as carriers of this gene may be more than normally prone to zinc deficiency in situations of marginal zinc intake or in periods of high zinc requirement such as pregnancy, adolescence, or wound healing.

The second type of AE is characterized by normal plasma zinc levels in infants but with other signs of zinc deficiency, such as alopecia and dermatitis (Garretts and Molokhia, 1977; Krieger and Evans, 1980). Like AE patients with hypozincemia, these individuals respond to zinc supplementation. It appears that the biochemical lesion(s) in AE patients with hypozincemia is different from that in AE patients with normal plasma zinc levels. In the first type, zinc transport acrosss the intestine is believed to be abnormal, but this may not be the case in those AE patients described as having normal zinc levels. AE patients with hypozincemia demonstrate reduced zinc absorption (up to 80% lower than normal) even when asymptomatic (Atherton *et al.*, 1979; Lombeck *et al.*, 1974, 1975; Walravens *et al.*, 1978). The biochemical mechanism of the reduced zinc absorption in patients with AE and hypozincemia has not yet been elucidated. The report that there are abnormal inclusions in the intestinal Paneth cells of AE patients, even when asymptomatic, suggests that these ultrastructural alterations are not a result of the zinc deficiency, but are in some way connected with the etiology of the condition. Experimental zinc deficiency does not result in similar ultrastructural abnormalities (Lombeck *et al.*, 1974; Wilson *et al.*, 1980), although it has been demonstrated by Elmes (1976) that there is an absence of zinc in duodenal cells of zinc deficient rats. Zinc absorption studies have not been reported for patients with the type of AE in which normal plasma zinc levels occur. If zinc absorption is normal in these patients, then the lesion in zinc metabolism may occur at receptor or transport sites in tissues other than the intestine. Fibroblast culture studies may be useful in discriminating between these two types of AE and in elucidating the differences between them.

3.1.2. "Acquired Acrodermatitis Enteropathica" – A Misnomer

There have been several reports of cases of AE which do not appear to have a genetic component involving zinc. Infants with this disease have usually consumed formulas low in zinc for prolonged periods of time or have been maintained for a long time with hyperalimentation preparations deficient in zinc (Weismann *et al.*, 1976; Katoh *et al.*, 1976; Bernstein and Leyden, 1978; Brazin *et al.*, 1979; Morishima *et al.*, 1981). These patients display signs quite similar to those observed in patients with genetic AE. Correction of the symptoms follows rapidly the introduction of zinc supplementation. Although these cases are interesting in that they demonstrate that dietary zinc deficiency can occur in normal infants (i.e., infants without the AE lesion), this disorder is not genetic. There are no data which suggest that zinc transport across the intestine of infants with "acquired AE" is impaired. Furthermore, the skin lesions typical of AE do not recur following the cessation of zinc therapy. Thus, it is totally inappropriate to call this type of disorder AE. Rather, it should be referred to simply as dietary zinc deficiency. The term AE should be restricted to cases in which zinc deficiency is caused by a genetic lesion.

3.2. Animal Models

There is still much to be learned regarding the metabolic defect in AE and the method of treatment for maximal therapeutic gain. The utilization of certain genetic mutants as animal models for the disease may provide valuable experimental information.

3.2.1. Lethal Milk in Mice

A potential model for studying some aspects of the human genetic disorder AE is the murine mutant *lethal milk (lm,* chromosome 2). The recessive mutant gene *lm* was discovered by Dickie (1969) in the C57BL/6J(B6) mouse strain. Phenotypic characteristics of this genotype are similar to some of the signs of AE. Offspring which suckle *lm/lm* dams exhibit stunted growth, alopecia, dermatitis, immunological incompetence, and rarely survive past weaning (Piletz and Ganschow, 1978a, 1978b; Beach *et al., 1980).* The *lm* lesion has been reported to be the inability of the suckling to obtain sufficient zinc from the milk of the mother (Piletz and Ganschow, 1978a, 1978b). However, in AE the metabolic lesion responsible for poor zinc nutriture is present in the infant; in the *lm* mutation, the defect resulting in neonatal lethality appears to be in the mother's milk. This conclusion is based on the observation that *lm/lm* pups cross-fostered onto normal dams do not display signs of the disorder until they are several months old; furthermore, normal pups cross-fostered onto *lm/lm* mothers quickly develop some of the phenotypic characteristics of the *lm* mutants (Green and Sweet, 1973; Dickie, 1969).

The zinc content of milk from *lm/lm* dams is significantly lower (35%) than that of normal mice (C57B1/6J) (Piletz and Ganschow, 1978a). This deficiency of zinc in the milk of the *lm/lm* is apparently compounded by decreased milk output by the *lm* mouse; thus zinc delivery is reduced to the suckling as a result both of consumption of milk low in zinc content and of a lower than normal intake (Piletz and Ganschow, 1978b). The decrease in milk production in the *lm* mutant may be a secondary effect of the *lm/lm* lesion, as it has been reported that lactating rats fed zinc deficient diets have lower than normal milk output (Mutch and Hurley, 1974, 1980). The pups of zinc deficient lactating rats showed several abnormal characteristics like those of *lm/lm* sucklings (Mutch and Hurley, 1974). It would be of interest to know if female AE carriers produce milk low in zinc content. Further evidence that it is the low zinc content of *lm/lm* milk that is responsible for many of the abnormalities in the suckling was provided by supplementation studies. When newborn *lm/lm* pups were injected with zinc glycinate, there was a significant improvement in their survival, weight, and coat growth (Piletz and Ganschow, 1978a, 1978b; Erway *et al.*, 1979).

While it is clear that the milk of the *lm/lm* mouse contains less zinc than that of a normal mouse, it is not known if this difference is exacerbated by a further difference in the molecular localization of the zinc. If there is a difference between *lm/lm* and normal mice in the localization of zinc in milk, this mutant may be a valuable model for studying *in vivo* the effect of zinc complexes in milk on zinc bioavailability.

Evidence that the normal mouse may also be a good model for studies on milk zinc is provided by the findings of Nishimura (1953). Nishimura reported that if normal newborn mice were deprived of zinc-rich colostrum and fostered by dams in mid- or late lactation, they developed severe zinc deficiency. In contrast, when fostering experiments similar to those of Nishimura were carried out in the rat, signs of zinc deficiency were not observed (Luckey *et al.*, 1954). This difference between the two species may be due in part to differences in liver zinc concentrations. Liver zinc concentration in the mouse does not change dramatically with age, while in the rat it declines with age (Keen and Hurley, 1980; Dungan *et al.*, 1980). Studies of changes with age in molecular localization of zinc in these two species suggest that the rat is born with a pool of liver zinc which is utilized during postnatal development, while the mouse lacks this pool and thus has a higher requirement for zinc during the early postnatal period (Keen *et al.*, 1981).

The deleterious effects of fostering newborn mice upon mice in mid- or late lactation may be related to changes in either the concentration or the localization of zinc in milk. If there is a change in the localization of zinc in mouse milk during the course of lactation, the normal mouse, as well as the *lm/lm* mouse, may be a good model for studying AE, particularly if a developmental change in the ability of the neonatal mouse to absorb the different zinc complexes can be demonstrated. Such an observation would be consistent with the hypothesis by Hurley *et al.* (1977) that the lesion produced by the AE gene is a failure in the maturation process of intestinal zinc absorption.

3.2.2. Adema Disease in Cattle

Another genetic lesion which is similar to AE is Adema disease in cattle (lethal trait A46, congenital thymus hypoplasia, congenital parakeratosis, hereditary zinc deficiency). This congenital disorder occurs in Black Pied cattle of Friesian descent. Genetic studies have shown that the disease is autosomal recessive (Andresen *et al.*, 1970). Affected calves are born "normal" but the signs of the disease usually appear 30-60 days postpartum. The signs include development of a dry scaly coat, alopecia, the formation of a hyperkeratotic crust around muscles, ears, and eyes, stomatitis, conjunctivitis, diarrhea, poor weight gain, immunological dysfunction (in particular severe thymic atrophy), and death occurring at three to four months of age (McPherson *et al.*, 1964; Grønborg-Pedersen, 1970; Stöber, 1971; Brummerstedt *et al.*, 1971, 1974; Kroneman *et al.*, 1975; Andresen *et al.*, 1970, 1974; Weismann and Flagstad, 1976; Bosma and Kroneman, 1979). Daily oral supplementation with zinc alleviates the phenotypic expression of the genotype. If supplementation with zinc is discontinued, the lesions reappear (Brummerstedt *et al.*, 1971; Stöber, 1971; Kroneman *et al.*, 1975; Flagstad, 1977).

As in AE patients, zinc absorption is impaired in cattle with Adema disease (Flagstad, 1976). The dosage used to treat Adema disease may be higher than 200 times the normal dietary zinc intake (Flagstad, 1976). However, the amount of zinc that must be supplemented can be reduced if chelating agents such as halogenated oxyquinolines are provided in the diet (Flagstad, 1977). It is not known whether the lesion is due to (1) a decreased ability by Adema cattle to catabolize zinc-containing proteins in the lumen, (2) failure to produce a LMW zinc binding ligand which would enhance zinc absorption from the lumen, or (3) a mutant protein in the mucosa of Adema cattle with high avidity for zinc, which may be by-passed by some, but not all, LMW zinc chelates. Interestingly, the removal rate of radiozinc from plasma is identical between normal and Adema calves, and the time of appearance of radiozinc in feces is similar (Flagstad, 1976). This suggests that the excretion of zinc into the intestine is not affected in Adema disease.

It is evident, therefore, that Adema disease and AE are similar disorders with similar signs that can be attributed to zinc deficiency. Thus, calves with Adema disease may be excellent models for studying the biochemical lesion in AE.

3.2.3.   Alaskan Malamute Chondrodysplasia in Dogs

Chondrodysplasia (short-limbed dwarfism) in Alaskan malamutes was initially described by Smart and Fletch (1971). While the disease causes several deformities, it is not life threatening. The mutant dogs have a severe bowing of the forelimbs with gross changes in the size and shape of the humerus, radius, and ulna. There is no difference between dwarf and normal dogs in the mineral content of long bones and forelimbs, but in some regions the calcium is more soluble than normal (as measured by extraction with EDTA). In addition to the apparent abnormality of calcium binding, the dwarfs have high levels of urinary acid mucopolysaccharides, suggesting a disturbance in normal bone maturation (Hoag *et al.*, 1976a, 1976b). Brown *et al.* (1978) have suggested that the abnormalities may be due in part to abnormal zinc metabolism, as there is some evidence that this element is required for normal connective tissue (McClain *et al.*, 1973; Pories *et al.*, 1976).

Dogs with this genetic lesion may also have hemolytic anemia. The red cells are macrocytic and hypochromic without concomitant reticulocytosis. Erythrocytes are fragile and have higher than normal intracellular cadmium and potassium and lower than normal glutathione. The specific defect that predisposes the cells to premature destruction is not known (Feldman, 1981). An additional sign of the disease is delayed sexual maturation. Mature dwarfs produce spermatozoa with 45% acrosomal defects, compared to 5% in controls. Significantly, this defect in spermatozoa can be corrected by dietary zinc supplementation (Brown *et al.*, 1978). Unfortunately, the effect of zinc supplementation on anemia, or on the occurrence of the bone abnormalities, has not been published. Although the correction of the spermatozoa defect by zinc supplementation suggests the presence of zinc deficiency, the zinc concentrations of liver, kidney, and pancreas are normal in the dwarfs, but heart zinc level is significantly lower (Brown *et al.*, 1977). Thus the defect in zinc metabolism may be tissue specific.

Recent *in vivo* studies by Brown *et al.* (1978) have shown that zinc absorption, as measured by the change in serum radioactivity after dogs were given an oral dose of ^{65}Zn, is significantly lower in dwarfs than in controls ($\leq 25\%$ of normal). Using an *in vitro* system, these investigators reported that the lesion in zinc absorption is a failure to release zinc from the mucosal cell to the circulation. In normal dogs absorbed zinc was first associated with a (presumably) protein fraction, then transferred to a nonprotein fraction, and finally to the circulation. In dwarfs, as in the normal dogs, the zinc was initially bound to a "protein" fraction, but there was no subsequent release to a nonprotein fraction. Unfortunately, the "protein" and nonprotein fractions discussed in this study were not purified. It is likely that the "protein" fraction is the 6000-8000 MW divalent cation binding protein metallothionein.

The identity of the nonprotein fraction is more difficult to predict. While it is tempting to suggest that this fraction represents an *in vivo* LMW ligand or complex involved in zinc transport out of the mucosal cell, it must be pointed out that in the *in vitro* study by Brown *et al.* (1978) tissues were incubated at 37°C for prolonged periods of time. Other investigators have shown that the chromatographic pattern of intestinal zinc in the rat is considerably altered by incubation of intestinal homogenates at 37°C; zinc is initially complexed to a protein with a MW of 6000-8000, but shifts to a LMW fraction (≤ 1500) during incubation.

This LMW zinc fraction is not observed if intestinal samples are prepared rapidly and chromatographed immediately; thus an *in vivo* role for this complex must be questioned (Cousins *et al.*, 1978; Lönnerdal *et al.*, 1980b). Whether or not the LMW zinc complex reported by Brown *et al.* (1978) has an *in vivo* role, their work does show that there is a difference between the dwarf and normal dogs in the metabolic handling of zinc at the site of the intestine.

It is interesting to note that if the lesion in the intestine of the dwarf involves a block at the site of metallothionein, because of either an abnormality in the protein itself or the absence of a transporting zinc ligand, this disease may be analogous to Menkes' syndrome, a genetic disorder of copper metabolism. In Menkes' disease, copper is absorbed, but not transported out of the mucosal cell (Danks *et al.*, 1978). Thus, the chondrodysplastic Alaskan malamute should be an excellent model for studying zinc absorption and the potential role of LMW ligands. Unfortunately for investigators in the field of zinc metabolism, attempts are now being made to eliminate heterozygous carriers of the dwarf gene from the breeding population (Harvey, 1980). It is hoped that at least a small number of these dogs will be maintained for further studies of zinc metabolism.

4.   Zinc Homeostasis

4.1.   Theories on Zinc Homeostasis

In the foregoing sections we have been concerned primarily with several genetic disorders of zinc metabolism. For each of these disorders it can be argued that part of the lesion may be due to a problem in zinc transport by LMW ligands and/or complexes. These lesions may be due either to an inadequate amount of the LMW ligands or to problems in transporting zinc to or from them. It is possible that under normal circumstances a maturational process occurs that eliminates the need for LMW ligands necessary or beneficial in the neonatal period. If this maturational shift is blocked, there may be a prolonged and enhanced requirement for LMW ligands to facilitate zinc absorption.

The concept that zinc metabolism is homeostatically controlled was introduced by Cotzias *et al.* (1962). Miller (1969) showed that this was true for ruminants. The idea that LMW ligands might function in normal zinc metabolism was initially suggested by the work of Starcher (1969), who reported the presence in mucosal cells of a LMW protein(s) (about 10,000 MW), that bound ^{65}Zn and ^{64}Cu given orally to chicks fed diets adequate in zinc. Later, VanCampen and Kowalski (1971) found that although some ^{65}Zn (either from an oral dose or added to rat intestinal homogenates) bound to a LMW fraction similar to that reported by Starcher (1969) and Evans *et al.* (1970), the majority of the ^{65}Zn was associated with high molecular weight (HMW) proteins. Kowarski *et al.* (1974) also found that the major zinc binding proteins of rat jejunal mucosa had high molecular weights (90-100,000). The potential role of these zinc species in absorption and subsequent release of zinc into the circulation has not been clarified. It has been demonstrated that there is relatively little induction of this class of proteins by high levels of zinc compared to that of the 10,000 MW species. The protein of the latter species, when purified, was found to have a MW of 6300-6600, with the chemical characteristics of metallothioneins (Richards and Cousins, 1977; Käi and Vallee, 1960).

Cousins (1979a, 1979b) has proposed that the amount of zinc absorbed from the gut is regulated via induction of metallothionein. According to this hypothesis, when the mucosal cell is exposed to excess zinc, metallothionein is induced, binds zinc, and is lost from the body by subsequent desquamation of the cell. Cousins has suggested that the metallothionein-bound zinc in the mucosal cell is not released to the circulation. It is interesting to note that this concept is in contrast to the known high turnover rate of induced metallothioneins in liver. Furthermore, zinc metallothionein does not appear to be a major zinc binding ligand in the intestinal cell unless a large amount of zinc is fed (Richards and Cousins, 1976). This observation suggests that the role of metallothionein may be in blocking the transport of zinc rather than in transporting it to the circulation, thereby acting as a homeostatic regulator. Another group of investigators (Evans *et al.*, 1979a) has argued against a role for metallothionein in zinc homeostasis, suggesting that zinc homeostasis is maintained through excretion. The idea that zinc homoestasis is maintained by secretion of zinc into the intestine was originally proposed by Weigand and Kirchgessner (1976a, 1977b).

Evans *et al.* (1976b) have suggested that there is a conflict between these two hypotheses: namely, that zinc status is controlled by (1) the amount of zinc absorbed being blocked by metallothionein or (2) secretion of zinc into the intestine. In our opinion, both of these factors may be involved in zinc homeostasis. Metallothionein may act as a fast "dampener" on meal to meal variations in zinc content, while excretion of zinc into the intestine may be a slower long range mechanism. Thus, if a single meal is high in zinc, part of it would be absorbed into the circulation and part of it would be bound to the metallothionein fraction. The metallothionein-bound zinc could be passed back into the intestinal lumen as a result of either the bidirectional flow of zinc which can occur across the brush border of the intact mucosal cell or the degradation of metallo-

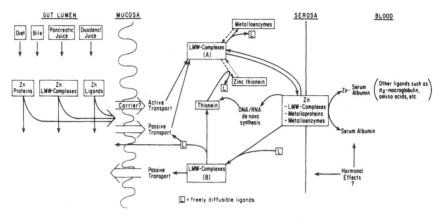

Fig. 1. A model of zinc absorption.

thionein following desquamation of the intestinal cell and its digestion by luminal enzymes. If the subsequent meal is low in zinc, the zinc released by this proposed process then could be absorbed and transferred to the circulation. If, however, the subsequent meal is high in zinc, the released zinc would be less likely to be absorbed. In addition, if the prior meal was also high in zinc, so that "zinc status" (tissue zinc concentration) was high, the amount of zinc absorbed would be even further reduced. In this manner the two proposed mechanisms of zinc homeostasis could be acting in concert.

In conclusion, we believe three different roles may be suggested for metallothioneins in the intestinal mucosal cell: (1) they act as detoxifying agents for divalent cations; (2) zinc metallothionein may provide zinc for zinc-requiring proteins or processes in the mucosal cell, and (3) they play a role in zinc homeostasis. Dietary zinc may bind to metallothionein, subsequently releasing zinc to another ligand which is in, or can pass across, the basolateral membrane; from here it can be released into the circulation if ligands that bind zinc are present (Fig. 1).

4.2. Zinc Binding Ligands in Bile and Pancreatic Fluid

A role for LMW zinc ligands in normal zinc homeostasis involving facilitation of zinc uptake by the mucosal cell was proposed by Evans and his colleagues (1975). They suggested that one factor in zinc homeostasis is a LMW zinc binding ligand secreted by the pancreas which binds zinc in the intestinal lumen; this complex is then transported through the intestinal microvilli into the epithelial cell. According to this hypothesis, the zinc then is transferred to binding sites in the basolateral membrane, with subsequent transfer to plasma albumin. It was suggested that the amount of metal-free albumin in the plasma is the major determinant of the amount of zinc transported out of the cell. While this hypothesis is attractive, methodological problems in its development may have led to inappropriate conclusions. One problem was that the pancreatic LMW zinc binding ligand discussed by Evans and coworkers was isolated from homogenized rat pancreas and not from pancreatic secretions. When pancreatic juice is collected by direct cannulation of the pancreatic duct in the rat, it can be shown that it does not contain either LMW complexes or an appreciable amount of LMW ligands. Addition of zinc to pancreatic fluid resulted in zinc binding by HMW rather than LMW zinc binding ligands (Lönnerdal *et al.*, 1980c). This finding argues against the idea that there are LMW ligands in pancreatic secretions that facilitate zinc absorption.

Lack of a role for pancreatic secretions in zinc absorption in the rat has been reported (Antonson *et al.*, 1979). However, rat bile contains LMW zinc binding ligands and zinc complexes. Over 90% of the zinc in bile occurs in complexes with molecular weights less than 2000. Two LMW peaks have been found (Cikrt *et al.*, 1974; Lönnerdal *et al.*, 1980c), but they have not been characterized. The observation that zinc added to bile binds to the two LMW fractions (Lönnerdal *et al.*, 1980c) is consistent with the idea that the LMW ligands in bile may facilitate dietary zinc absorption. While it is not known if biliary zinc complexes undergo enterohepatic circulation, it has been reported that other divalent cations (manganese, copper, and mercury) are reabsorbed (35%, 17%, and 21%, respectively) after biliary excretion (Cikrt, 1973). It is thought, at least in the case of copper, that the LMW metal complexes in bile are important for copper reabsorption and therefore for copper homeostasis (Cikrt *et al.*, (1974). The finding that zinc added to bile is associated with LMW compounds stresses the need to clarify the importance of this compound(s) in zinc metabolism, since there may be a similar mechanism for other cations. Methfessel and Spencer (1973) have presented data that support the idea that biliary zinc may be reabsorbed, although their own conclusion concerns pancreatic zinc. Also, Birnstiegl *et al.* (1956) have shown that there is a considerable amount of zinc secreted in dog bile. We conclude that the bioavailability of exogenous zinc may be affected by biliary secretions through the complexing of dietary zinc by LMW compounds in bile (Fig. 1). These zinc complexes could then be absorbed by the microvilli in a manner analogous to the hypothesis earlier suggested by Evans *et al.* (1975).

4.3. Zinc Binding Ligands and Complexes in Duodenum

Another possible origin of zinc binding ligands could be duodenal secretions exclusive of pancreatic and biliary fluids. Casey *et al.* (1978) have examined zinc binding in human duodenal secretions and reported that zinc was associated with complexes of 20,000, 18,000, and 6-8000 MW. These workers did not report the presence of LMW zinc complexes in duodenal secretions. However, in the same paper it was noted that the 6000-8000 MW peak found in duodenal secretions eluted in the same position as the LMW complex present in human milk. The authors cited the apparent MW of 6000-8000 as being in agreement with the MW reported by Eckhert *et al.* (1977) for the LMW zinc complex in human milk. We have since shown, however, that this MW was incorrect and that the true MW is approximately 600 (Lönnerdal *et al.*, 1980a, 1980b). Thus, it is possible that the actual MW of the LMW zinc complex reported by Casey *et al.* (1978) in human duodenal secretions was less than 2000. This point needs to be resolved, particularly since Casey *et al.* (1978) found that the 6000-8000 MW zinc complex was abnormal in AE patients. Similar investigations of duodenal and biliary secretions from the *lm* mouse, Alaskan chondrodysplastic malamute, and Adema calf, along with their respective nonmutant controls, are needed.

4.4. Methodological Considerations

It is evident that the mechanism of zinc entry into mucosal cells has not been elucidated. Electron probe studies of the epithelial membrane, as well as histochemical studies with antibodies to different zinc complexes, are needed. The homeostasis hypothesis as proposed by Evans *et al.* (1975) suggests that LMW zinc complexes would be found in the mucosal cell. In support of the hypothesis, several investigators have reported the presence of LMW zinc complexes in rat intestinal mucosa (Hahn and Evans, 1973; Johnson and Evans, 1978a; Hurley *et al.*, 1977b; Duncan and Hurley, 1978; Schricker and Forbes, 1978). A developmental aspect to the occurrence of this LMW zinc complex in rat intestine has been proposed (Duncan and Hurley, 1978). However, Cousins *et al.* (1978) reported that such LMW zinc complexes were not found in rat intestine if the samples were prepared rapidly and chromatographed immediately, and they suggested that these complexes were degradation products. This conclusion was based on the observation that a 4 hr incubation at 37°C resulted in a shift of zinc from high molecular weight compounds to LMW compounds. This observation has been confirmed and extended by Lönnerdal *et al.* (1980b), who found a significant shift in zinc localization in rat intestine after only 30 min of incubation.

An additional problem which may contribute to discrepancies in results among laboratories may be the use of gels with residual charges. Although, in our opinion, earlier experimenters have overestimated the amount of LMW (≤ 2000) zinc present in mucosal cells, these complexes can be detected in some situations even when charge-free gels are used and extreme precautions are taken to process tissues immediately at low temperature and in a nitrogen atmosphere (Keen *et al.*, 1981). The half-life of this compound(s) appears to be extremely short, and techniques such as freeze clamping and pulse labeling may be needed to ascertain the function, if any, of these complexes. In rat and mouse intestine that is handled carefully, with precautions that reduce shifts in localization, zinc is present in two major molecular classes with MW of $\geq 30{,}000$ and $\simeq 8000$, with most of the zinc in the former peak. The latter class of protein (MW $\simeq 8000$) probably consists of metallothioneins. We have found that there are pronounced age-related changes in the amount of zinc and the proportion associated with this class of proteins (Keen *et al.*, 1981).

4.5. A Proposed Model of Zinc Absorption

The role(s) of LMW zinc ligands and complexes in zinc homeostasis is still poorly understood. In Fig. 1, using zinc as a model, several potential loci at which LMW ligands may have a role are shown schematically. In this figure, zinc in the intestine is shown to be derived from the diet, pancreatic juice, bile, and duodenal secretions. As discussed earlier, zinc in the diet, bile, and duodenal secretions may be present in both HMW and LMW complexes, while zinc in pancreatic juice is found only in HMW form (Lönnerdal *et al.*, 1980c). Free LMW ligands capable of binding zinc may be present in all four of these zinc sources. Interchange of zinc among zinc proteins, zinc complexes, and zinc ligands occurs according to their relative concentrations and affinity constants and gastrointestinal conditions. In addition, gut transit time must be considered. Some of the zinc proteins may be absorbed by pinocytosis and others may be degraded by pancreatic proteases, resulting in redistribution of zinc among the different ligands. The bioavailability of the various LMW Zn-complexes will be dependent on the partition coefficients and on transport sites in the mucosal cell. As Fig. 1 shows, zinc transport may be either active (ATP-dependent) or passive. It is not known whether the transport site involves a carrier, and the fate of zinc once it has crossed the membrane has not been delineated.

We propose that after zinc enters the cytoplasmic pool of intestinal cells it forms LMW complexes (A) with short half-lives. The zinc from these complexes then can be transferred to metalloenzymes, to thionein, or across the serosal or mucosal border. The metalloenzyme (HMW) pool would appear to have first priority for zinc, since the HMW zinc pool is fairly constant regardless of dietary zinc status (Cousins, 1979a). The secondary priority for zinc would be its transport across the serosal border where it is then transferred to plasma proteins, albumin being the major receptor, or to a smaller extent to LMW ligands such as amino acids. Transport through the serosal border may occur through diffusion of

LMW complexes or may involve a carrier. If the transport site(s) of zinc on the serosal membrane becomes saturated either as a result of high intracellular zinc or high zinc binding of plasma proteins and ligands, more zinc will bind to the LMW intracellular zinc complexes, increasing their concentration. One result of such an increase could be the binding of zinc by other LMW ligands to form another type of LMW complex (B). If the concentration of A and/or B is high enough, *de novo* synthesis of thionein is increased. This mechanism would work particularly well if the ligand free of zinc acted as a restrictor of thionein synthesis; thus a ligand with sulfhydryl groups may bind to RNA, preventing its transcription/ translation. When the intracellular concentration of zinc is high enough, the ligand would form a complex, thereby removing the suppressor. Synthesis of thionein subsequently will result in increased zinc thionein, decreased LMW intracellular zinc and thus the suppression of thionein synthesis by the putative ligand. We propose that the fate of the zinc bound to thionein (zinc thionein) depends on subsequent cellular uptake of zinc. Thus, if the dietary intake of zinc or plasma zinc is reduced, zinc from zinc thionein can be released and complexed with free ligands and transported to the circulation. On the other hand, if dietary zinc intake remains high, thionein zinc will not be released. By this mechanism, thionein acts as a "dampener" in zinc absorption in a manner similar to that proposed by Cousins (1979a, 1979b). We also propose that this scheme of zinc absorption has a developmental aspect. Maturational changes may occur at any or all of the sites described above. LMW zinc complexes such as those found in human milk may have an important role during this period as they may allow the bypassing of one of these steps, most likely the one of membrane transport.

It should be emphasized that this diagram is a simplistic linear approach to zinc absorption and its control. Other factors such as hormonal input or prostaglandins and their effect on membrane permeability may alter this flow scheme.

5.   Concluding Remarks

The focus of this review has been on the process of zinc absorption and its potential facilitation by LMW ligands. It is apparent that the study of genetic lesions of zinc metabolism, particularly those involving absorption, can provide considerable information about normal zinc metabolism. Future work is needed to study the individual loci of zinc absorption and metabolism in the intestinal cell and, in our opinion, special emphasis should be directed to developmental changes in these steps as well as in dietary supply and/or endogenous secretions. Finally, studies on the identification of optimal ligands for enhancing zinc absorption could be of great benefit, in particular in their use in infant formulas and therapy.

References

Aggett, P.J., Fenwick, P.K., and Kirk, H., 1982, A study *in vitro* of the effect of picolinic acid on metal translocation across model membranes, *Proc. Nutr. Soc.* **41**:68A.

Andresen, E., Flagstad, T., Basse, A., and Brummerstedt, E., 1970, Evidence of a lethal trait, A46, in Black Pied Danish Cattle of Friesian descent, *Nord. Vet. Med.* **22**:473.

Andresen, E., Basse, A., Brummerstedt, E., and Flagstad, T., 1974, Lethal trait A46 in cattle. Additional genetic investigations, *Nord. Vet. Med.* **26**:275.

Antonson, D.L., Barak, A.J., and Vanderhoof, J.A., 1979, Determination of the site of zinc absorption in rat small intestine, *J. Nutr.* **109**:142.

Atherton, D.J., Muller, D.P.R., Aggett, P.J., and Harries, J.T., 1979, A defect in zinc uptake by jejunal biopsies in acrodermatitis enteropathica, *Clin. Sci.* **56**:505.

Beach, R.S., Gershwin, M.E., and Hurley, L.S., 1980, T cell function in the lethal milk *(lm/lm)* mutant mouse, in: *Immunology 80: Fourth International Congress of Immunology* (M. Fougereau and J. Dausset, eds.), Academic Press, New York.

Berggren, L., and Hansson, O., 1966, Treating acrodermatitis enteropathica, *Lancet* **1**:52.

Bergsma, D., 1973, in: *Birth Defects,* Williams and Wilkins, Baltimore.

Bernstein, B., and Leyden, J.J., 1978, Zinc deficiency and acrodermatitis after intravenous hyperalimentation, *Arch. Dermatol.* **114**:1070.

Birnstiegl, M., Stone, B., and Richards, V., 1956, Excretion of radioactive zinc (Zn^{65}) in bile, pancreatic and duodenal secretions of the dog, *Am. J. Physiol.* **186**:377.

Blakeborough, P., Salter, D.N., and Gurr, W.I., 1981, Zinc binding in human and cow's milk, in: *Nutrition in Health and Disease and International Development:* symposia from the 12th International Congress of Nutrition (A.E. Harper and G.K. Davis, eds.), no. 957, Liss, New York.

Bosma, A.A., and Kroneman, J., 1979, Chromosome studies in cattle with hereditary zinc deficiency (lethal trait A46), *Vet. Quart.* **1**:121.

Brandt, T., 1936, Dermatitis in children with disturbances of the general condition and the absorption of food elements, *Acta Derm. Venereol.* **17**:513.

Brazin, S.A., Johnson, W.T., and Abramson, L.J., 1979, The acrodermatitis enteropathica-like syndrome, *Arch. Dermatol.* **115**:597.

Brise, H., and Hallberg, L., 1962, Absorbability of different iron compounds, *Acta Med. Scand.* **171**:Suppl. 376, 23.

Brown, R.G., Hoag, G.N., Smart, M.E., Boechner, G., and Subden, R.E., 1977, Alaskan malamute chondrodysplasia. IV. Concentrations of zinc, copper, and iron in various tissues, *Growth* **41**:215.

Brown, R.G., Hoag, G.N., Smart, M.E., and Mitchell, L.H., 1978, Alaskan malamute chondrodysplasia. V. Decreased gut zinc absorption, *Growth* **42**:1.

Brummerstedt, E., Flagstad, T., Basse, A., and Andresen, E., 1971, The effect of zinc on calves with hereditary thymus hypoplasia (lethal trait A46), *Acta Path. Microbiol. Scand., section A* **79**:686.

Brummerstedt, E., Andresen, E., Basse, A., and Flagstad, T., 1974, Lethal trait A46 in cattle. Immunological investigations, *Nord Vet-Med.* **26**:279.

Canfield, W., Lykken, G., Milne, D., and Sandstead, H., 1982, Effect of oral picolinic acid on ^{65}Zn retention in normal men, *Am. J. Clin. Nutr.* **35**:843.

Casey, C.E., Hambidge, K.M., Walravens, P.A., Silverman, A., and Neldner, K.H., 1978, Zinc binding in human duodenal secretions, *Lancet* **2**:423.

Casey, C.E., Walravens, P.A., and Hambidge, K.M., 1981, Availability of zinc: loading tests with human milk, cow's milk, and infant formulas, *Pediatrics* **68**:394.

Cavell, P.A., and Widdowson, E.M., 1964, Intakes and excretions of iron, copper and zinc in the neonatal period, *Arch. Dis. Child.* **39**:496.

Chandra, R.K., 1980, Acrodermatitis enteropathica: zinc levels and cell-mediated immunity, *Pediatrics* **66**:789.

Chvapil, M., 1973, New aspects in the biological role of zinc: a stabilizer of macromolecules and biological membranes, *Life Sci.* **13**:1041.

Cikrt, M., 1973, Enterohepatic circulation of ^{64}Cu, ^{52}Mn, and ^{203}Hg in rats, *Arch. Toxicol.* **31**:51.

Cikrt, M., Havrdova, J., and Tichy, M., 1974, Changes in the binding of copper and zinc in the rat bile during 24 hours after application, *Arch. Toxicol.* **32**:321.

Cotzias, G.C., Bong, C., and Selleck, B., 1962, Specificity of zinc pathway through the body: turnover of ^{65}Zn in the mouse, *Am. J. Physiol.* **202**:359.

Cousins, R.J., 1979a, Regulation of zinc absorption: role of intracellular ligands, *Am. J. Clin. Nutr.* **32**:339.

Cousins, R.J., 1979b, Regulatory aspects of zinc metabolism in liver and intestine, *Nutr. Rev.* **37**:97.

Cousins, R.J., Smith, K.T., Failla, M.L., and Markowitz, L.A., 1978, Origin of low molecular weight zinc binding ligands in rat intestine, *Life Sci.* **23**:1819.

Danbolt, N., and Closs, K., 1943, Acrodermatitis enteropathica, *Acta Derm. Venereol.* **23**:127.

Danks, D.M., Camakaris, J., and Stevens, B.J., 1978, The cellular defect in Menkes' syndrome and in mottled mice, in: *Trace Element Metabolism in Man and Animals - 3* (M. Kirchgessner, ed.), pp. 401-404, Technische Universität München, Freising-Weihenstephan.

Dauncey, M.J., Shaw, J.C.L., and Urman, J., 1977, The absorption and retention of magnesium, zinc and copper by low birth weight infants fed pasteurized breast milk, *Pediatr. Res.* **11**:911.

Dickie, M.M., 1969, Lethal-milk: a new mutation in the mouse, *Mouse News Letter* **41**:30.

Dillaha, C.J., Lorincz, A.L., and Aavik, O.R., 1953, Acrodermatitis enteropathica. Review of the literature and report of a case successfully treated with diodoquin, *J. Am. Med. Assn.* **152**:509.

Dreosti, I.E., and Hurley, L.S., 1975, Depressed thymidine kinase activity in zinc-deficient rat embryos, *Proc. Soc. Exp. Biol. Med.* **150**:161.

Dreosti, I.E., Grey, P.C., and Wilkins, P.J., 1972, Deoxyribonucleic acid synthesis protein synthesis and teratogenesis in zinc-deficient rats, *S. Afr. Med. J.* **46**:1585.

Duncan, J.R, and Hurley, L.S., 1978, Intestinal absorption of zinc: a role for a zinc binding ligand in milk, *Am. J. Physiol.* **235**:E556.

Dungan, D.D., Keen, C.L., Lönnerdal, B., and Hurley, L.S., 1980, Developmental changes in concentrations of iron, copper and zinc in mouse and rat tissues, *Fed. Proc.* **39**:903.

Eckhert, C.D., and Hurley, L.S., 1977, Reduced DNA synthesis in zinc deficiency: regional differences in embryonic rats, *J. Nutr.* **107**:855.

Eckhert, C.D., Sloan, M.V., Duncan, J.R., and Hurley, L.S., 1977, Zinc binding: a difference between human and bovine milk, *Science* **195**:789.

Eichhorn, G.L., Berger, N.A., Butsow, J.J., Clark, P., Hein, J., Pitha, I., Richardson, C., Rifkind, R.M., Shin, Y., and Tarien, E., 1973, in: *Metal Ions in Biologial Systems* (S.K. Dhar, ed.), p. 43, Plenum Press, New York.

Elmes, M.E., 1976, The Paneth cell population of the small intestine of the rat — effects of fasting and zinc deficiency on total count and on dithizone-reactive count, *J. Path.* **118**:183.

Entwisle, B.R., 1965, Acrodermatitis enteropathica. Report of a case in a twin with dramatic response to expressed human milk, *Austral. J. Derm.* **8**:13.

Erway, L.C., Piletz, J.E., and Ganschow, R.E., 1979, Zinc supplementation prevents mortality in lethal-milk mice, *Mouse News Letter* **60**:41.

Evans, G.W., 1980, Normal and abnormal zinc absorption in man and animals: the tryptophan connection, *Nutr. Rev.* **38**:137.

Evans, G.W., and Johnson, P.E., 1977, Defective prostaglandin synthesis in acrodermatitis enteropathica, *Lancet* **1**:52.

Evans, G.W., and Johnson, E.C., 1980a, Zinc concentration of liver and kidneys from rat pups nursing dams fed supplemental zinc dipicolinate or zinc acetate, *J. Nutr.* **110**:2121.

Evans, G.W., and Johnson, P.E., 1980b, Characterization and quantitation of a zinc binding ligand in human milk, *Pediatr. Res.* **14**:876.

Evans, G.W., Majors, P.F., and Cornatzer, W.E., 1970, Mechanism of cadmium and zinc antagonism of copper metabolism, *Biochem. Biophys. Res. Commun.* **40**:1142.

Evans, G.W., Grace, C.I., and Votava, H.J., 1975, A proposed mechanism for zinc absorption in the rat, *Am. J. Physiol.* **228**:501.

Evans, G.W., Johnson, E.C., and Johnson, P.E., 1979a, Zinc absorption in the rat determined by radioisotope dilution, *J. Nutr.* **109**:1258.

Evans, G.W., Johnson, P.E., Brushmiller, J.G., and Ames, R.W., 1979b, Detection of labile zinc binding ligands in biological fluids by modified gel filtration chromatography, *Anal. Chem.* **51**:839.

Feldman, B.F., 1981, Anemias associated with blood loss and hemolysis, *Vet. Clin. North America: Small Animal Practice* **11**:265.

Flagstad, T., 1976, Lethal trait A46 in cattle, *Nord. Vet.-Med.* **28**:160.

Flagstad, T., 1977, Intestinal absorption of ^{65}zinc in A46 (Adema disease) after treatment with oxychinolines, *Nord. Vet.-Med.* **29**:96.

Flagstad, T., 1981, Zinc absorption in cattle with a dietary picolinic acid supplement, *J. Nutr.* **40**:65.

Fomon, S.J., 1974, in: *Infant Nutrition,* pp. 370-371, W.B. Saunders Co., Philadelphia.

Freier, S., Faber, J., Goldstein, R., and Mayer, M., 1973, Treatment of acrodermatitis enteropathica by intravenous amino acid hydrolysate, *J. Pediatr.* **82**:109.

Garretts, M., and Molokhia, M., 1977, Acrodermatitis enteropathica without hypozincemia, *J. Pediatr.* **91**:492.

Giroux, E., and Prakash, N.J., 1977, Influence of zinc-ligand mixtures on serum zinc levels in rats, *J. Pharm. Sci.* **66**:391.

Graves, K., Kestenbaum, T., and Kalivas, J., 1980, Hereditary acrodermatitis enteropathica in an adult, *Arch. Dermatol.* **116**:562.

Green, M.C., and Sweet, H.O., 1973, The Jackson Laboratory—Linkages and Chromosomes, *Mouse News Letter* **48**:35.

Grønborg-Pedersen, H., 1970, Morbus ademae, *Medlemsbl. Danske Dyrlaege forening.* **53**:143.

Hahn, C.J., and Evans, G.W., 1973, Identification of a low molecular weight ^{65}Zn complex in rat intestine, *Proc. Soc. Exp. Biol. Med.* **144**:793.

Hambidge, K.M., Hambidge, C., Jacobs, M., and Baum, J.D., 1972, Low levels of zinc in hair, anorexia, poor growth and hypogeusia in children, *Pediatr. Res.* **6**:868.

Hambidge, K.M., Walravens, P.A., Brown, R.M., Webster, J., White, S., Anthony, M., and Roth, M.L., 1976, Zinc nutrition of preschool children in the Denver Head Start program, *Am. J. Clin. Nutr.* **29**:734.

Hambidge, K.M., Walravens, P.A., Casey, C.E., Brown, R.M., and Bender, C., 1979, Plasma zinc concentrations of breast-fed infants, *J. Pediatr.* **94**:607.

Hambraeus, L., 1977, Proprietary milk versus human breast milk in infant feeding, *Pediatr. Clin. N. Amer.* **24**:17.

Harvey, J.W., 1980, Canine hemolytic anemias, *J. Am. Vet. Med. Assoc.* **176**:970.

Henderson, L.M., and Swan, P.B., 1971, Picolinic acid carboxylase, in: *Methods in Enzymology* Vol. 18B (D.B. McCormick and L.D. Wright, eds.), pp. 175-180, Academic Press, New York.

Henkin, R.I., Schulman, J.D., Schulman, C.B., and Bronzert, D.A., 1973, Changes in total, nondiffusible and diffusible plasma zinc and copper during infancy, *J. Pediatr.* **82**:831.

Hirsch, F.S., Michel, B., and Strain, W.H., 1976, Gluconate zinc in acrodermatitis enteropathica, *Arch. Dermatol.* **112**:475.

Hoag, G.W., Brown, R.G., Smart, M.E., and Subden, R.E., 1976a, Alaskan malamute chondrodysplasia. I. Bone composition studies, *Growth* **40**:3.

Hoag, G.W., Brown, R.G., Smart, M.E., and Subden, R.E., 1976b, Alaskan malamute chondrodysplasia. II. Urinary excretion of hydroxyproline, uronic acid and acid mucopolysaccharides, *Growth* **40**:13.

Hurley, L.S., 1976, Interaction of genes and metals in development, *Fed. Proc.* **35**:2271.

Hurley, L.S., 1980, in: *Developmental Nutrition,* 335 pp., Prentice Hall, Englewood Cliffs, NJ.

Hurley, L.S., and Lönnerdal, B., 1980, Tryptophan, picolinic acid and zinc absorption: an unconvincing case, *J. Nutr.* **110**:2536.

Hurley, L.S., and Lönnerdal, B., 1981, Picolinic acid as a zinc–binding ligand in human milk: an unconvincing case, *Pediatr. Res.* **15**:166.

Hurley, L.S., and Lönnerdal, B., 1982, Zinc binding in human milk: citrate versus picolinate, *Nutr. Rev.* **40**:65.

Hurley, L.S., Eckhert, C.D., Duncan, J.R., and Sloan, M.V., 1977a, Acrodermatitis enteropathica and human breast milk, *Lancet* **1**:195.

Hurley, L.S., Duncan, J.R., Sloan, M.V., and Eckhert, C.D., 1977b, Zinc-binding ligands in milk and intestine: A role in neonatal nutrition? *Proc. Nat. Acad. Sci. USA* **74**:3547.

Hurley, L.S., Keen, C.L., Young, H.M., and Lönnerdal, B., 1982, Effect of chelates on zinc-concentration in rat maternal and pup tissues, *Fed. Proc.* **41**:781.

Jackson, M.J., 1977, Zinc and diodohydroxyquinoline therapy in acrodermatitis enteropathica, *J. Clin. Pathol.* **30**:284.

Jackson, M.J., Jones, D.A., and Edwards, R.H.T., 1981, Zinc absorption in the rat, *Br. J. Nutr.* **46**:15.

Jameson, S., 1976, Effects of zinc deficiency in human reproduction, *Acta Med. Scand.* Suppl. 593.

Jameson, S., 1981, Zinc nutrition and pregnancy in humans, in: *Fourth International Symposium on Trace Elements in Man and Animals* (TEMA-4) (J. McHowell, J.M. Gawthorne, and C.L. White, eds.), pp. 243-248, Griffin Press Ltd., Netley, South Australia.

Jenness, R., 1974, The composition of milk: salts, in: *Lactation* Vol. III (B.L. Larson and V.R. Smiths, eds.), pp. 43-44, Academic Press, New York and London.

Johnson, P.E., and Evans, G.W., 1978a, Identification of a prostaglandin E_2-zinc complex in human breast milk, and porcine and rat duodenum, *Fed. Proc.* **37**:889.

Johnson, P.E., and Evans, G.W., 1978b, Relative zinc availability in human breast milk, infant formulas, and cow's milk, *Am. J. Clin. Nutr.* **31**:416.

Käi, J.H.R., and Vallee, B.L., 1960, Metallothionein: a cadmium and zinc containing protein from equine renal cortex, *J. Biol. Chem.* **235**:3460.

Katoh, T., Igarashi, M., Okhashi, E., Ohi, R., Hebiguchi, T., and Seiji, M., 1976, Acrodermatitis enteropathica-like eruption association with parenteral nutrition, *Dermatologica* **152**:119.

Keen, C.L., and Hurley, L.S., 1980, Developmental changes in concentrations of iron, copper, and zinc in mouse tissues, *Mech. Ageing Dev.* **13**:161.

Keen, C.L., Lönnerdal, B., and Hurley, L.S., 1981, Developmental changes in zinc and copper in mouse and rat tissues, in: *Fourth International Symposium on Trace Elements in Man and Animals* (TEMA-4) (J. McHowell, J.B. Gawthorne, and C.L. White, eds.), pp. 287-290, Griffin Press Ltd., Netley, South Australia.

Kowarski, S., Blair-Stanek, C.S., and Schachter, D., 1974, Active transport of zinc and identification of zinc binding protein in rat jejunal mucosa, *Am. J. Physiol.* **226**:401.

Krieger, I., 1980, Picolinic acid in the treatment of disorders requiring zinc supplementation, *Nutr. Rev.* **38**:148.

Krieger, I., and Evans, G.W., 1980, Acrodermatitis enteropathica without hypozincemia. Therapeutic effect of a pancreatic enzyme preparation due to a zinc binding ligand, *J. Pediatr.* **96**:32.

Kroneman, J., Mey, G.J.W. van der, and Helder, A., 1975, Hereditary zinc deficiency in Dutch Friesian cattle, *Zbl. Vet. Med. A* **22**:201.

Lombeck, I., Von Bassewitz, D.B., Becker, K., Tinschmann, P., and Kastner, H., 1974, Ultrastructural findings on acrodermatitis enteropathica, *Pediatr. Res.* **8**:82.

Lombeck, I., Schnippering, H.G., Ritzl, F., Feiendegen, L.E., and Bremer, H.J., 1975, Absorption of zinc in acrodermatitis enteropathica, *Lancet* **1**:855.

Lönnerdal, B., and Fransson, G.B., 1981, Distribution of copper, zinc, calcium and magnesium in human milk, in: *Nutrition in Health and Disease and International Development: symposia from the 12th International Congress of Nutrition* (A.E. Harper and G.K. Davis, eds.), no. 220, Liss, New York.

Lönnerdal, B., and Hoffman, B., 1981, Alkaline reduction of dextran gels and cross-linked agarose to overcome nonspecific binding of trace elements, *Biol. Trace Element Res.* **3**:301.

Lönnerdal, B., Stanislowski, A.G., and Hurley, L.S., 1980a, Isolation of a low molecular weight zinc binding ligand from human milk, *J. Inorg. Biochem.* **12**:71.

Lönnerdal, B., Keen, C.L., Sloan, M.V., and Hurley, L.S., 1980b, Molecular localization of zinc in rat milk and neonatal intestine, *J. Nutr.* **110**:2414.

Lönnerdal, B., Schneeman, B.O., Keen, C.L., and Hurley, L.S., 1980c, Molecular distribution of zinc in biliary and pancreatic secretions, *Biol. Trace Element Res.* **2**:149.

Lönnerdal, B., Keen, C.L., and Hurley, L.S., 1981a, Iron, copper, zinc and manganese in milk, *Ann. Rev. Nutr.* **1**:149.

Lönnerdal, B., Keen, C.L., and Hurley, L.S., 1981b, Trace elements in milk from various species in: *Fourth International Symposium on Trace Elements in Man and Animals* (TEMA-4) (J. McHowell, J.M. Gawthorne, and C.L. White, eds.), pp. 249-252, Griffin Press, Netley, South Australia.

Lönnerdal, B., Hoffman, B., and Hurley, L.S., 1982, Zinc- and copper-binding proteins in human milk, *Am. J. Clin. Nutr.* (in press).

Luckey, T.D., Mende, T.J., and Pleasants, J., 1954, The physical and chemical characterization of rat's milk, *J. Nutr.* **54**:345.

MacDonald, L., and Gibson, R.S., 1981, The zinc status of breast and formula fed infants, in: *Fourth International Symposium on Trace Elements in Man and Animals* (TEMA-4) (J. McHowell, J.M. Gawthorne, and C.L. White, eds.), pp. 121-124, Griffin Press Ltd., Netley, South Australia.

Martin, M.T., Licklider, K.F., Brushmiller, J.G., and Jacobs, F.A., 1981, Detection of low molecular weight copper (II) and zinc (II) binding ligands in ultrafiltered milks —the citrate connection, *J. Inorg. Biochem.* **15**:55.

McClain, P.E., Wiley, E.R., Beecher, R.G., Anthony, W.L., and Hsu, J.M., 1973, Influence of zinc deficiency on synthesis and crosslinking of rat skin collagen, *Biochim. Biophys. Acta* **304**:45.

McMaster, D., Steel, L., and Love, A.H.G., 1981, Zinc absorption by vascularly perfused small intestine, in: *Fourth International Symposium on Trace Elements in Man and Animals* (TEMA-4) (J. McHowell, J.M. Gawthorne, and C.L. White, eds.), pp. 121-124, Griffin Press Ltd., Netley, South Australia.

McPherson, E.A., Beattie, I.S., and Young, G.B., 1964, An inherited defect in Friesian calves, *Nord. Vet. Med. Suppl.* **1**:533.

Menard, M.P., and Cousins, R.J., 1982, Zinc transport by isolated brush border membrane vehicles from rat intestine, *Fed. Proc.* **41**:779.

Methfessel, A.H., and Spencer, H., 1973, Zinc metabolism in the rat. I. Intestinal absorption of zinc, *J. Appl. Physiol.* **34**:58.

Michäelsson, G., 1974, Zinc therapy in acrodermatitis enteropathica, *Acta Dermatovener. (Stockholm)* **54**:377.

Miller, W.J., 1969, Absorption, tissue distribution, endogenous excretion, and homeostatic control of zinc in ruminants, *Am. J. Clin. Nutr.* **22**:1323.

Morishima, T., Yaji, S., and Takemura, T., 1981, An acquired form of acrodermatitis enteropathica due to long-term lactose-free milk alimentation, in: *Fourth International Symposium on Trace Element Metabolism in Man and Animals* (TEMA-4) (J. McHowell, J.M. Gawthorne, and C.L. White, eds.), pp. 487-490, Griffin Press Ltd., Netley, South Australia.

Moynahan, E.J., 1966, Acrodermatitis enteropathica with secondary lactose intolerance, and tertiary deficiency state, probably due to chelation of essential nutrients by di-iodohydroxyquinoline, *Proc. Roy. Soc. Med.* **59**:445.

Moynahan, E.J., 1974, Acrodermatitis enteropathica: a lethal inherited human zinc-deficiency disorder, *Lancet* **2**:399.

Moynahan, E.J., and Barnes, P.M., 1973, Zinc deficiency and a synthetic diet for lactose intolerance, *Lancet* **1**:676.

Mutch, P.B., and Hurley, L.S., 1974, Effect of zinc deficiency during lactation on postnatal growth and development, *J. Nutr.* **104**:828.

Mutch, P.B., and Hurley, L.S., 1980, Mammary gland function and development: effect of zinc deficiency in rat, *J. Physiol.* **238**:E26.

Neldner, K.H., and Hambidge, K.M., 1975, Zinc therapy of acrodermatitis enteropathica, *N. Engl. J. Med.* **292**:879.

Nishimura, H., 1953, Zinc deficiency in suckling mice deprived of colostrum, *J. Nutr.* **49**:79.

O'Dell, B.L., 1981, Metabolic functions of zinc—a new look, in: *Fourth International Symposium on Trace Elements in Man and Animals* (TEMA-4) (J. McHowell, J.M. Gawthorne, and C.L. White, eds.), Griffin Press Ltd., Netley, South Australia.

Ohlsson, A., 1981, Case Report: acrodermatitis enteropathica: reversibility of cerebral atrophy with zinc therapy, *Acta. Paediatr. Scand.* **70**:269.

Ohtake, M., 1977, Serum zinc and copper levels in healthy Japanese infants, *Tohoku J. Exp. Med.* **123**:265.

Ølholm-Larsen, P., 1978, Untreated acrodermatitis enteropathica in adults, *Dermatologica* **156**:155.

Ølholm-Larsen, P., 1979, Serum zinc levels in heterozygous carriers of the gene for acrodermatitis enteropathica, *Hum. Genet.* **46**:65.

Parisi, A.F., and Vallee, B.L., 1969, Zinc metalloenzymes: characteristics and significance in biology and medicine, *Am. J. Clin. Nutr.* **22**:1222.

Piletz, J.E., and Ganschow, R.E., 1978a, Zinc deficiency in murine milk underlies expression of the lethal milk (lm) mutation, *Science* **199**:181.

Piletz, J.E., and Ganschow, R.E., 1978b, Lethal-milk mutation results in dietary zinc deficiency in nursing mice, *Am. J. Clin. Nutr.* **31**:560.

Pories, W.J., Henzel, J.H., Rob, G.G., and Strain, W.H., 1976, Acceleration of wound healing in man with zinc sulfate given by mouth, *Lancet* **1**:121.

Portnoy, B., and Molokhia, M., 1974, Zinc in acrodermatitis enteropathica, *Lancet* **2**:663.

Prasad, A.S., 1976, in: *Trace Elements in Human Health and Disease,* Academic Press, New York.

Rebello, T., Lönnerdal, B., and Hurley, L.S., 1982, Picolinic acid in milk, pancreatic juice and intestine: inadequate for role in zinc absorption, *Am. J. Clin. Nutr.* **35**:1.

Richards, M.P., and Cousins, R.J., 1976, Zinc binding protein: relationship to short-term changes in zinc metabolism, *Proc. Soc. Exp. Biol. Med.* **153**:52.

Richards, M.P., and Cousins, R.J., 1977, Isolation of an intestinal metallothionein induced by parenteral zinc, *Biochem. Biophys. Res. Commun.* **75**:286.

Roberts, J.A.F., 1970, in: *An Introduction to Medical Genetics,* Oxford University Press, London.

Robertson, A.F., and Sotos, J., 1975, Treatment of acrodermatitis enteropathica with zinc sulfate, *Pediatrics* **55**:738.

Rodin, A.E., and Goldman, A.S., 1969, Autopsy findings in acrodermatitis enteropathica, *J. Am. Clin. Path.* **51**:315.

Schricker, B.R., and Forbes, R.M., 1978, Studies on the chemical nature of a low molecular weight zinc binding ligand in rat intestine, *Nutr. Rep. Int.* **18**:159.

Siegel, H., and McCormick, D.B., 1970, On the discriminating behavior of metal ions and ligands with regard to their biological significance, *Acc. Chem. Res.* **3**:201.

Smart, M.E., and Fletch, S., 1971, A hereditary growth defect in purebred Alaskan malamutes, *Can. Vet. J.* **12**:31.

Sobel, A.E., Goldenberg, M., and Schmerzler, E., 1954, Calcification. XI. Studies on the incorporation of citrate in calcification *in vitro, J. Dent. Res.* **33**:492.

Song, M.K., and Adham, N.F., 1978, Role of prostaglandin E_2 in zinc absorption in the rat, *Am. J. Physiol.* **234**:E99.

Song, M.K., and Adham,N.F., 1979, Evidence for an important role of prostaglandins E_2 and F_2 in the regulation of zinc transport in the rat, *J. Nutr.* **109**:2152.

Starcher, B.C., 1969, Studies on the mechanism of copper absorption, *J. Nutr.* **97**:321.

Stober, M., 1971, Parakeratose beim schwarzbunten Niederungskalb. I. Klinisches Bild und Atiologie, *Dtsch. Tierarztl. Wschr.* **78**:257.

Sunderman, F.W., Jr., 1975, Current status of zinc deficiency in the pathogenesis of neurological, dermatological and musculoskeletal disorders, *Ann. Clin. Lab. Sci.* **5**:132.

Thunberg, T., 1953, Occurrence and significance of citric acid in the animal organism, *Physiol. Rev.* **33**:1.

Underwood, E.J., 1977, in: *Trace Elements in Human and Animal Nutrition,* pp. 545, Academic Press, New York.

Van Campen, D.R., and Kowalski, T.J., 1971, Studies on zinc absorption: ^{65}Zn binding by homogenates of rat intestinal mucosa, *Proc. Soc. Exp. Biol. Med.* **136**:294.

Vohra, P., and Kratzer, F.H., 1966, Influence of various phosphates and other complexing agents on the availability of zinc for turkey poults, *J. Nutr.* **89**:106.

Walravens, P.A., and Hambidge, K.M., 1976, Growth of infants fed a zinc supplemented formula, *Am. J. Clin. Nutr.* **29**:1114.

Walravens, P.A., Hambidge, K.M., Neldner, K.H., Silverman, A., Van Doorninck, W.J., Mierau, G., and Favara, B., 1978, Zinc metabolism in acrodermatitis enteropathica, *J. Pediatr.* **93**:71.

Wapnir, R.A., Wang, J. Exeni, R.A., and McVicar, M., 1981, Experimental evaluation of ligands for the intestinal absorption of zinc *in vivo, Am. J. Clin. Nutr.* **34**:651.

Weigand, E., and Kirchgessner, M., 1976a, Radioisotope dilution technique for determination of zinc absorption *in vivo, Nutr. Metab.* **20**:307.

Weigand, E., and Kirchgessner, M., 1976b, ^{65}Zn-labeled tissue zinc for determination of endogenous fecal zinc excretion in growing rats, *Nutr. Metabol.* **20**:314.

Weigand, E., and Kirchgessner, M., 1980, Total true efficiency of zinc utilization: determination and homeostatic dependence upon the zinc supply status in young rats, *J. Nutr.* **110**:469.

Weismann, K., and Flagstad, T., 1976, Hereditary zinc deficiency (Adema disease) in cattle, and animal parallel to acrodermatitis enteropathica, *Acta Dermatovener. (Stockholm)* **56**:151.

Weismann, K. Hjorth, N., and Fischer, A., 1976, Zinc depletion syndrome with acrodermatitis during long-term intravenous feeding, *Clin. Exp. Dermatol.* **1**:237.

Wilson, I.D., McClain, C.J., and Erlandsen, S.L., 1980, Ileal paneth cells and IgA system in rats with severe zinc deficiency: an immunohistochemical and morphological study, *Histochemical J.* **12**:457.

Yoza, N., 1977, Determining the stability constant of a metal complex by gel chromatography, *J. Chem. Ed.* **54**:284.

Chapter 7

The Clinical Implications of Dietary Fiber

David J.A. Jenkins and Alexandra L. Jenkins

1. Introduction

In the decade since Trowell and Burkitt (1975) hypothesized that many diseases seen in the West were due to overconsumption of refined carbohydrate foods, there has been a rekindling of interest in many hitherto neglected aspects of human nutrition. This in turn has been reflected in the pronouncements of official bodies concerned with health and public policy (American Diabetes Association, 1979; Canadian Diabetes Association, 1981; British Diabetic Association, 1981; U.S. Senate Select Committee on Nutrition and Human Needs, 1977; American Heart Association, 1982) and by the generous coverage given in the popular press to some of the issues involved (Alen, 1982; Engel, 1982; Fellman, 1982; Gillie, 1977). However, interest in the possible value of plant foods is not solely the result of the "fiber hypothesis." The drive to substitute unsaturated for saturated fat during previous decades was a force acting in the same direction. Both moves implicitly favored a reduction in the intake of animal products and an increased consumption of plant foods.

The impact of the fiber hypothesis no doubt owes much to the emphasis which it placed on the effects of fiber in modifying gastrointestinal events. It was through these means that the hypothesis sought to explain the mechanism of action of fiber in the prophylaxis and treatment

David J. A. Jenkins and Alexandra L. Jenkins • Department of Nutritional Sciences, Faculty of Medicine, University of Toronto, Toronto, Ontario M5S 1A8.

of disease. As a result, it has stimulated research and the development of concepts not only in clinical nutrition but also in gastrointestinal physiology and metabolism.

2.   Dietary Fiber Hypothesis

On returning to Britain after a lifetime of medical experience in Uganda, Burkitt and Trowell were impressed by the great differences which existed in the pattern and nature of diseases affecting the affluent West as opposed to less prosperous communities. As a result, their "fiber hypothesis" was formulated, in which they suggested that the consumption of unrefined, high fiber carbohydrate foods protected against many Western ailments, including colon cancer, diverticular disease, appendicitis, constipation, hemorrhoids, hiatus hernia, varicose veins, diabetes, heart disease, gallstones, obesity and many others. On the face of it these claims appeared too extensive to be realistic. Detractors suggested that the incidence of these diseases could be related as well to food availability, exercise patterns, motor cars, televisions and those variables which divide the more affluent from the less developed communities throughout the world. However, with the passage of time, laboratory and clinical evidence has been gathered to support much of the original hypothesis. The disorders about which the hypothesis has had greatest impact on medical thinking include constipation (Eastwood *et al.,* 1973; Cummings *et al.,* 1976), diverticular disease (Brodribb and Humphries, 1976; Painter *et al.,* 1972), diabetes (Anderson and Chen, 1979; Anderson and Ward, 1979; Doi *et al.,* 1979; Jenkins *et al.,* 1976a, 1980c, 1980h; Kiehm *et al.,* 1976) and hyperlipidemia (Kritchevsky, 1978; Miettinen and Tarpila, 1977; Jenkins *et al.,* 1980b; Thiffault *et al.,* 1970) and, to some extent, obesity (Van Itallie, 1978). In addition, evidence has also been adduced to suggest a use for high fiber diets in the treatment of Crohn's disease (Heaton *et al.,* 1979).

Nevertheless, in view of the many nutritional differences which exist between Western and underdeveloped communities, the exact importance of dietary fiber *per se* continues to be debated. It may be that in the long term the greatest value of this hypothesis will be to increase awareness of the possible differences between foods, their degree of processing, the form in which they are eaten and their content of non- or anti-nutrient components. All these factors may influence digestibility and alter metabolic events.

To illustrate these features, special attention will be paid in this review to the development of the fiber hypothesis in relation to carbohydrate and lipid metabolism.

3.   Diabetes

In this area the fiber hypothesis and its implications have helped to usher in a radical change in thinking in the nutritional management of disease. For several decades diabetics in Western communities have tended to be managed on 40% fat, 20% protein, and 40% carbohydrate diets. In many instances the fat content has been higher and the carbohydrate lower. Over a three-year period this policy has been reversed, first by the American Diabetes Association (1979), then by the Canadian Diabetes Association (1981) and most recently by the British Diabetic Association (1981). These changes in practice owed much to the epidemiological leads which formed the basis of the fiber hypothesis and to the experimental studies which followed. Such studies were of two sorts: trials involving purified fiber and tests of high fiber, high carbohydrate diets derived from unprocessed foods.

3.1. Purified Fiber and Diabetic Management

In early studies, purified fiber preparations added to the diet (Jenkins *et al.*, 1977c; Miranda and Horwitz, 1978) or to individual meals (Jenkins *et al.*, 1976a; Morgan *et al.*, 1979), with no alteration in carbohydrate intake, were shown to reduce urinary glucose loss and flatten postprandial glucose responses. Tests involving the "artificial pancreas" have shown that the dietary fiber guar (a galactomannan from the Indian cluster bean) reduced the insulin requirement of insulin-dependent diabetics (Christiansen *et al.*, 1980). In 5-day metabolic ward studies, addition of 25 g guar to prepared foods reduced the urinary glucose output of a group of insulin-dependent diabetics to 50% of that seen during the control period (Jenkins *et al.*, 1977c). Similar improvements in diabetic control have been seen in outpatient trials. Development of a crispbread containing 1 g guar per slice (Speywood Laboratories, Bingham, Nottingham, England) allowed longer term studies to be undertaken than were possible with previous preparations of this viscous material. By one year, when guar crispbread had been taken by a group of eight diabetics, urinary glucose concentrations were still reduced by over 50% below pretreatment values and, in addition, there was a 20% reduction in insulin requirement (Jenkins *et al.*, 1980h) (Fig. 1).

In metabolic and outpatient studies, non-viscous fiber-enriched breads were also shown to reduce the postprandial glucose response of diabetics throughout the day. This suggested that viscosity was not the only attribute of purified fiber necessary for its action (Miranda and Horwitz, 1978). The effects, however, did not appear to be as marked as those seen with viscous fiber.

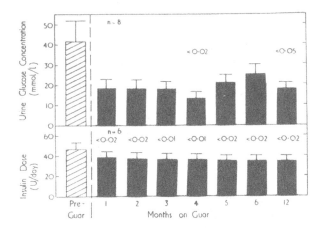

Fig. 1. Urine glucose concentration and insulin dose of eight diabetics (7 taking insulin) before and during the first year of taking guar crispbread (14-26 g guar depending on the individual). Jenkins *et al.* (1980).

Experiments with purified fiber also indicated how fiber in unprocessed foods might be acting. It was found to be of great importance that the purified fiber be mixed intimately with the food. This was demonstrated in a study in which guar was taken 2 min before a glucose drink; in this situation it had little effect on glucose tolerance in normal volunteers (Jenkins *et al.*, 1979b). Also, in a long-term study in which guar was given in gelatin-coated capsules to obese "brittle" diabetics, no significant improvement in control was achieved (Cohen *et al.*, 1980). Incidentally, this last study also indicates the value of performing preliminary test meal studies before embarking on full-scale long-term clinical trials.

The therapeutic use of fiber supplements or supplemented foods (the latter being preferable due to adequacy of mixing) is therefore in the developmental stage. At present, only guar in crispbread form has been found satisfactory for long-term use among those types of fiber with the greatest effect on glucose tolerance (Jenkins *et al.*, 1978). With other preparations of these viscous materials, compliance is a major problem.

3.2. Fiber in Unrefined Foods and Diabetic Management

The majority of therapeutic trials involving dietary fiber have been carried out using high carbohydrate diets in which the fiber has been derived from the foods eaten rather than given as a supplement. An example is the pioneer work of Anderson and his colleagues (Anderson and Chen,

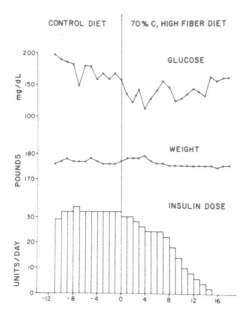

Fig. 2. Fasting blood glucose, body weight and insulin dose in a diabetic man on control carbohydrate (43%), low fiber and high carbohydrate (70%), high fiber diet. Modified from Anderson and Ward (1979).

1979; Anderson and Ward, 1979; Kiehm *et al.*, 1976) who demonstrated that diabetic control could be greatly improved with high fiber diets composed of whole grain cereals, leguminous seeds and generous portions of leafy vegetables in which 70% of the calories were supplied as carbohydrate. This work was of great importance in opening up a new approach to diabetic management. Furthermore, this work showed that high fiber diets must of necessity be high carbohydrate by virtue of the association of fiber and carbohydrate in the same foods. Acceptance of the possible value of fiber in the diabetic diet therefore has done much to enhance acceptance of the high carbohydrate diet, not only to replace fat but as a vehicle for fiber which has other possible attributes.

In these studies, diabetics on less than 20 units of insulin and 43% carbohydrate diets (once the norm for diabetic management) invariably were well controlled without insulin at the end of the hospital period on high carbohydrate, high fiber intakes (Fig. 2). Many patients then were given high fiber, 60% carbohydrate diets out of hospital and remained in good control without insulin.

Similar results have been obtained in both non-insulin dependent diabetes management (NIDDM) and insulin dependent diabetes management (IDDM) outpatients using fiber largely of cereal origin (Simpson *et al.*,

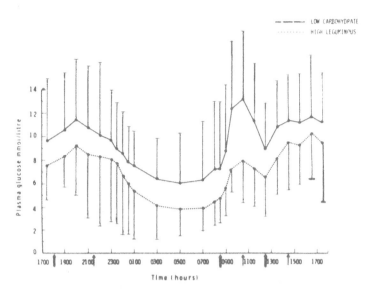

Fig. 3. Plasma glucose values during 24 hr profiles of IDDM patients on high legume (HL) and low carbohydrate (LC) diets. Simpson *et al.* (1981).

1979a, 1979b), fiber from legumes and non-cereal vegetables (Rivellese *et al.*, 1980) and fiber from legumes and cereals (Simpson *et al.*, 1981). In these outpatient situations the most dramatic effects were seen when large amounts (over 100 g dry weight/d) of cooked legumes were used (Fig. 3). Such studies have been criticized because both fiber and carbohydrate intakes were increased simultaneously. However, the two must be taken together if unprocessed foods are to provide the source of fiber.

Nevertheless, Brunzell and coworkers (Brunzell *et al.*, 1971) demonstrated that high carbohydrate intakes *per se* may improve diabetic control. Since cereal fiber as such has a negligible effect on carbohydrate tolerance (Jenkins *et al.*, 1981b), the high carbohydrate intake may explain the improved diabetic control seen on the high carbohydrate, high cereal fiber diets.

In support of an effect associated with chronic cereal fiber intake, bran supplementation for six months, with no apparent increase in carbohydrate intake, improved the glucose tolerance test of middle-aged patients with diverticular disease (Brodribb and Humphreys, 1976). It is possible that such changes may produce subtle alterations in food selection or perhaps long-term small intestinal adaptation either in terms of morphology (Tasman Jones, 1980) or function.

It has been shown that, with no alterations in carbohydrate intake, simple addition of certain fibers improves diabetic control. Moreover, experience with purified fiber preparations indicates that they are most effective when taken against a background of higher carbohydrate intake (over 43% of calories) (Jenkins et al., 1980c) and it therefore seems justifiable to advise high carbohydrate diets as the vehicle for providing fiber from unprocessed foods. This is especially so in view of the possible advantages of a low fat diet in the prevention of raised blood lipids and vascular disease.

Thus, from the standpoint of official recommendations the impact of the fiber hypothesis and the subsequent work it engendered has been considerable. In terms of unprocessed carbohydrate foods, the question raised by fiber studies is: which carbohydrate foods are both acceptable and therapeutically worthwhile to incorporate into the diabetic diet (British Diabetic Association, 1981)? Possible answers to this question are beginning to emerge but they are extending the field far beyond the confines of the dietary fiber hypothesis and into the realm of nutrient interactions, antinutrients in general and the determinants of food form.

3.3. Mechanism of Action: Lente Carbohydrate

3.3.1. Reduced Postprandial Blood Glucose Response

A wide range of dietary fibers and fiber analogues has been shown to flatten the postprandial blood glucose response when added to glucose drinks and mixed meals taken by normal and diabetic individuals (Gold et al., 1980; Holt et al., 1979; Jeffrys, 1974; Jenkins et al., 1976a, 1977b, 1978, 1979b; Levitt et al., 1980; Miranda and Horwitz, 1978; Morgan et al., 1979; O'Connor et al., 1981). In general, the more viscous the fiber, the greater the reduction in postprandial glycemia. In a comparative study of six fibers and fiber analogues taken with 50 g glucose, it was the most viscous types, guar and tragacanth, which produced the greatest flattening in the blood glucose response (Fig. 4) (Jenkins et al., 1978). It is therefore likely that viscosity is an important factor determining glycemic response. Although a study comparing the viscosity of various guar preparations with the glycemic response failed to show a linear relationship, the nonviscous preparation had minimal effects (O'Connor et al., 1981).

The fact that simple addition of a variety of fibers, for example to a 50 g glucose tolerance test (GTT), can substantially flatten the blood glucose response challenges the relevance of the so-called "carbohydrate exchange" principle (Jenkins et al., 1978). This concept maintains that equal portions of available (or "absorbable") carbohydrate from one food, as estimated by chemical analysis of the food, may be exchanged for the same amount in another food in diets, such as those for diabetics, in which it is considered important that daily carbohydrate intake be held

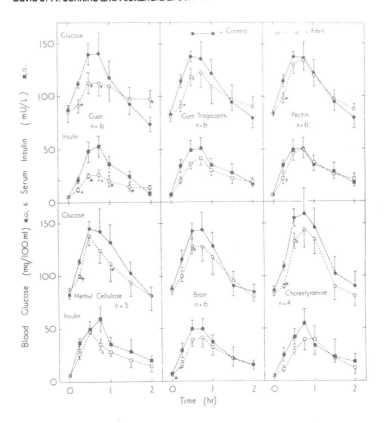

Fig. 4. Mean blood glucose and serum insulin levels of volunteers after taking 50 g glucose tolerance tests (control) to which had been added the equivalent of 12 g of a range of dietary fibers and fiber analogs (test). Jenkins *et al.* (1980).

constant. Yet fiber studies suggest that the biological response to the carbohydrate in a food may be greatly modified by the other food constituents.

3.3.2. Reduced Postprandial Endocrine Response

No increase in insulin levels has been seen which could account for the flattened postprandial blood glucose responses (Brodribb and Humphreys, 1976; Gold *et al.*, 1980; Holt *et al.*, 1979; Levitt *et al.*, 1980; Morgan *et al.*, 1979; O'Connor *et al.*, 1981). Rather, the converse was observed. Although fiber preparations have induced substantial reductions in the postprandial glycemia in normal subjects (Gold *et al.*, 1980; Holt *et al.*, 1979; Jenkins *et al.*, 1978; O'Connor *et al.*, 1981), diabetics (Jenkins *et al.*, 1976a; Morgan *et al.*, 1979) and postgastric surgery patients (Jenkins *et al.*, 1980a; Leeds *et al.*, 1978; Taylor, 1979), even larger reductions

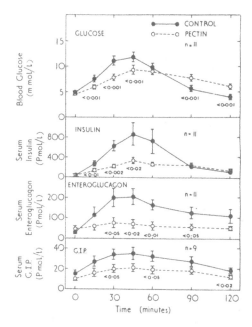

Fig. 5. The mean blood glucose and plasma insulin, enteroglucagon and GIP in postgastric surgery patients who took 50 g glucose (control) alone and with 14.5 g pectin on separate occasions after overnight fasts. Jenkins *et al.* (1980).

were seen in insulin and other hormone responses, especially those of the enteroinsular axis including gastric inhibitory peptide (Jenkins *et al.*, 1980a, Morgan *et al.*, 1979) and enteroglucagon (Jenkins *et al.*, 1980a) (Fig. 5). Such changes throw light on the mechanism of action of fiber, indicating that it may reduce the effective concentration of nutrients at the enterocyte surface.

3.3.3.   Reduced Rate of Absorption, Not Malabsorption

It is possible that substantial malabsorption may occur after fiber administration and that this may explain the flattened blood glucose response. To test this hypothesis, xylose was added to fiber (guar) test meals as a marker of carbohydrate absorption (Jenkins *et al.*, 1978). After the fiber meal, blood xylose levels and urinary xylose output were significantly lower than in fiber-free controls for 2 hr. However, over the following 6 hr significantly more xylose was excreted than in the controls, so that by 8 hr the total xylose excreted was the same (Jenkins *et al.*, 1978). Similar results have been found with pectin and guar, both viscous forms of fiber, when paracetamol (Holt *et al.*, 1979) and digoxin (Kasper *et al.*, 1979) were used as markers of absorption. Again, no evidence was found for

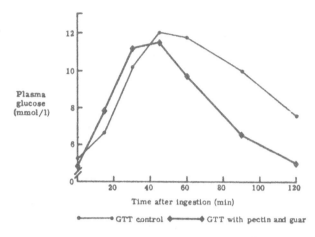

Fig. 6. Effect of guar and pectin on the glucose tolerance curve of a patient who had had total gastrectomy. Holt *et al.* (1979).

malabsorption and it appeared that the rate of absorption was slower.

Breath hydrogen evolution is another test which has been used as a marker of carbohydrate malabsorption. Hydrogen is produced by bacteria which ferment unabsorbed sugars reaching the colon. When lactulose, an unabsorbable sugar, was mixed with guar, very substantial hydrogen evolution was seen within 3 hr. The absence of hydrogen evolution by 5 hr after guar and glucose were taken together (Jenkins *et al.*, 1977b) therefore indicated that guar does not cause glucose malabsorption even though it flattens the blood glucose response curve.

3.3.4.   Gastric Emptying and Mouth to Cecum Transit Time

Delayed gastric emptying (Holt *et al.*, 1979; Leeds *et al.*, 1978) and a reduced rate of small intestinal absorption (Elsenhans *et al.*, 1980; Jenkins *et al.*, 1978) are factors which may explain the action of viscous fiber in reducing postprandial glycemia. However, it is unlikely that either effect provides the whole explanation. In gastric emptying studies, addition of pectin to test meals of normal individuals and patients who had undergone gastric surgery failed to demonstrate a relationship between inhibition of postprandial glycemia and the slower gastric emptying (Leeds *et al.*, 1978). In addition, after pectin and guar, the area under the blood glucose response curve of a completely gastrectomized individual was still reduced by over 20% (Taylor, 1979) (Fig. 6). Thus, although gastric emptying may be one factor, these observations suggest that events in the small intestine may be important in determining the shape of the glucose response curve.

In recent dialysis studies, guar prevented the outward diffusion of glucose (Khan *et al.*, 1979). The same effect may be produced in the small intestine by the action of fiber in increasing the thickness of the unstirred water layer adjacent to the brush border of the enterocyte which may form an important barrier to substrate diffusion (Thomson and Dietschy, 1980).

These possible mechanisms of action of viscous fiber may be of considerable importance, since it has been shown that in diabetics with autonomic neuropathy and gastroparesis, addition of guar to test meals was without effect (Levitt *et al.*, 1980). It may be that, in the absence of normal gastrointestinal motility, the effective thickness of the small intestinal unstirred layer is already increased to the extent that viscous fiber can produce no further increase. Together with slower gastric emptying due to autonomic neuropathy, this would minimize the effectiveness of guar since these are also the likely mechanisms of its action.

Delayed gastric emptying may be one reason for the slower upper gastrointestinal transit time observed with viscous fiber. Lactulose added to test meals has been used to demonstrate that viscous fibers delay the appearance of hydrogen in the breath while wheat bran shortens the time of first appearance (Jenkins *et al.*, 1978). Since the appearance of hydrogen reflects the mouth to caecum transit time, such studies provide evidence for prolonged transit time with viscous fiber and decreased transit time with bran. In earlier studies, McCance and colleagues also found that barium-impregnated wholemeal bread left the stomach and traversed the small intestine more rapidly than did white bread (McCance *et al.*, 1953).

3.3.5.   Diversity of Responses to Carbohydrate from Different Foods

Studies have indicated that addition of purified fiber to a food profoundly influences the glycemic response, thereby challenging the concept of carbohydrate exchange. Similar conclusions were arrived at by using carbohydrate from different food sources and comparing processed and unprocessed foods. The factors responsible, however, are likely to be more diverse than in the purified fiber studies.

The early work of Campbell (1971) indicated that the blood glucose response to carbohydrate in foods was less than to an equivalent amount of carbohydrate given as glucose. This work has been extended using apples, apple puree, and apple juice to show that both disruption and removal of fiber enhance the insulin response to a food (Haber *et al.*, 1977). The importance of food form, of which fiber may be only one aspect, has been demonstrated by the greater blood glucose and insulin response to ground as opposed to unground rice (O'Dea *et al.*, 1980).

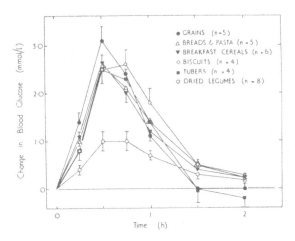

Fig. 7. Change in blood glucose concentration after eating 50 g carbohydrate portions of individual grains, breads and pasta, breakfast cereals, biscuits, tubers, and dried legumes. Jenkins *et al.* (1980).

Further work has demonstrated that equal carbohydrate portions from different foods raise the blood glucose and insulin levels to different extents (Crapo *et al.*, 1977; Jenkins *et al.*, 1980d, 1980e, 1980f, 1980g; Schauberger *et al.*, 1978). Possible explanations have included fiber, simple sugar, the nature of the starch, and the protein and fat content of the foods. In a study in which over 60 foods and sugars were tested in healthy volunteers, only fat and protein showed significant negative correlations with blood glucose rise (Jenkins *et al.*, 1980e). Obviously, many unidentified factors, such as enzyme inhibitors, lectins, etc., may have played a part. It is therefore of special interest that dried beans, which as a class are particularly rich in such materials, caused one of the smallest rises in blood glucose of the foods tested on an equivalent carbohydrate content basis (Jenkins *et al.*, 1980d) (Fig. 7).

As with the purified fiber studies, emphasis has been placed on the rate at which foods release their products of digestion in the gastrointestinal tract as a possible determinant of the glycemic response (Jenkins *et al.*, 1980g). This factor may explain the flatter insulin and GIP responses, resembling those observed after purified fiber, seen after ingestion of some foods (Jenkins *et al.*, 1982d).

3.3.6.   *In Vitro* Digestion - Lente Carbohydrate

Studies in which foods have been digested with human enzymes in dialysis bags (Jenkins *et al.*, 1982c, 1980g) have shown that a food which pro-

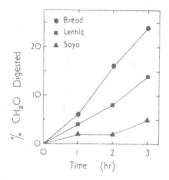

Fig. 8. Estimated percentage of bread, lentil and soya bean carbohydrate digested in an *in vitro* dialysis system over 3 hours.
Jenkins *et al.* (1980).

duces a large rise in blood glucose releases its carbohydrate products of digestion rapidly. On the other hand, the leguminous seeds such as lentils or soybeans, which result in small or negligible postprandial rises in blood glucose, release their carbohydrate products of digestion into the dialysate slowly (Jenkins *et al.*, 1980g) (Fig. 8). The results, therefore, are very similar to those observed when using fiber-glucose mixtures in dialysis bags.

Recently a similar relationship, emphasizing the importance of food form, has been found for ground and unground cooked rice, the latter being digested more slowly *in vitro* and producing a flatter blood glucose response on feeding (O'Dea *et al.*, 1981). Finally, the relationship between the rate of digestion *in vitro* over 1 hr and the area under the 2 hr blood glucose response curve to 14 foods indicated that digestibility was a major determinant of the glycemic response (Jenkins *et al.*, 1982b). There was, in fact, only a marginally significant relationship with food fiber content. Other factors therefore must be considered.

3.3.7.   Factors Other Than Fiber

These factors include the starch-protein and starch-fat interactions and the level and type of enzyme inhibitors, lectins, phytates, tannins, and antigenic proteins present in the food. Fat and protein content have both been shown to have negative relationships with the glycemic response to foods (Jenkins *et al.*, 1980e). Evidence for an effect of protein in reducing digestibility comes from a breath hydrogen study on carbohydrate malabsorption after feeding a variety of breads (Anderson *et al.*, 1981). It was estimated that 10-20% of the starch in bread was malabsorbed. Removal of the gluten from the flour reduced malabsorption to zero but subsequent addition of gluten to the bread mix was without effect (Anderson *et al.*, 1981). It was concluded that the natural relationship of starch and protein was an important determinant of digestibility. If this finding is borne out it will have important implications for patients with diabetes and gastrointestinal disease.

The therapeutic use of fiber in clinical medicine has been accompanied by the development of enzyme inhibitors. Inhibitors form part of the unabsorbable material within the gut lumen which limits the rate of nutrient digestion and absorption, and they may be important factors to consider when ascribing to fiber the function of determining digestibility of unprocessed foods. Enzyme inhibitors have been found in high concentrations in many plant seeds. Their function presumably is to stop autodigestion of the plant during storage and perhaps to limit attack by moulds and other agents (Salunkhe and Wu, 1977).

In addition to protease inhibitors, on which much work has been done (Liener, 1979), there are also inhibitors of carbohydrate digestion. Notable among these are the α-amylase inhibitors (Jaffe *et al.*, 1973; Kneen and Sandstedt, 1946) found in wheat (Kneen and Sandstedt, 1946; Militzer *et al.*, 1946), rye and beans (Hernandez and Jaffe, 1968; Jaffe and Vega Lette, 1968), taro root (Miller and Kneen, 1947), unripe mango (Mattoo and Modi, 1970) and soybeans (Narayana Rao *et al.*, 1967). The inhibitor in wheat is a non-dialysable protein denatured at 95°C in 10-15 min at alkaline pH but more stable at acid pH, which is able to withstand baking in bread (Kasper *et al.*, 1979). The amylase inhibitor in *Faseolus vulgaris* beans was found to be destroyed by heating at 100°C for 15 min (Militzer *et al.*, 1946). Nevertheless, it is possible that, as with bread, under certain cooking conditions appreciable amounts of this inhibitor may be present in cooked bean dishes.

During the last decade there has been considerable interest in the possibility that inhibition of carbohydrate digestion and absorption may be of benefit to the diabetic, the overweight, and to individuals with carbohydrate-induced hypertriglyceridemia or with the dumping syndrome. This spectrum of disease bears a surprising similarity to those which may be amenable to dietary fiber therapy.

The first inhibitor to be tested in man was a heat labile α-glycosidase inhibitor in wheat which limited the breakdown of starch (Puls and Keup, 1973). It was shown to flatten the glycemic response to a mixed meal in healthy volunteers. The potential clinical use of such agents has been confirmed using a similar substance (Acarbose, Bayer Company) isolated from bacteria. Acarbose has predominantly anti-sucrose and anti-amylase activity, plus some anti-maltase activity, but it is ineffective against lactose or monosaccharides (Jenkins *et al.*, 1981a). Such an inhibitor, given at a dose of 100 mg (Caspary, 1978) to 200 mg (Jenkins *et al.*, 1979c) causes 50% to almost 100% malabsorption of a 50-100 g sucrose load. At

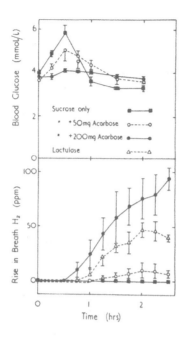

Fig. 9. Blood glucose response to sucrose with acarbose 50 mg, acarbose 200 mg, and placebo. Breath hydrogen responses to sucrose and lactulose. Jenkins *et al.* (1978).

lower doses (e.g., 50 mg) (Jenkins *et al.*, 1979c) carbohydrate malabsorption is minimal, judging from symptoms and breath hydrogen (Fig. 9). Nevertheless, large percentage reductions are seen in the area under the glucose response curve following a carbohydrate load. This suggests that, as with dietary fiber, slow absorption rather than malabsorption is all that is required to produce the reduction in glycemia. In a study in which Acarbose and guar were used in combination, their flattening effects on the glycemic response were additive (Jenkins *et al.*, 1979c) (Fig. 10). There are now a number of short- and long-term studies indicating that this form of enzyme inhibition improves diabetic control (Fig. 10) (Gerard *et al.*, 1982; Hillebrand *et al.*, 1979).

The effects of other factors (lectins, phytates, tannins, etc.) which may limit absorption or digestion, have not yet been studied in man. Nevertheless, the fiber hypothesis and the studies which it prompted have already resulted in a new approach to the nutritional therapy of diabetes. Fat intakes have been reduced. New foods are likely to be introduced into the diet and in the future purified fiber supplements and enzyme inhibitors may also be used when diet, with or without insulin, has not achieved the therapeutic goal.

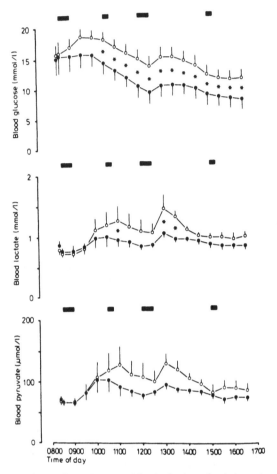

Fig. 10. Effect of acarbose on blood glucose, lactate and pyruvate concentrations on control (O) and test (O) days (Mean ± SEM). = Meals and snacks. Significant difference (P<0.05) between mean control and test day values. Walton *et al.* (1979).

4.   Dumping Syndrome

Following logically from the physiological effects of fiber already described, in the dumping syndrome the dietary fiber pectin has been shown to reduce associated abnormalities including hypoglycemia, hyper-insulinemia, hypovolemia, carbohydrate malabsorption and the rate of gastric emptying after a glucose challenge (Jenkins *et al.*, 1977a, 1980a; Leeds *et al.*, 1981). There is also a suggestion that fiber supplementation may be of use in the long term (Jenkins *et al.*, 1977a). As with diabetes, success in slowing carbohydrate absorption through enzyme inhibition,

Table 1. Binding In Vitro of Bile Acids and Salts to Fiber[a]

Bile acid or salt	Fiber			
	Alfalfa[b]	Bran	Cellulose	Lignin
Cholic	1.00	0.51	0.15	2.20
Chenodeoxycholic	1.25	0.91	0.10	1.17
Deoxycholic	0.52	0.27	0.01	0.87
Taurocholic	1.00	0.20	0.15	3.20
Taurochenodeoxycholic	2.18	1.42	0	3.68
Taurodeoxycholic	1.54	0.49	0.10	4.48
Glycocholic	1.00	0.33	0.10	1.96
Glycochenodeoxycholic	1.30	1.86	0.02	2.19
Glycodeoxycholic	2.42	0.68	0.41	4.57

[a]After Story and Kritchevsky(1976).

[b]For each group alfalfa-cholic acid or salt is set as 1.00. Actual percent binding to alfalfa: cholic, 19.9; taurocholic, 6.9; glycocholic, 11.5.

using the α-glucoside hydrolase inhibitor Acarbose, has given this approach to therapy credibility both in test meal (McLoughlin *et al.*, 1979) and one-week studies (Jenkins *et al.*, 1982a). The therapeutic role of fiber and agents such as enzyme inhibitors may be similar in the dumping syndrome and in diabetics.

5.   Dietary Fiber and Hyperlipidemia

In terms of official recommendations or therapeutic measures in the management of hyperlipidemia the fiber hypothesis has, as yet, made less impact than it has in diabetes. The evidence for a hypocholesterolemic effect of dietary fiber predated the genesis of the dietary fiber hypothesis. Guar, pectin and a wide range of unabsorbable viscous plant polysaccharides were shown in the 1960's to have hypocholesterolemic properties (Fahrenbach *et al.*, 1965; Palmer and Dixon, 1966). Lignin was shown to increase fecal bile acid losses (Eastwood and Hamilton, 1968) and, as one of the first major conceptual advances of the "fiber era," Kritchevsky and his colleagues demonstrated that a range of indigestible plant materials were capable of binding bile salts *in vitro* (Kritchevsky and Story, 1974,

1975; Story and Kritchevsky, 1976; Vahouny *et al.*, 1980) (Table I). This did much to provide a rationale for further work, allowing an analogy to be drawn between some of the actions of dietary fiber and those of synthetic hypocholesterolemic agents, the bile salt binding anion exchange resins (e.g., cholestyramine) used in clinical practice.

As yet, extensive studies on groups of patients have not been possible due to lack of available fiber preparations acceptable for long-term use. Nevertheless, the above approach has stimulated much interest in the therapeutic (Connor and Connor, 1982) as well as prophylactic attributes of dietary fiber.

5.1. Purified Fiber and Hypolipidemic Effects

Guar (Jenkins *et al.*, 1979a, 1980b), pectin (Miettinen and Tarpila, 1977), lignin (Thiffault *et al.*, 1970), oat bran (Kirby *et al.*, 1981) and locust bean gum (Zavoral *et al.*, in press) have been used in trials involving type II and IV hyperlipidemic patients. The viscous materials guar (10-25 g/d) and pectin (40 g/d) (Miettinen and Tarpila, 1977) have been found to lower serum cholesterol levels by 10-16%, the effect being more marked with guar. Guar also has been shown to be hypocholesterolemic in diabetics (Jenkins *et al.*, 1980h). Reports on the effect of lignin have been conflicting (Lindner and Möller, 1973). Decreases on fiber administration have been in LDL cholesterol (Jenkins *et al.*, 1980b; Miettinen and Tarpila, 1977) with no reduction in HDL cholesterol (Fig. 11). Mean serum triglyceride levels have tended to fall; however, the decreases have not been significant. The modes of administration of the supplements have included a raw powder (Fahrenbach *et al.*, 1965; Miettinen and Tarpila, 1977; Jenkins *et al.*, 1979a, 1980b), a canned soup (Jenkins *et al.*, 1980b), breads and biscuits (Zavoral *et al.*, in press) guar crispbread (Jenkins *et al.*, 1980b) and, more recently, a granulate (Speywood Laboratories, Bingham, Nottingham, England) which was mixed with fruit drinks. Mixing with the food did not seem to be as important for the hypocholesterolemic effect as it was for the flattening of postprandial glycemia in diabetes, and thus the granulate, which is very acceptable, is particularly suitable for long-term treatment.

5.2. High Fiber Foods and Hypolipidemic Effects

The presence of other potentially hypocholesterolemic factors in high fiber foods makes it difficult to assess the role played by fiber in unprocessed foods. The majority of studies using foods have been carried out on diabetics who, for the most part, were normolipidemic. In such studies, decreases in serum cholesterol have been observed (Anderson and Ward, 1979; Kiehm *et al.*, 1976; Rivellese *et al.*, 1980; Simpson *et al.*, 1979a,

1979b, 1981), but since the diets were high in both fiber and carbohydrate, the reduced fat intake may have contributed to the lower cholesterol levels. Nevertheless, these studies are remarkable in that high carbohydrate diets high in fiber failed to raise serum triglyceride levels. On the contrary, in hypertriglyceridemic individuals reductions in serum triglyceride levels were seen (Anderson and Chen, 1979; Simpson et al., 1981). The fiber foods included cereals, non-cereal vegetables and beans. Recent studies on Trappist monks fed diets in which the fiber came from legumes, cereals, and vegetables demonstrated significant reductions in serum cholesterol (Lewis et al., 1981). In addition, a recent prospective study suggested that increasing fiber intake may be a useful dietary manipulation in preventing cardiovascular disease (Hjermann et al., 1981).

Only in a retrospective study by Morris and coworkers (1977) was there a suggestion that cereal fiber may be protective against cardiovascular disease. However, cereal fiber, for the most part, has had little effect on blood lipids. The majority of studies therefore have emphasized the possible value of legumes or legume fiber in lowering plasma lipids.

In the 1960's, the work of Mathur in India (Mathur et al., 1968) demonstrated that Bengal gram (chick peas, Cicer arietanum), fed to healthy volunteers whose serum cholesterol levels had been raised by increasing their butterfat intake, dramatically lowered their cholesterol levels after a latent period of three months. This fall was maintained over the remaining three months of the study, at which time there was a large mean increase in fecal bile acid loss. Earlier work of Keys and coworkers (Grande et al., 1965) established the hypolipidemic effect of beans but the effect was attributed to the carbohydrate.

A recent report from Sishuan province indicates that supplementing a Chinese diet with 30 g dried beans or more daily resulted in a 10% fall in serum cholesterol in a group of hyperlipidemic individuals over a three-month period (Bingwen et al., 1981). This may be the first report of supplementation of a hyperlipidemic group with a high fiber food. No change was seen in serum triglyceride.

5.3. Mechanism of Action of Fiber on Lipid Metabolism

5.3.1. Purified Fiber

Lower serum cholesterol levels (Durrington et al., 1976; Jenkins et al., 1976b, 1979a, 1979d; Kay and Truswell, 1977) after feeding purified fiber have been associated with increases in fecal excretion of acidic and neutral sterols and, in some studies, in fecal fat (Jenkins et al., 1976b; Kay and Truswell, 1977). One mechanism of action may be the binding of bile acids by fiber (Kritchevsky and Story, 1974). However, although bile

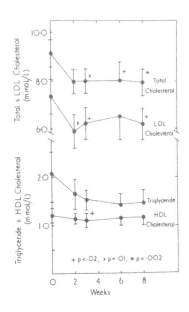

Fig. 11. Hypolipidemic effect of guar (10-15 g) taken in crispbread over 8 weeks by hyperlipidemic patients. Jenkins *et al.*
(1980).

salt and cholesterol losses are probable factors in the lowering of serum
cholesterol, it is likely that other factors also are involved, since these
losses are increased by only 40% after pectin administration while after
guar the loss is 80% for a fall in cholesterol of 12-16% (Jenkins *et al.*,
1979d). In addition, bile acid losses of the same magnitude as seen with
pectin have been seen with soft white wheat bran which, unlike hard red
spring wheat bran (Munoz *et al.*, 1979), has no effect on serum cholesterol (Jenkins *et al.*, 1975a).

   In general, particulate fiber preparations – e.g., soft white wheat bran
(Jenkins *et al.*, 1975a; Munoz *et al.*, 1979; Truswell and Kay, 1976), carrot and cabbage (Jenkins *et al.*, 1979d) – have for the most part been
ineffective in reducing serum cholesterol. Unlike the viscous fibers, the
particulate fiber preparations may cause large increases in fecal bulk
(Cummings *et al.*, 1978). The ability of a dietary fiber to increase fecal
bulk is not necessarily related to its ability to lower serum cholesterol
(Jenkins *et al.*, 1979d) (Fig. 11).

5.3.2.   High Fiber Foods and Associated Hypolipidemic Constituents

As mentioned in relation to high fiber foods, chick peas increase bile acid
losses (Mathur *et al.*, 1968). Legumes cause little rise in blood glucose
(Jenkins *et al.*, 1980d) and hence stimulate insulin secretion to a lesser

extent than other foods of equivalent carbohydrate content (Jenkins *et al.,* 1982d). This latter effect may reduce the stimulus to hepatic cholesterol synthesis and, together with increased fecal sterol losses, may reduce serum cholesterol. However, other factors may be involved as well as fiber.

Work carried out in India on commonly consumed leguminous seeds (Saraswathy Devi and Kurup, 1979) (black gram, *Phaseolus mungo* red gram, *Cajanus cajan* horse gram, *Dolichos biflorus)* indicated that the hypocholesterolemic activity of beans lay in the insoluble carbohydrate and protein fractions. Although the protein fraction was more effective, both supplements increased fecal losses of neutral and acidic sterols (Saraswathy Devi and Kurup, 1973) and this may have been part of the explanation of their hypocholesterolemic effect.

Such studies have indicated a possible role for vegetable protein, as well as fiber, in lowering serum cholesterol. There is much work indicating a difference between animal and vegetable protein (Carroll, 1978; Carroll *et al.,* 1978; Carroll and Hamilton, 1975; Gatti and Sirtori, 1977; Sirtori *et al.,* 1977; Kritchevsky *et al.,* 1978), the latter producing lower cholesterol levels. Studies by Sirtori *et al.* (1977) on hyperlipidemic patients indicated that substitution of soy for meat protein in an Italian diet significantly reduced serum cholesterol levels and that a protein isolate of a leguminous seed had therapeutic potential. This effect was not produced by supplementation with methionine, a nutritionally limiting amino acid in legumes (Gatti and Sirtori, 1977). It has been suggested that the higher ratio of arginine to lysine or the lower lysine content of plant proteins may be an important factor (Kritchevsky *et al.,* 1978). It has been hypothesized that lysine may inhibit hepatic arginase activity, making more arginine available for synthesis of arginine-rich apolipoprotein.

Apart from their protein and fiber components, a number of other factors may be involved in the hypolipidemic effect of whole foods. Many leafy vegetables and seeds contain appreciable amounts of saponin. Alfalfa (Malinow *et al.,* 1979) and soy (Topping *et al.,* 1980) saponins have a hypocholesterolemic effect in experimental animals. Only one study has been carried out on the effect of saponin in humans, and this was at the low level of 1.1 g/man/d (Potter *et al.,* 1980) compared to the hypocholesterolemic dose of .21 g/d in rats (Topping *et al.,* 1980). Despite the low dose used, an increased fecal bile acid loss was noted even though no hypocholesterolemic effect was seen. Higher doses of saponins also may have hypocholesterolemic effects.

Lectins are glycoproteins which are found in high concentrations in certain vegetable foods. Small amounts survive normal cooking procedures in a wide range of common foods (Nachbar and Oppenheim, 1980). Lectins bind to cell surfaces causing agglutination of cells in suspension and may block absorption sites in the small intestine. Large amounts, fed as uncooked bean meal to experimental animals, are toxic to the small intestinal mucosa (Puzstai et al., 1979). While the amount of lectins which survives normal cooking is small, it may exert an inhibitory effect on small intestinal absorption. This may be another way in which ileal absorption of bile salts is reduced. The rate of nutrient absorption along the small intestine also may be reduced, and the resulting blunted insulin response may lessen the stimulus to hepatic lipid synthesis. In this respect, it is of interest that certain high fiber diets result in reduced villous height in the small intestine of rodents (Tasman Jones, 1980), an effect which also may be induced by lectins (Puzstai et al., 1979).

As discussed in relation to diabetes, enzyme inhibitors, found in large amounts in plant storage organs (grains, beans, roots, etc.) are currently receiving consideration in the treatment of a range of metabolic disorders. The α-glucoside hydrolase inhibitor with anti-amylase, anti-maltase, and anti-sucrase activity (Jenkins et al., 1981a) produced by the Bayer Company was developed for the control of diabetes (Walton et al., 1979) as well as hypertriglyceridemia. In hypertriglyceridemia, it is proposed that the flattened insulin and glucose responses observed when the inhibitor is taken with carbohydrate-containing meals will reduce the stimulus to triglyceride synthesis. The same may be true of enzyme inhibitors in food.

Plant sterols are another group of food constituents which are found in high fiber foods. They reduce the absorption of cholesterol but are absorbed only to a negligible degree themselves (Gould, 1955; Salen et al., 1970). Addition of β-sitosterol to the diet has been found to produce small but significant decreases in serum cholesterol in healthy volunteers (Salen et al., 1970).

The above food constituents, possessing the spectrum of possible hypolipidemic effects ascribed to fiber, make it difficult to define the action of fiber in whole foods. In view of their occurrence and mechanism of action it may be appropriate to consider them along with fiber. Together with fiber in purified form, they may be useful in the pharmacological management of hyperlipidemia and related disorders.

6.   Weight Control

As part of the "fiber hypothesis," it was pointed out that obesity was not seen in those parts of the world where large amounts of fiber were eaten. However, in many communities the consumption of unprocessed foods is associated with decreased caloric intake and thus a causal relationship with fiber is difficult to establish. Methylcellulose, cellulose and other non-absorbable materials have been used to induce satiety as non-caloric "fillers" (Van Itallie, 1978). Small (1.5-2.5 kg) weight losses were seen in diabetics (Jenkins *et al.*, 1980b) and hyperlipidemics (Jenkins *et al.*, 1980b) after long-term guar supplementation, and similar small weight losses have been reported in diabetics fed high fiber, high carbohydrate diets (Kiehm *et al.*, 1976). Increased fecal fat losses have been found in association with diets supplemented with bran (Cummings *et al.*, 1976), guar (Jenkins *et al.*, 1976b) or pectin (Jenkins *et al.*, 1976b; Kay and Truswell, 1977), but these losses seldom have been more than 7 g per day greater than those seen on the control diet.

It is tempting to speculate that the weight loss induced by viscous fiber may be due to its ability to prolong absorption and so to prevent large postprandial excursions above and below the fasting blood glucose level. Together with associated alterations in amino acid metabolism (Anderson, 1979), this may increase satiety. In addition, reduction in endogenous insulin production or a decreased requirement in diabetics for exogenous insulin may be an important factor favoring weight loss.

Not only the type of fiber but also the effect it has on food form and food selection must be considered. This effect has been clearly shown in studies in which rats became obese when fed human "junk" foods (biscuits and chocolate coated pastries, etc.) as opposed to Chow[R] (Van Itallie, 1978). Further studies must be undertaken before the role of fiber *per se* in the prevention and treatment of obesity can be defined.

7.   Constipation

The laxative effect of wholemeal bread was noted in the writings of Hippocrates. However, despite systematic studies on the mechanism carried out in the 1930's (Williams and Olmsted, 1936) and 1940's (Werch and Ivy, 1941), and the use of fiber preparations as laxatives, general interest in this aspect of colonic physiology has been greatly stimulated by the fiber hypothesis. In the past decade, studies on purified fiber have demonstrated large differences in the fecal bulking ability of different fibers (Brodribb and Groves, 1978; Cummings *et al.*, 1976, 1978, 1979; Eastwood *et al.*, 1973; Kirwan *et al.*, 1974) and have further suggested that pentosan content may be an important determinant (Cummings *et al.*, 1978). Wheat bran, which is rich in pentosans, has been found to be the most effective fiber in increasing fecal bulk (Cummings *et al.*, 1978). The importance of particle size has also been demonstrated (Brodribb and Groves, 1978; Kirwan *et al.*, 1974), bran of coarse particle size being more effective than fine bran. Fibers, such as the pectins, which are largely metabolized in the colon, contribute little to fecal output (Werch and Ivy, 1941; Cummings *et al.*, 1979; Kay and Truswell, 1977) and it is likely that a greater proportion of the fine bran than of coarse bran is metabolized.

Volatile fatty acids (VFA) formed from fiber and other dietary carbohydrates are no longer thought to be major cathartics (Cummings *et al.*, 1981). An exception is lactate (Cummings *et al.*, 1981), formed from lactose, which is associated with infant diarrhea. In view of the substantial amounts of acetate, proprionate and butyrate formed from carbohydrates entering the colon and the small amounts of VFA which are passed in the stool, it appears that the colon has a substantial ability to absorb VFA (Cummings *et al.*, 1981). These acids may contribute up to 10% of the energy requirements of individuals consuming a Western diet and appreciably more (20%) in subjects consuming the high fiber diets of some other communities (Cummings *et al.*, 1981).

It has been shown that the bulking effect of different fibers is achieved through different mechanisms (Cummings *et al.*, 1981). Thus, with cabbage fiber, the increase was due to an increase in fecal bacteria, whereas with bran it was due to the fiber, salts, and water content of the stool.

Investigations into these aspects of colonic function did much to arouse interest in many facets of colonic metabolism and bacteriology. It was demonstrated that a negative relationship exists between fecal weight and transit time (Cummings *et al.*, 1976; Eastwood *et al.*, 1973). This finding, together with the possibility that fiber alters stool bacteriology (Hill, 1982), suggested a means by which fiber may prevent colonic cancer, i.e., by diluting carcinogens, speeding their elimination and modifying their synthesis. At present, the role of fiber in colonic cancer prophylaxis is speculative. Nevertheless, current evidence suggests that the initial advice given to all individuals who suffer from constipation should include a recommendation to eat wholemeal bread, breakfast cereals, and other unrefined wheat and cereal products. Fiber intake may be further increased by the use of a high bran breakfast cereal, bran muffins, and biscuits. If this is ineffective, additional bran may be taken as such on breakfast cereals, etc., or in high bran foods prepared by the pharmaceutical industry. Much simple constipation may be improved in this way.

8.   Diverticular Disease

Until relatively recently, diverticular disease was treated in the long term with low residue diets in order to "rest" the bowel (Littlewood *et al.*, 1981). The use of high fiber diets in the outpatient management of diverticular disease therefore represents a dramatic reversal of the established clinical practice. The early work of Painter and colleagues demonstrated significant improvement in symptoms in patients with diverticular disease fed bran supplements (Painter and Burkitt, 1971, 1975; Painter *et al.*, 1972). Further work indicated that raised intraluminal pressures could be reduced by coarse bran, and bowel function normalized and symptomatic relief induced by bran supplements (Brodribb and Humphreys, 1976). It has also been suggested that individuals with diverticular disease have a history of consuming diets which contained lower levels of fiber than matched controls from the general population (Brodribb and Humphreys, 1976). Although the value of fiber in the treatment of relatively mild symptoms of diverticular disease recently has been disputed (Ornstein *et al.*, 1981), the weight of published evidence remains in support of high fiber diets in the long-term outpatient management of diverticular disease, i.e., a treatment regimen similar to that advocated for the prevention of constipation (Heaton, 1981).

9.   Crohn's Disease

It has been reported that, as with diverticular disease, patients with Crohn's consume less fiber and more sugar prior to diagnosis than do controls (Kasper and Sommer, 1979; Thornton *et al.*, 1979). Such retrospective evidence prompted a prospective trial of dietary modification which included a reduction in sugar intake and an increase in fiber consumption in Crohn's disease patients (Heaton *et al.*, 1979). Over a five-year period the test group required one-fifth the time in hospital that the control group did and were subject to fewer operations (Heaton *et al.*, 1979). This trial, which was undertaken irrespective of the presence of strictures, suggests that the role of fiber in the management of Crohn's warrants further attention.

10.   Gallstones, Hiatus Hernia, and Varicose Veins

Comparison of the diet histories of test and control groups has suggested that not only patients with diverticular disease but also sufferers from the remaining two diseases of Saint's triad, gallstones and hiatus hernia, consumed significantly less fiber in their diets (Capron *et al.*, 1978). The original fiber hypothesis suggested that both varicose veins and hiatus hernia may be the result of increased intra-abdominal pressure arising from attempts to pass constipated feces (Burkitt and Meisner, 1979). Fiber, by regularizing bowel habit, may prevent this increase. The rationale for fiber in the prevention of gallstones was not clearly stated initially, but subsequent studies have indicated that bran supplements increase the proportion of chenodeoxycholic acid in the circulating bile salt pool. The decrease in lithogenicity of the bile may provide a rationale for freedom from gallstones (Kay *et al.*, 1979; Pomare *et al.*, 1976).

11.   Fiber, Antinutrients and Neoplastic Disease

It has been reported that in those areas where a large amount of dietary fiber is eaten there is a lower incidence of some forms of neoplastic disease (e.g., colonic cancer) than in Western communities (Burkitt, 1971). Strong positive correlations exist between cancer incidence and total caloric intake and the intakes of fat and protein (Wynder, 1979). It is possible that fiber and antinutrients in general play a part: fiber, by decreasing the absorption of potential toxins, diluting their concentration in the colon and altering colonic metabolism in general; lectins and immunologically active plant proteins by enhancing the immune response, and antinutrients by decreasing effective caloric intake while reducing the rate of absorption of nutrients. Of the antinutrients, only fiber has attracted interest, and clear-cut experimental data on humans are lacking. However, a recent Dutch study suggests that, within a given population, a high intake of fiber is associated with a lower incidence of both colonic cancer and cardiovascular disease (Kromhout *et al.,* 1982).

12.   Possible Adverse Effects of Fiber

Originally, fiber (as crude fiber) was at best thought to be inert and at worst an antinutrient which limited the nutritional value of food. Since the dietary fiber concept of Trowell and Burkitt (1975), emphasis has shifted from the demerits to the possible benefits of the consumption of fiber-rich foods. There is now a reappraisal of the merits and demerits of the so-called "antinutrients" in general.

Nevertheless, if an increased intake of these substances is advocated, the possibility of adverse effects must be considered. At present there is no clear indication that dietary fiber *per se* compromises mineral and trace element status in the long term to the extent of producing clinical deficiency states in man. There are studies, however, which indicate that high cereal fiber (Heaton and Pomare, 1974; Jenkins *et al.,* 1975a) or cellulose diets (Pak *et al.,* 1974) may lower blood levels or increase fecal losses of calcium and of iron. Only when associated with unleavened breads, possibly caused by an increased intake of phytate normally destroyed during leavening and cooking, was a zinc deficiency seen (Reinhold *et al.,* 1976). The relevance of these findings to fiber supplementation of a Western diet is not clear. Certain groups, such as the malnourished and elderly, may require special consideration, but there is no evidence for general concern. Long term studies on guar given to diabetics for six months failed to show any reduction in serum Ca^{++}, Zn^{++}, or Cu^{++} levels (Jenkins *et al.,* 1980b). Similar results have been found using the Anderson diet (Anderson *et al.,* 1980).

13. Conclusion

Many diseases have been related to dietary fiber. The treatment of diverticular disease and diabetes has already been markedly influenced by the fiber hypothesis. Much has been learned of the role of fiber in gastrointestinal physiology and its links with metabolic events, blood glucose and endocrine responses to meals, bile acid and lipid metabolism, colonic metabolism, volatile fatty acid generation and absorption, and metal ion and trace element metabolism. Much more remains to be learned about the relevance of these effects to disease states. But in the long term, perhaps the most important contribution the fiber hypothesis has made to nutritional and medical knowledge comes from the impetus it has given to exploration of the links between nutrition and disease.

14. Acknowledgments

Supported by funds from the Canadian Diabetes Association, the Natural Sciences and Engineering Research Council, and the Ileitis and Colitis Foundation.

References

Alen, C., 1982, The food that fights diabetes, heart disease and high blood pressure, *National Enquirer* January 5, p. 2.

American Diabetes Association, Committee on Food and Nutrition, 1979, Special report: principles of nutrition and dietary recommendations for individuals with diabetes mellitus, *Diabetes Care* **2**:520.

American Heart Association, 1982, Report of Nutrition Committee: Rationale of the diet-heart statement of the American Heart Association, *Circulation* **65**:839A.

Anderson, G.H., 1979, Control of protein and energy intake: role of plasma amino acids and brain neurotransmitters, *Can. J. Physiol. Pharmacol.* **57**:1043.

Anderson, I.H., Levine, A.S., and Levitt, M.D., 1981, Incomplete absorption of the carbohydrate in all-purpose wheat flour, *N. Engl. J. Med.* **304**:891.

Anderson, J.W., and Chen, W.L., 1979, Plant fiber: carbohydrate and lipid metabolism, *Am. J. Clin. Nutr.* **32**:346.

Anderson, J.W., and Ward, K., 1979, High-carbohydrate, high-fiber diets for insulin-treated men with diabetes mellitus, *Am. J. Clin. Nutr.* **32**:2312.

Anderson, J.W., Ferguson, S.K., Karounos, D., O'Malley, L., Seiling, B., and Chen, W.J.L., 1980, Mineral and vitamin status on high-fiber diets: long-term studies of diabetic patients, *Diabetes Care* **3**:38.

Bingwen, L., Zhaofeny, W., Wanshen, L., and Rongjue, Z., 1981, Effects of bean meal on serum cholesterol and triglycerides, *Chin. Med. J.* **94**:455.

British Diabetic Association, Nutrition Subcommittee of the Medical Advisory Committee, Dietary Recommendations for Diabetes for the 1980's, final draft (July 1981).

Brodribb, A.J.M., and Groves, C., 1978, Effect of bran particle size on stool weight, *Gut* **19**:60.

Brodribb, A.J.M., and Humphreys, D.M., 1976, Diverticular disease: three studies, *Br. Med. J.* **1**:424.

Brunzell, J.D., Lerner, R.L., and Hazzard, W.R., 1971, Improved glucose tolerance with high carbohydrate feeding in mild diabetes, *N. Engl. J. Med.* **284**:521.

Burkitt, D.P., 1971, Epidemiology of cancer of the colon and rectum, *Cancer* **28**:3.

Burkitt, D.P., and Meisner, P., 1979, How to manage constipation with a high fiber diet, *Medical Report* **24**:30.

Campbell, G.D., 1971, Frequency of diabetes with special respect to diet, in: *Diabetes*, Proc. 7th Congress Int. Diab. Fedn., (R.R. Rodrigues and J. Vallance-Owen, eds.) Excerpta Medica, Amsterdam, (Int. Cong. Ser., 231) 1971.

Canadian Diabetes Association, Special Report of Committee on Guidelines for the Nutritional Management of Diabetes Mellitus, 1981, *J. Can. Diet. Assoc.* **42**:110.

Capron, J.P., Pajenneville, H., Dumont, M., Dupas, J.L., and Loniaug, A., 1978, Evidence for an association between cholelithiasis and hiatus hernia, *Lancet* **2**:329.

Carroll, K.K., 1978, The role of dietary protein in hypercholesterolemia and atherosclerosis, *Lipids* **13**:360.

Carroll, K.K., and Hamilton, R.M.G., 1975, Effects of dietary protein and carbohydrate on plasma cholesterol levels in relation to atherosclerosis, *J. Food Sci.* **40**:18.

Carroll, K.K., Giovannetti, P.M., Huff, M.W., Moase, O., Roberts, D.C.K., and Wolfe, B.M., 1978, Hypocholesterolemic effect of substituting soybean protein for animal protein in the diet of healthy young women, *Am. J. Clin. Nutr.* **31**:1312.

Caspary, W.F., 1978, Sucrose malabsorption in man after ingestion of alpha-glucosidehydrolase inhibitor, *Lancet* **1**:1231.

Christiansen, J.W., Bonnevie-Nielsen, V., Svendsen, P.A., Rubin, P., Ronn, B., and Nerup, J., 1980, Effect of guar gum on 24-hour insulin requirements of insulin-dependent diabetic subjects as assessed by an artificial pancreas, *Diabetes Care* **3**:659.

Cohen, M., Leong, V.W., Salmon, E., and Martin, F.I.R., 1980, Role of guar and dietary fibre in the management of diabetes mellitus, *Med. J. Aust.* **1**:59.

Connor, W.E., and Connor, S.L., 1982, The dietary treatment of hyperlipidemia: rationale, technique and efficacy, *Med. Clin. N. Am.* **66**:485.

Crapo, P.A., Reaven, G., and Olefsky, J., 1977, Postprandial plasma-glucose and -insulin responses to different complex carbohydrates, *Diabetes* **26**:1178.

Cummings, J.H., 1981, Progress report: Short chain fatty acids in the human colon, *Gut* **22**:763.

Cummings, J.H., Hill, M.J., Jenkins, D.J.A., Pearson, J.R., and Wiggins, H.S., 1976, Changes in fecal composition and colonic function due to cereal fiber, *Am. J. Clin. Nutr.* **29**:1468.

Cummings, J.H., Branch, W., Jenkins, D.J.A., Southgate, D.A.T., Houston, H., and James, W.P.T., 1978, Colonic response to dietary fibre from carrot, cabbage, apple, bran, and guar gum, *Lancet* **1**:5.

Cummings, J.H., Southgate, D.A.T., Branch, W.J., Wiggins, H.S., Houston, H., Jenkins, D.J.A., Jivraj, T., and Hill, M.J., 1979, The digestion of pectin in the human gut and its effect on calcium absorption and large bowel function, *Br. J. Nutr.* **41**:477.

Doi, K., Matsuuram, M., Kawara, A., and Baba, S., 1979, Treatment of diabetes with glucomannan (Konjac mannan), *Lancet* **1**:987.

Durrington, P.N., Manning, A.P., Bolton, C.H., and Hartog, M., 1976, Effect of pectin on serum lipids and lipoproteins, whole-gut transit-time and stool weight, *Lancet* **2**:394.

Eastwood, M.A., and Hamilton, D., 1968, Studies on the adsorption of bile salts to nonabsorbed components of the diet, *Biochim. Biophys. Acta* **152**:165.

Eastwood, M.A., Kirkpatrick, J.R., Michell, W.D., Bone, A., and Hamilton, T., 1973, Effects of dietary supplements of wheat bran and cellulose on faeces and bowel function, *Br. Med. J.* **4**:392.

Elsenhans, B., Sufke, U., Blume, R., and Caspary, W.F., 1980, The influence of carbohydrate gelling agents on rat intestinal transport of monosaccharides and neutral amino acids *in vitro, Clin. Sci.* **59**:373.

Engel, J., 1982, Is there enough fibre in your diet? *Chatelaine* **55(6)**:24.

Fahrenbach, M.J., Riccardi, B.A., Saunders, J.L., Lowrie, I.N., and Heider, J.G., 1965, Comparative effects of guar gum and pectin on human cholesterol levels, *Circulation* **31/32**:Suppl.2:11.

Fellman, B., 1982, Put pectin to work for you, *Prevention* **4**:109.

Gatti, E., and Sirtori, C.R., 1977, Soybean-protein diet and plasma-cholesterol, *Lancet* **1**:805.

Gerard, J., Luyckx, A.S., and Lefebvre, P.J., 1982, Improvement of metabolic control in insulin dependent diabetics treated with the alpha-glucosidase inhibitor Acarbose for two months, *Diabetologia* **21**:446.

Gillie, O., 1977, Indian bean gum helps diabetics and slimmers, *The Sunday Times,* Oct. 23.

Gold, L.A., McCourt, J.P., and Merimee, T.J., 1980, Pectin: an examination in normal subjects, *Diabetes Care* **3**:50.

Gould, R.G., 1955, Absorbability of beta-sitosterol, *Trans. N.Y. Acad. Sci.* **18**:129.

Grande, F., Anderson, J.T., and Keys, A., 1965, Effect of carbohydrates of leguminous seeds, wheat and potatoes on serum cholesterol concentration in man, *J. Nutr.* **86**:313.

Haber, G.B., Heaton, K.W., Murphy, D., and Burroughs, L.F., 1977, Depletion and disruption of dietary fibre: effects on satiety, plasma-glucose, and serum-insulin, *Lancet* **2**:679.

Heaton, K.W., 1981, Is bran useful in diverticular disease? *Br. Med. J.* **2**:1523.

Heaton, K.W., and Pomare, E.W., 1974, Effect of bran on blood lipids and calcium, *Lancet* **1**:49.

Heaton, K.W., Thornton, J.R., and Emmett, P.M., 1979, Treatment of Crohn's disease with an unrefined-carbohydrate, fibre-rich diet, *Br. Med. J.* **2**:764.

Hernandez, A., and Jaffe, W.G., 1968, Inhibitor of pancreatic amylase from beans *(Phaseolus vulgaris), Acta. Cient. Venez.* **19**:183.

Hill, M.J., 1982, Colonic bacterial activity: effect of fiber on substrate concentration and on enzyme action, in: *Dietary Fiber in Health and Disease* (G.V. Vahouny and D. Kritchevsky, eds.), Plenum Press, New York (in press).

Hillebrand, I., Boehme, K., Frank, G., Fink, H., and Berchtold, P., 1979, The effects of the alpha-glucosidase inhibitor Bay g 5421 (Acarbose) on meal-stimulated elevations of circulating glucose, insulin, and triglyceride levels in man, *Res. Exp. Med. (Berl.)* **175**:81.

Hjermann, I., Byre, K.V., Holme, I., and Lerer, P., 1981, Effect of diet and smoking intervention on the incidence of coronary heart disease: Report from the Oslo Study Group of a randomised trial of healthy men, *Lancet* **2**:1303.

Holt, S., Heading, R.C., Carter, D.C., Prescott, L.F., and Tothill, P., 1979, Effect of gel fibre on gastric emptying and absorption of glucose and paracetamol, *Lancet* **1**:636.

Jaffe, W.G., and Vega Lette, C.L., 1968, Heat-labile growth inhibiting factors in beans (Phaseolus vulgaris), *J. Nutr.* **94**:203.

Jaffe, W.G., Moreno, R., and Wallis, V., 1973, Amylase inhibitors in legume seeds, *Nutr. Rep. Int.* **7**:169.

Jeffrys, D.B., 1974, The effect of dietary fibre on the response to orally administered glucose, *Proc. Nutr. Soc.* **33**:11A.

Jenkins, D.J.A., Hill, M.S., and Cummings, J.H., 1975a, Effect of wheat fiber on blood lipids, fecal steroid excretion and serum iron, *Am. J. Clin. Nutr.* **28**:1408.

Jenkins, D.J.A., Leeds, A.R., Newton, C., and Cummings, J.H., 1975b, Effect of pectin, guar gum, and wheat fibre on serum-cholesterol, *Lancet* **1**:1116.

Jenkins, D.J.A., Leeds, A.R., Gassull, M.A., Wolever, T.M.S., Goff, D.V., Alberti, K.G.M.M., and Hockaday, T.D.R., 1976a, Unabsorbable carbohydrates and diabetes: decreased post-prandial hyperglycaemia, *Lancet* **2**:172.

Jenkins, D.J.A., Leeds, A.R., Gassull, M.A., Houston, H., Goff, D.V., and Hill, M.J., 1976b, The cholesterol lowering properties of guar and pectin, *Clin. Sci. Mol. Med.* **51**:8.

Jenkins, D.J.A., Gassull, M.A., Leeds, A.R., Metz, G., Dilawari, J.B., Slavin, B., and Blendis, L.M., 1977a, Effect of dietary fiber on complications of gastric surgery: prevention of postprandial hypoglycemia by pectin, *Gastroenterology* **73**:215.

Jenkins, D.J.A., Leeds, A.R., Gassull, M.A., Cochet, B., and Alberti, K.G.M.M., 1977b, Decrease in postprandial insulin and glucose concentrations by guar and pectin, *Ann. Intern. Med.* **86**:20.

Jenkins, D.J.A., Wolever, T.M.S., Hockaday, T.D.R., Leeds, A.R., Howarth, R., Bacon, S., Apling, E.C., and Dilawari, J., 1977c, Treatment of diabetes with guar gum, *Lancet* **2**:779.

Jenkins, D.J.A., Wolever, T.M.S., Leeds, A.R., Gassull, M.A., Haisman, P., Dilawari, J., Goff, D.V., Metz, G.L., and Alberti, K.G.M.M., 1978, Dietary fibres, fibre analogues, and glucose tolerance: importance of viscosity, *Br. Med. J.* **1**:1392.

Jenkins, D.J.A., Leeds, A.R., Slavin, B., Mann, J., and Jepson, E.M., 1979a, Dietary fiber and blood lipids: reduction of serum cholesterol in type II hyperlipidemia by guar gum, *Am. J. Clin. Nutr.* **32**:16.

Jenkins, D.J.A., Nineham, R., Craddock, C., Craig-McFeely, P., Donaldson, K., Leigh, T., and Snook, J., 1979b, Fibre and diabetes, *Lancet* **1**:434.

Jenkins, D.J.A., Taylor, R.H., Nineham, R., Goff, D.V., Bloom, S.R., Sarson, D., and Alberti, K.G.M.M., 1979c, Combined use of guar and acarbose in reduction of postprandial glycaemia, *Lancet* **2**:924.

Jenkins, D.J.A., Reynolds, D., Leeds, A.R., Waller, A.L., and Cummings, J.H., 1979d, Hypocholesterolemic action of dietary fiber unrelated to fecal bulking effect, *Am. J. Clin. Nutr.* **32**:2430.

Jenkins, D.J.A., Bloom, S.R., Albuquerque, R.H., Leeds, A.R., Sarson, D.L., Metz, G.L., and Alberti, K.G.M.M., 1980a, Pectin and complications after gastric surgery: normalisation of postprandial glucose and endocrine responses, *Gut* **21**:574.

Jenkins, D.J.A., Reynolds, D., Slavin, B., Leeds, A.R., Jenkins, A.L., and Jepson, E.M., 1980b, Dietary fiber and blood lipids: treatment of hypercholesterolemia with guar crispbread, *Am. J. Clin. Nutr.* **33**:575.

Jenkins, D.J.A. Wolever, T.M.S., Bacon, S., Nineham, R., Lees, R., Rowden, R., Love, M., and Hockaday, T.D.R., 1980c, Diabetic diets: high carbohydrate combined with high fiber, *Am. J. Clin. Nutr.* **33**:1729.

Jenkins, D.J.A., Wolever, T.M.S., Taylor, R.H., Barker, H., and Fielden, H., 1980d, Exceptionally low blood glucose response to dried beans: comparison with other carbohydrate foods, *Br. Med. J.* **2**:578.

Jenkins, D.J.A., Wolever, T.M.S., Taylor, R.H., Barker, H., Fielden, H., Baldwin, J.M., Bowling, A.C., Newman, H.C., Jenkins, A.L., and Goff, D.V., 1980e, Glycemic index of foods: a physiological basis for carbohydrate exchange, *Am. J. Clin. Nutr.* **34**:362.

Jenkins, D.J.A., Wolever, T.M.S., Taylor, R.H., Barker, H., Fielden, H., and Jenkins, A.L., 1980f, Effect of guar crispbread with cereal products and leguminous seeds on blood glucose concentrations of diabetes, *Br. Med. J.* **2**:1248.

Jenkins, D.J.A., Wolever, T.M.S., Taylor, R.H., Ghafari, H., Jenkins, A.L., Barker, H., and Jenkins, M.J.A., 1980g, Rate of digestion of foods and postprandial glycaemia in normal and diabetic subjects, *Br. Med. J.* **2**:14.

Jenkins, D.J.A., Wolever, T.M.S., Taylor, R.H., Reynolds, D., Nineham, R., and Hockaday, T.D.R., 1980h, Diabetic glucose control, lipids, and trace elements on long-term guar, *Br. Med. J.* **1**:1353.

Jenkins, D.J.A., Taylor, R.H., Goff, D.V., Fielden, H., Misiewicz, J.J., Sarson, D.L., Bloom, S.R., and Alberti, K.G.M.M., 1981a, Scope and specificity of Acarbose in slowing carbohydrate absorption in man, *Diabetes* **30**:951.

Jenkins, D.J.A., Wolever, T.M.S., Taylor, R.H., Barker, H.M., Fielden, H., and Gassull, M.A., 1981b, Lack of effect of refining on the glycemic response to cereals, *Diabetes Care* **4**:509.

Jenkins, D.J.A., Barker, H.M., Taylor, R.H., and Fielden, H., 1982a, Low dose Acarbose without symptoms of malabsorption in the dumping syndrome, *Lancet* **1**:109.

Jenkins, D.J.A., Ghafari, H., Wolever, T.M.S., Taylor, R.H., Jenkins, A.L., Barker, H.M., Fielden, H., and Bowling, A.C., 1982b, Relationship between rate of digestion of foods and post-prandial glycaemia, *Diabetologia* **22**:450.

Jenkins, D.J.A., Thorne, M.J., Camelon, K., Jenkins, A.L., Rao, A.V., Taylor, R.H., Thompson, L.U., Kalmusky, J., Reichert, R., and Francis, T., 1982c, Effect of processing on digestibility and the blood glucose response: a study of lentils, *Am. J. Clin. Nutr.* **36**:1093.

Jenkins, D.J.A., Wolever, T.M.S., Taylor, R.H., Griffiths, C., Krzeminska, K., Lawrie, J.A., Bennett, C.M., Goff, D.V., Sarson, D.L., and Bloom, S.R., 1982d, Slow release dietary carbohydrate improves second meal tolerance, *Am. J. Clin. Nutr.* **35**:1339.

Kasper, H., and Sommer, H., 1979, Dietary fiber and nutrient intake in Crohn's disease, *Am. J. Clin. Nutr.* **32**:1898.

Kasper, H., Zilly, W., Fassl, H., and Fehle, F., 1979, The effect of dietary fiber on post-prandial serum digoxin concentration in man, *Am. J. Clin. Nutr.* **32**:2436.

Kay, R.M., and Truswell, A.S., 1977, Effects of citrus pectin on blood lipids and fecal steroid excretion in man, *Am. J. Clin. Nutr.* **30**:171.

Kay, R.M., Wayman, M., and Strasberg, S.M., 1979, Effect of autohydrolysed lignin and lactulose on gallbladder bile (GB) composition in the hampster, *Gastroenterology* **76**:1167 (abstract).

Keys, A., Grande, R., and Anderson, J.T., 1961, Fiber and pectin in the diet and serum cholesterol concentration in man, *Proc. Soc. Exp. Biol. Med.* **106**:555.

Khan, P., Macrae, R., and Robinson, R.K., 1979, The novel use of a chromatography refractive index detector for monitoring model dialysis experiments, *Lab. Pract.* **28**:260.

Kiehm, T.G., Anderson, J.W., and Ward, K., 1976, Beneficial effects of a high carbohydrate, high fiber diet on hyperglvcemic diabetic men, *Am. J. Clin. Nutr.* **29**:895.

Kirby, R.W., Anderson, J.W., Sieling, B., Rees, E.D., Chen, W.L., Miller, R.E., and Kay, R.M., 1981, Oat-bran intake selectively lowers serum low-density lipoprotein concentrations, *Am. J. Clin. Nutr.* **34**:824.

Kirwan, W.O., Smith, A.N., McConnell, A.A., Mitchell, W.D., and Eastwood, M.A., 1974, Action of different bran preparations on colonic function, *Br. Med. J.* **4**:187.

Kneen, E., and Sandstedt, R.M., 1946, Distribution and general properties of an amylase inhibitor in cereals, *Arch. Biochem.* **9**:235.

Kritchevsky, D., 1978, Fiber, lipids, and atherosclerosis, *Am. J. Clin. Nutr.* **31**:S65.

Kritchevsky, D., and Story, J.A., 1974, Binding of bile salts *in vitro* by nonnutritive fiber, *J. Nutr.* **104**:458.

Kritchevsky, D., and Story, J.A., 1975, In vitro binding of bile acids and bile salts, *Am. J. Clin. Nutr.* **28**:305.

Kritchevsky, D., Tepper, S.A., and Story, J.A., 1978, Influence of soy protein and casein on atherosclerosis in rabbits, *Fed. Proc.* **37**:747.

Kromhout, D., Bosschieter, E.B., and de Lezenne Coulander, C., 1982, Dietary fibre and 10-year mortality from coronary heart disease, cancer, and all causes, *Lancet* **2**:518.

Leeds, A.R., Ralphs, D.N., Boulos, P., Ebied, R., Metz, G.L., Dilawari, J.B., Elliot, A., and Jenkins, D.J.A., 1978, Pectin and gastric emptying in the dumping syndrome, *Proc. Nutr. Soc.* **37**:23A.

Leeds, A.R., Ralphs, D.N.L., Ebied, F., Metz, G., and Dilawari, J.B., 1981, Pectin in the dumping syndrome: Reduction of symptoms and plasma volume changes, *Lancet* **1**:1075.

Levitt, N.S., Vinik, A.I., Sive, A.A., Child, P.T., and Jackson, W.P., 1980, The effect of dietary fiber on glucose and hormone responses to a mixed meal in normal subjects and in diabetic subjects with and without autonomic neuropathy, *Diabetes Care* 3:515.

Lewis, B., Hammett, F., Katen, M., Kay, R.M., Merkx, I., Nobels, A., Miller, N.E., and Swan, A.V., 1981, Towards an improved lipid-lowering diet: Additive effects of changes in nutrient intake, *Lancet* 2:1310.

Liener, I.E., 1979, The nutritional significance of plant protease inhibitors, *Proc. Nutr. Soc.* 38:109.

Lindner, P., and Möller, B., 1973, Lignin: a cholesterol lowering agent? *Lancet* 2:1259.

Littlewood, E.R., Ornstein, M.H., McLean Baird, I., and Cox, A.G., 1981, Doubts about diverticular disease, *Br. Med. J.* 2:1524.

Malinow, M.R., McLaughlin, P., Stafford, C., Livingstone, A.L., Kohler, G.O., and Cheeke, P.R., 1979, Comparative effects of alfalfa saponins and alfalfa fiber on cholesterol absorption in rats, *Am. J. Clin. Nutr.* 32:1810.

Mathur, K.S., Khan, M.A., and Sharma, R.D., 1968, Hypocholesterolaemic effect of bengal gram: a long-term study in man, *Br. Med. J.* 1:30.

Mattoo, A.K., and Modi, V.V., 1970, Partial purification and properties of enzyme inhibitors from unripe mangos, *Enzymologia* 39:237.

McCance, R.A., Prior, K.M., and Widdowson, E.M., 1953, A radiological study of the rate of passage of brown and white bread through the digestive tract of man, *Br. J. Nutr.* 7:98.

McLoughlin, J.C., Buchanan, K.D., and Alam, M.J., 1979, A glucoside-hydrolase inhibitor in treatment of dumping syndrome, *Lancet* 2:603.

Miettinen, T.A., and Tarpila, S., 1977, Effect of pectin on serum cholesterol, fecal bile acids, and biliary lipids in normolipidemic and hyperlipidemic individuals, *Clin. Chim. Acta* 79:471.

Militzer, W., Ikeda, C., and Kneen, E., 1946, The preparation and properties of an amylase inhibitor of wheat, *Arch. Biochem.* 9:30.

Miller, B.S. and Kneen, E., 1947, The amylase inhibitor of Leoti sorghum, *Arch. Biochem. Biophys.* 15:251.

Miranda, P.M., and Horwitz, D.L., 1978, High fiber diets in the treatment of diabetes mellitus, *Ann. Intern. Med.* 88:482.

Morgan, L.M., Goulder, T.J., Tsiolakis, D., Marks, V., and Alberti, K.G.M.M., 1979, The effect of unabsorbable carbohydrate on gut hormones: modification of postprandial GIP secretion by guar, *Diabetologia* 17:85.

Morris, J.N., Marr, J.W., and Clayton, D.G., 1977, Diet and heart: a postscript, *Br. Med. J.* 2:1307.

Munoz, J.M., Sandstead, H.H., Jacob, R.A., Logan, G.M. Jr., Reck, S.J., Klevay, L.M., Dintzis, F.R., Inglett, G.E., and Shuey, W.C., 1979, Effects of some cereal brans and textured vegetable protein on plasma lipids, *Am. J. Clin Nutr.* 32:580.

Nachbar, M.S., and Oppenheim, J.D., 1980, Lectins in the United States diet: a survey of lectins in commonly consumed foods and a review of the literature, *Am. J. Clin. Nutr.* 33:2338.

Narayana Rao, M., Shurpalekar, K.S., and Sundaravalli, O.E., 1967, An amylase inhibitor in *Colocasia esculenta, Indian J. Biochem.* 4:185.

O'Connor, N., Tredger, J., and Morgan, L., 1981, Viscosity differences between various guar gums, *Diabetologia* 20:612.

O'Dea, K., Nestel, P.J., and Antonoff, L., 1980, Physical factors influencing postprandial glucose and insulin responses to starch, *Am. J. Clin. Nutr.* 33:760.

O'Dea, K., Snow, P., and Nestel, P., 1981, Rate of starch hydrolysis in vitro as a predictor of metabolic responses to complex carbohydrate in vivo *Am. J. Clin. Nutr.* 34:1991.

Ornstein, M.H., Littlewood, E.R., McLean Baird, I., Fowler, J., North, W.R.S., and Cox, A.G., 1981, Are fibre supplements really necessary in diverticular disease of the colon? A controlled clinical trial, *Br. Med. J.* 1:1353.

Painter, N.S., and Burkitt, D.P., 1971, Diverticular disease of the colon: a deficiency disease of Western civilisation, *Br. Med. J.* 2:450.

Painter, N.S., and Burkitt, D.P., 1975, Diverticular disease of the colon, a 20th Century problem, *Clin. Gastroenterol.* 4:3.

Painter, N.S., Almeida, A.Z., and Colebourne, K.W., 1972, Unprocessed bran in treatment of diverticular disease of the colon, *Br. Med. J.* 1:137.

Pak, C.W., Delea, C.S., and Bartter, F.C., 1974, Treatment of recurrent nephrolithiasis with cellulose phosphate, *N. Engl. J. Med.* 290:175.

Palmer, G.H., and Dixon, D.G., 1966, Effect of pectin dose on serum cholesterol levels, *Am. J. Clin. Nutr.* 18:437.

Pomare, E.W., Heaton, K.W., Low-Beer, T.S., and Espiner, H.J., 1976, The effect of wheat bran upon bile salt metabolism and upon the lipid composition of bile in gallstone patients, *Am. J. Dig. Dis.* 21:521.

Potter, J.D., Illman, R.J., Calvert, G.D., Oakenfull, D.G., and Topping, O.C., 1980, Soya saponins, plasma lipids, lipoproteins and fecal bile acids: a double blind cross-over study, *Nutr. Rep. Int.* 22:521.

Puls, W., and Keup, U., 1973, Influence of an α-amylase inhibitor (Bay d 7791) on blood glucose, serum insulin and NEFA in starch loading tests in rats, dogs and man, *Diabetologia* 9:97.

Puzstai, A., Clarke, E.M.W., and King, T.P., 1979, The nutritional toxicity of *Phaseolus vulgaris* lectins, *Proc. Nutr. Soc.* 38:115.

Reinhold, J.G., Faradji, B., Abadi, P., and Ismail-Beigi, F., 1976, Decreased absorption of calcium, magnesium, zinc and phosphorus by humans due to increased fiber and phosphorus consumption as wheat bread, *J. Nutr.* 106:493.

Rivellese, A., Riccardi, G., Giacco, A., Pacioni, D., Genovese, S., Mattioli, P.L., and Mancini, M., 1980, Effect of dietary fibre on glucose control and serum lipoproteins in diabetic patients, *Lancet* 2:447.

Salen, G. Ahrens, E.H. Jr., and Grundy, S.M., 1970, Metabolism of β-sitosterol in man, *J. Clin. Invest.* 49:952.

Salunkhe, D.K., and Wu, M.T., 1977, Toxicants in plants and plant products, *CRC Critical Reviews in Food Science and Nutrition* 9:265.

Saraswathy Devi, K., and Kurup, P.A., 1973, Hypolipidaemic activity of the protein and polysaccharide fraction from *Phaseolus Mungo* (blackgram) in rats fed a high-fat high-cholesterol diet, *Atherosclerosis* 18:389.

Saraswathy Devi, K., and Kurup, P.A., 1979, Effects of certain Indian pulses on the serum, liver and aortic lipid levels in rats fed a hypercholesterolaemic diet, *Atherosclerosis* 11:479.

Schauberger, G., Brinck, U.C., Suldner, G., Spaethe, R., Niklas, L., and Otto, H., 1978, Exchange of carbohydrates according to their effect on blood glucose, *Diabetes* 26:415 (abstract).

Simpson, R.W., Mann, J.I., Eaton, J., Carter, R., and Hockaday, T.D.R., 1979a, High carbohydrate diets in insulin-dependent diabetes, *Br. Med. J.* 2:523.

Simpson, R.W., Mann, J.I., Eaton, J., Moore, R.A., Carter, R., and Hockaday, T.D.R., 1979b, Improved glucose control in maturity-onset diabetes treated with high-carbohydrate-modified fat diet, *Br. Med. J.* 1:1753.

Simpson, H.C.R., Simpson, R.W., Lousley, S., Carter, R.D., Geekie, M., Hockaday, T.D.R., and Mann, J.I., 1981, A high carbohydrate leguminous fibre diet improves all aspects of diabetic control, *Lancet* 1:1.

Sirtori, C.R., Agradi, E., Conti, F., Mantero, O., and Gatti, E., 1977, Soybean-protein diet in the treatment of type-II hyperlipipoproteinaemia, *Lancet* **1**:275.

Story, J.A., and Kritchevsky, D., 1976, Comparison of the binding of various bile acids and bile salts *in vitro* by several types of fiber, *J. Nutr.* **106**:1292.

Tasman Jones, C., 1980, Effects of dietary fiber on the structure and function of the small intestine, in: *Medical Aspects of Dietary Fiber* (G.A. Spiller and R.M. Kay, eds.), pp. 67-74, Plenum Press, New York.

Taylor, R.H., 1979, Gastric emptying, fibre, and absorption, *Lancet* **1**:872.

Thiffault, C., Belanger, M., and Pouliot, M., 1970, Traitment de l'hyperlipoproteinemie essentielle de type II par un nouvel agent therapeutique, la Celluline, *Can. Med. Assoc. J.* **103**:165.

Thomson, A.B.R., and Dietschy, J.M., 1980, Experimental demonstration of the effect of unstirred water layer on the kinetic constants of the membrane transport of D-glucose in rabbit jejunum, *J. Membr. Biol.* **54**:221.

Thornton, J.R., Emmett, P.M., and Heaton, K.W., 1979, Diet and Crohn's disease: characteristics of the pre-illness diet, *Br. Med. J.* **2**:762.

Topping, D.C., Trimble, R.P., Illman, R.J., Potter, J.D., and Oakenfull, D.G., 1980, Prevention of dietary hypercholesterolemia in the rat by soy flour high and low in saponins, *Nutr. Rep. Int.* **22**:315.

Trowell, H.C., and Burkitt, D.P., 1975, Concluding considerations, in: *Refined Carbohydrate Foods and Disease* (D.P. Burkitt and H.C. Trowell, eds.), pp. 333-345, Academic Press, London.

Truswell, A.S., and Kay, R.M., 1976, Bran and blood-lipids, *Lancet* **1**:367.

U.S. Senate Select Committee on Nutrition and Human Needs, Dietary Goals for The United States, U.S. Government Printing Office, Washington, D.C., 1977.

Vahouny, G.V., Tombes, R., Cassidy, M.M., Kritchevsky, D., and Gallo, L.L., 1980, Dietary fibers: V. Binding of bile salts, phospholipids and cholesterol from mixed micelles by bile sequestrants and dietary fibers, *Lipids* **15**:1012.

Van Itallie, T.B., Dietary fiber and obesity, 1978, *Am. J. Clin. Nutr.* **31**:S43.

Walton, R.J., Sherif, I.T., Noy, G.A., and Alberti, K.G.M.M., 1979, Improved metabolic profiles in insulin-treated diabetic patients given an alpha-glucosidehydrolase inhibitor, *Br. Med. J.* **1**:220.

Werch, S.C., and Ivy, A.C., 1941, On the fate of ingested pectin, *Am. J. Dig. Dis.* **8**:101.

Williams, R.D., and Olmsted, W.H., 1936, The effect of cellulose, hemicellulose, and lignin on the weight of the stool: a contribution to the study of laxation in man, *J. Nutr.* **11**:433.

Wynder, E.L., 1979, Dietary habits and cancer epidemiology, *Cancer* **43**:1955.

Zavoral, J.H., Smith, C.M., Hedlund, B.E., Hanson, M., Kuba, K., Frantz, I.D., and Jacobs, D.R., Locust bean gum in food products fed to familial hypercholesterolemic families, one adult with ileal bypass and one adult with type IV hyperlipidemia, ACS Symposium Series (in press).

Chapter 8

The Role of Selenium in Keshan Disease

Guangqi Yang, Junshi Chen, Zhimei Wen,
Keyou Ge, Lianzhen Zhu, Xuecun Chen,
and Xiaoshu Chen

1. Introduction

Since the discovery of the essentiality of selenium (Se) for rats by Schwarz and Foltz (1957) and for chicks by Patterson et al. (1957), several naturally occurring animal diseases related to Se and vitamin E (VE) deficiency have been reported (Muth et al., 1958; Hartley and Grant, 1961). Since the recognition of Se as an essential nutrient for animals, suggestions that Se may also be essential in human nutrition have been based upon three kinds of findings. One was the demonstration of certain uncomplicated Se-deficiency diseases produced in the laboratory chick (Thompson and Scott, 1970) and rat (McCoy and Weswig, 1969; Wu et al., 1979), with the assumption that a similar specific requirement should occur in humans. Second was the isolation of glutathione peroxidase (GSHpx) from human erythrocytes (Awasthi et al., 1975) and human placenta (Awasthi and Dao, 1978). There was also the finding that Se is required for the growth of human fibroblasts in culture (McKeehan et al., 1976). It was, therefore, reasonable to assume that some human disorders may be related to nutritional Se deficiency. The association of Se with

Guangqi Yang, Junshi Chen, Zhimei Wen, Keyou Ge, Lianzhen Zhu, Xuecun Chen, and Xiaoshu Chen • Institute of Health, Chinese Academy of Medical Sciences, Beijing, People's Republic of China.

protein-rich foods prompted the early investigation of its possible relation to the pathogenesis of protein-calorie malnutrition (Schwarz, 1965a; Burk *et al.*, 1967). Moreover, the detrimental effects of Se deficiency on heart function observed in animals (Godwin, 1965; Godwin and Frash, 1966) led some researchers in this field to speculate that there may be a certain relationship between Se deficiency and human cardiovascular disease. For example, epidemiological evidence (Shamberger and Willis, 1976) suggested an inverse relationship between human heart disease mortality and nutritional Se status. Andrews *et al.* (1981) suggested that the high mortality rate from cardiovascular disease in the coastal plain of Georgia, U.S.A., may be related to Se deficiency. Frost and Lish (1975) also suggested that insufficient Se and VE may contribute to human heart disease.

This article reports on our research on the possible relationship between Se deficiency in humans and Keshan disease, an endemic cardiomyopathy of unknown cause in China. In view of the endemic occurrence of white muscle disease in lambs and of diarrhea in young asses with the same endemic distribution as Keshan disease, and in view of some similarities between the human and animal symptoms as well as the epidemiological characteristics between these two diseases, in 1965, the Institute of Veterinary Medicine, Shanxi (unpublished), suggested that the human and animal diseases could both be responsive to selenium. Small scale human intervention studies and Se analysis of blood and hair from local residents and of local foods were therefore carried out by our group from 1969. Until the large-scale study by the Keshan Disease Research Group of the Chinese Academy of Medical Sciences (1979) was carried out in Mianning County in 1974, no conclusive evidence could be drawn as to the effect of supplementation with Se either alone (Keshan Research Group, unpublished) or in combination with VE (Farm Hospital of Fu County, Shanxi and Xian Medical College, unpublished), although supplementation (particularly with Se alone) invariably seemed beneficial in each trial. An extensive comparative study of the level of Se in human blood and hair was carried out by the Keshan Research Group (unpublished). Data obtained in that year revealed that the Se status of people in the Keshan disease areas was extremely low. These results encouraged the continuation of the Se intervention study and finally led to its amazing success.

In order to evaluate more thoroughly the role of Se in the etiology of Keshan disease, the epidemiological relationship of this disease to Se status in susceptible populations was studied. Although the results indicate that Se deficiency plays an important role in the etiology of the disease, other factors, including microorganisms, in the local environment also may contribute to its occurrence. To our knowledge, this is the first time that a direct relationship between Se deficiency and a human disease, as well as

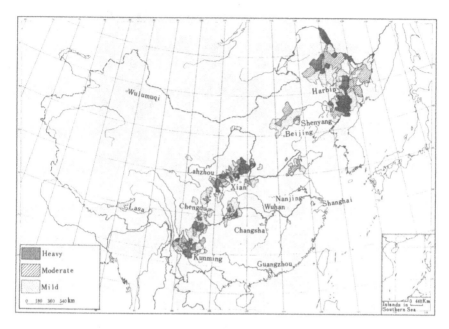

Fig. 1. Regional distribution of Keshan disease in China. Tan, J.A. *et al*. (1979).

the essentiality of Se in human nutrition, have been demonstrated. Since then, research groups working in other parts of China (Xian Medical College, unpublished; Institute of Geography, Chinese Academy of Sciences, unpublished) have confirmed these results.

2.   Brief Description of Keshan Disease

2.1. History and Epidemiology

In the winter of 1935, an outbreak of an unknown disease with sudden onset of precardial oppression and pain, nausea and vomiting (yellowish fluid), and fatal termination in severe cases occurred in Keshan County, Heilongjiang Province. It was called Keshan disease by the Japanese since its cause was not known. Later on, the disease was also discovered in other counties in the northeast as well as in other parts of China. It had a very high case-fatality, with more than 80% in the 1940's and around 30% since. It was at first suspected to be an acute infectious disease, and in 1936 was shown to be a myocardial disease with necrotic lesions of unknown nature.

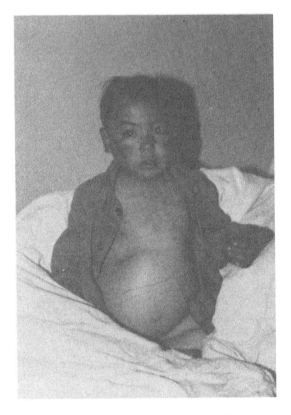

Fig. 2. A nine year-old boy with chronic Keshan disease who suffered from congestive heart failure for three years, showing slight protrusion of the precordial region, significant enlargement of liver, edema of extremities and ascites.

The distribution of Keshan disease has its own characteristics. It shows a belt-like regional distribution throughout the mainland of the country from northeast to southwest (Fig. 1) with specific focal distributions within the wide belt. The disease is generally found in hilly and mountainous areas in 14 provinces. The geographic features of affected areas are "eroded," in contrast to that of non-affected areas which are "precipitated." The epidemics are irregular from year to year. There is marked seasonal fluctuation in its prevalence. The peak season is winter in the northeast and summer in the south. The most susceptible populations are two- to seven-year-old children and women of child-bearing age. The disease is more prevalent in farmers' families than in the families of factory workers and other professionals. New cases have been found in families within three months after moving from non-affected areas into affected areas. Several cases of cardiomyopathy in fetuses from affected mothers have been reported.

2.2. Clinical Manifestations and Course

The outstanding feature of Keshan disease is the acute or chronic episode of a diseased heart characterized by cardiogenic shock, cardiac enlargement, congestive heart failure, cardiac arrhythmias, and electrocardiographic changes. Embolic episodes from cardiac thromboses are sometimes seen.

The course and features of the disease vary in different patients. These are categorized into four types which have some characteristics in common.

2.2.1.   Acute Type

Acute Keshan disease occurs abruptly in otherwise healthy people without any preexistence of cardiac disorders or in patients with a chronic or insidious disease as a result of rapid changes in its clinical course. The most frequent symptoms include dizziness, malaise, loss of appetite, nausea, even projectile vomiting, chilly sensation, precardia or substernal discomfort and dyspnea. The principal physical signs of acute Keshan disease are those resulting from cardiogenic shock, such as pallor, constricted veins in the extremities and low arterial pressure (below 80/60 mm Hg). The features of congestive heart failure always become obvious after the shock is controlled. Adam-Stock syndrome as a result of A-V block is not uncommon while the heart beat is less than 40 per minute. The electrocardiographic changes include proximal tachycardia, diminished voltage of QRS waves, prolongation in the atrio-ventricular conduction time and Q-T intervals, A-V Block, right bundle branch block, changes in S-T segment and inversion of T wave. Multiple variable and sudden changes of the ECG are the characteristics of acute Keshan disease. Roentgenographic studies usually show evidence of slight to moderate enlargement of the heart with diminished cardiac pulsation.

2.2.2.   Chronic Type

The prominent characteristic of chronic Keshan disease is chronic congestive heart failure, which is the late stage of acute or subacute disease or a result of a long lasting cardiac disorder of an insidious onset (Fig. 2). The symptoms vary according to the degree of cardiac insufficiency. The patients always complain of palpitation with consciousness of their own heart beat at rest, shortness of breath (exacerbated on exertion), cough with hemoptysis, pain in the right upper quadrant, edema, and oliguria. The classic signs are significant enlargement of the heart, reduced intensity of heart sound, a relatively soft and changeable systolic murmur, gallop rhythm, rales on the base of the lungs, hepatomegaly and edema. The changes in electrocardiography are even more pronounced in chronic

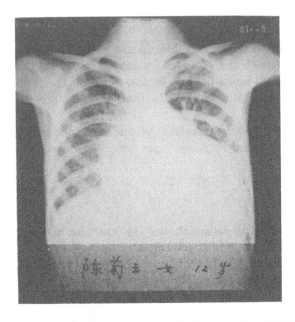

Fig. 3. Posterioanterio roentgenogram of the chest of a 12-year-old girl who had suffered from Keshan disease since five years of age, showing significant dilation and hypertrophy of the cardiac chambers (especially the left ventricle) and congestion of the lungs due to chronic congestive heart failure.

Keshan disease: proximal tachycardia (ventricular or supraventricular), arterial fibrillation, frequent ventricular premature beats, bundle branch block, A-V block, changes in ST-T segment and T wave. A significantly enlarged heart forming a flask shape with weak pulsation is characteristic under X-ray examination (Fig. 3).

2.2.3.   Subacute Type

This type is the most common, especially in children. Its clinical manifestations vary in proportion to the severity of heart failure. The onset is less sudden and the insidious period is about one to two weeks. With a mild attack or in the early stage there may be nothing more than malaise, restlessness, gallop rhythm and slight dilation of the heart. The clinical features of the subacute type are similar to those of the chronic type except that the severely affected patient seems to be on the verge of collapse and the disease has a more accelerated course.

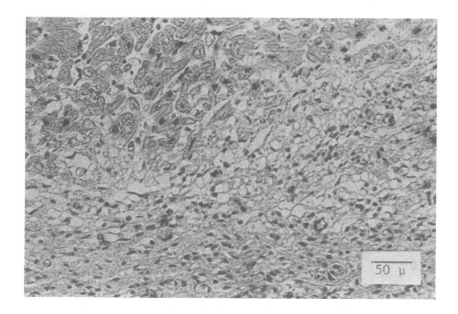

Fig. 4. A typical view of the myocardial lesion in subacute Keshan disease. Note the myocytolysis between the relatively intact myocardium (upper part) and the scar tissue (bottom).

2.2.4. Insidious Type

The patient may be unaware of the disease, which may be discovered only as an incidental finding upon routine autopsy or routine physical examination. Dizziness, fatigue, and palpitation after physical exercise or work are the most common complaints. These symptoms are associated with a mildly enlarged heart and abnormal ECG changes, including right bundle branch block, first or second degree A-V block, and infrequent premature ventricular contractions.

2.3. Morphological Observations

A great many autopsies have been carried out since the recognition of Keshan disease. These have shown the heart to be the main target organ. Moderate enlargement with dilation of all chambers represents the gross cardiac feature in most cases. Histopathologically, multifocal necrosis and fibrous replacement of myocardium is scattered throughout the heart muscle, the ventricles being more severely affected than the atria, and the

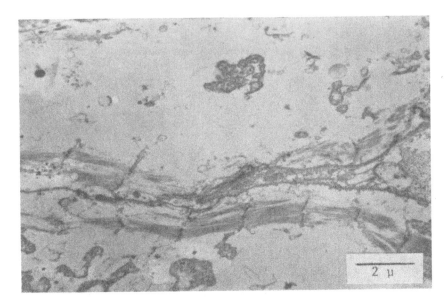

Fig. 5. Mitochondrial pattern of necrosis. Most organelles have
disappeared. The remaining sarcolemma outlines the damaged
myocytes.

left side more than the right side. Myocytolysis has been referred to as
the characteristic change (Fig. 4). Dense scar tissue, early fibrosis and
fresh necrosis may appear simultaneously in a single field. Subendocardial
patches of contraction band necrosis predominate in some acute cases,
while dense scars are frequently the main abnormalities in chronic cases.
Inflammatory cell infiltrates are variable. Lymphocytes and macrophages
are commonly seen within and around the necrotic foci. No substantial
lesions have been described for the endo- or epi-cardium. Changes in other
organs are usually not appreciable.

By electron microscopy, two necrotic patterns of myocardiopathy
have been demonstrated. One is characterized by mitochondrial disorgan-
ization and the other by myofibril segmentation. They are consistent with
the myocytolysis and the contraction band necrosis, respectively, found in
light microscopy. Mitochondrial swelling is the initial event in the mitochondria pattern. Balloon-like mitochondria are packed between the myofibrils or cluster around the nucleus. After most parts of the organelles
have disappeared, the remaining sarcolemma often outlines the location of
the affected myocytes (Fig. 5). This is the substantial feature observed in
most cases, and it may play the principal part in the development of the
heart lesion. The morphological appearance of myofibril patterns indicates
an over-contraction of the myofibrils. Irregular pieces or transversal bands

of contractile substance are seen interspersing among the relatively compacted mitochondria. This pattern has been considered as the consequence of the severe circulatory disorders usually observed in acute cases.

Histochemical enzymatic studies have been performed on Keshan hearts for comparison with controls. Acid phosphatase activity, shown as brown-black particles in the myoplasma by Holt's method (Holt, 1959) is sharply increased in myofibers surrounding the necrotic foci in the myocardium. Succinic dehydrogenase activity, measured by the Nachlas method (Nachlas *et al.,* 1957), appears as fine violet granules in normal heart muscle, but is greatly reduced or completely absent in the damaged myocardiocytes of Keshan disease patients.

2.4. Treatment

For acute patients, the case-fatality can be greatly reduced by early management, although there is no specific therapy. The patient should be kept warm and quiet in bed and a tranquilizer is recommended. Huge doses of ascorbic acid administered intravenously, first introduced by Xian Medical College (Xie, unpublished), are effective as a treatment of the cardiogenic shock. The first dose of 5-10 g with not less than 30 g within 24 hr is administered by slow intravenous injection to adults. Children less than ten years old are given up to half this dose. The dosage is repeated for two to three days and then is reduced slowly with the subsidence of symptoms. Patients with congestive heart failure require prompt and optimal digitalization, although the response in some cases is poor. Other treatments (antibiotics, oral diuretics, and moderate restriction of salt) are of value in the treatment of congestive heart failure.

2.5. Etiology

The etiology of Keshan disease is unknown. Although attempts had been made to relate experimental cardiomyopathy in animals to Keshan disease in man, only a few pathological changes of heart muscle resembling the changes in Keshan disease have been seen, and no experimental animal, to date, has been found to be an ideal model for research on the etiology of Keshan disease.

Several hypotheses on Keshan disease etiology have been suggested from epidemiological as well as laboratory studies.

2.5.1. Infection

Keshan disease is thought to be a kind of naturally occurring infection transmitted by unknown mediators or to be an infectious disease of viral origin. Extensive efforts have been made to grow suspect causative agents in microbiological media and in tissue cultures. Different strains of enterovirus, such as Echo and Coxsackie, have been isolated from the blood and organs of patients (Su *et al.,* 1979). A strain of Coxsackie B4 virus was isolated from a subacute patient, but could not be found in other patients. A viral agent is still considered to be the most probable factor, although no specific virus is yet known.

2.5.2.   Intoxication

It has been suggested that Keshan disease may occur through chronic poisoning by: (a) environmental toxicants including humin and other inorganic chemicals such as barium, nitrite, etc., in drinking water; (b) imbalanced intake of electrolytes and trace elements; and (c) biological products, such as mycotoxins, in moldy foods. No definitive evidence of these toxicants has yet been found, although extensive analyses have been conducted.

2.5.3.   Nutrient Deficiency

Protein and energy intakes in the affected areas are comparable to those in the non-affected areas. Most foods eaten by residents in the affected areas are grown locally on land which is "eroded" instead of "precipitated." There is a certain monotony in food selection in most affected areas, which are strictly rural. In these diets, deficiencies of molybdenum, magnesium, and thiamin have been suggested as causes of Keshan disease.

Unfortunately, none of these hypotheses has explained adequately the etiology of Keshan disease. Controversial reports are still appearing. Consequently, it has been suggested that Keshan disease is the result of a combination of several factors. The demonstration that Se supplementation can reduce the incidence of Keshan disease thus afforded important information on the etiology of this disease.

3.   Selenium Status and Keshan Disease

3.1.   Selenium Status of Residents in Keshan Disease Areas

Blood samples were analyzed by a modification of the fluorometric method of Watkinson (1966). A total of 173 and 111 blood samples were collected from affected and non-affected areas, respectively. The results, shown in Fig. 6, indicate that samples containing less than 0.02 ppm Se

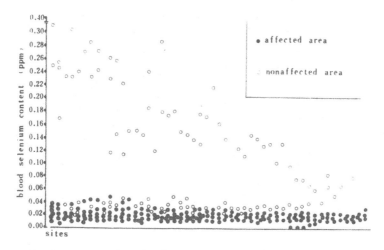

Fig. 6. Concentration of selenium in blood samples of residents in Keshan disease-affected and non-affected areas.

were almost exclusively from Keshan disease-affected areas, whereas all the samples from the non-affected areas contained more than 0.02 ppm of Se. The average blood Se content was 0.021 ± 0.001 ppm for the affected areas and 0.095 ± 0.088 ppm for non-affected areas. The blood Se levels of people in Keshan disease-affected areas are substantially lower than the lowest blood Se levels reported in free-living persons anywhere in the world, e.g., 0.048 ± 0.01 ppm for children in Tapanuin, New Zealand (Thomson and Robinson, 1980) and 0.054-0.079 ppm for persons in Egypt (WHO, 1973).

Concentrations of Se in the hair of residents in the Keshan disease-affected and non-affected areas were also studied. Because of the relative difficulty of sampling blood, it was preferable to use scalp hair for wide-scale screening of Se status. It should be noted that with carefully controlled methods of collecting scalp hair samples, there is an excellent correlation (r=0.81) between blood and hair Se concentrations within a practically important range (i.e., 0.002 to 0.266 ppm for blood and 0.023 to 0.890 ppm for hair) (Wang et al., 1979). As can be seen from Fig. 7, average hair Se concentrations in the affected areas were all less than 0.12 ppm, while those from non-affected areas were greater than 0.16 ppm, and those near the affected areas ranged from 0.12 to 0.20 ppm. Hair Se concentrations of residents at different sites of affected areas averaged 0.074 ± 0.050 ppm and those of non-affected areas 0.343 ± 0.173 ppm, the difference being highly significant (P<0.001). In parts of China located far from the wide belt of endemic Keshan disease, most hair Se concentrations fall within a range of 0.25 to 0.50 ppm. Hair Se concentrations of staff members in our Institute in Beijing averaged 0.80 ppm.

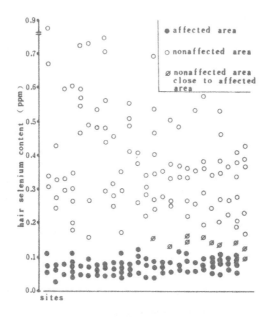

Fig. 7. Concentration of selenium in the hair of residents in Keshan disease-affected and non-affected areas.

Hair Se levels of people living in the seleniferous area of Enshi County, Hubei Province, averaged 32.2 ppm in regions with chronic human selenosis and 3.7 ppm in regions without human selenosis (Yang *et al.,* unpublished).

Taking the average of Se concentrations at individual sampling sites within a county as the representative figure for each county, a total of 128 representative values was obtained for both areas. As shown in Fig. 8, as these values are arranged in decreasing order it is evident that Keshan disease-affected counties are consistently low in Se status (see also Fig. 7), but the converse is not always true.

The urinary excretion of Se by residents of areas affected by Keshan disease and of non-affected areas was investigated. The amounts and concentrations of Se in urine were found to be strongly influenced by daily intakes of both Se and water. To minimize individual variation, the amount of Se excreted in 12-hr night urine, collected separately for three consecutive days, was measured. These data (Yin *et al.,* 1979) showed that the average urinary excretion of Se by rural children was 0.69 ± 0.18 μg in affected areas and 1.50 ± 0.13 μg in non-affected areas (P<0.05). Both values differed markedly from those obtained for children in Beijing (11.9 ± 1.34 μg), whose diets contain much more animal foods and are much higher in Se.

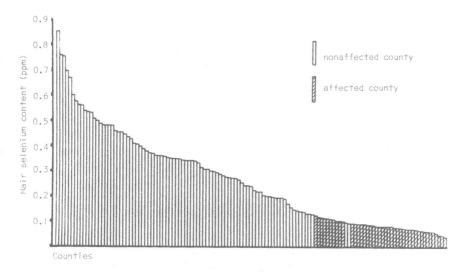

Fig. 8. Average hair selenium levels of residents in Keshan disease-affected and non-affected counties.

For further evaluation of Se status, in 1975 whole blood selenium-dependent GSHpx activity of residents of Keshan disease endemic and non-endemic areas in Sichuan Province were examined by the residual sulfhydryl method of Hafeman *et al.* (1974). The results are presented in Table I. The enzyme activities of children in the affected area were significantly lower than those in the non-affected area (P<0.001). The enzyme activities of patients suffering from Keshan disease were a little lower than those of normal children from the same areas, but the difference was not significant (P>0.05). After oral administration of sodium selenite for one year, the enzyme activities of children in affected areas had increased to levels comparable to those of children in the non-affected areas. Significant correlations were found between the Se concentrations in blood and hair and blood GSHpx activity (0.57 and 0.64, respectively) (Zhu *et al.*, unpublished).

3.2.   Concentrations of Selenium in Foods

The Se content of staple foods, such as maize and rice, in affected and non-affected areas was determined as described above. The data are shown in Fig. 9 and Fig. 10. It is clear that all maize and most rice samples from affected areas contained less than 0.01 ppm Se, whereas all the samples from non-affected areas contained more than 0.01 ppm Se. A few rice samples from the affected areas contained between 0.01 and 0.02 ppm Se. On the basis of the apparent levels of Se intake in rural areas

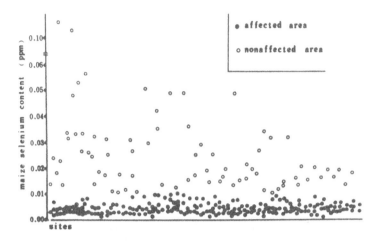

Fig. 9. Concentration of selenium in maize grown in Keshan
disease-affected and non-affected areas.

Table I. Blood glutathione peroxidase (SeGSHpx) activities and
selenium status of children from Keshan disease-affected
and non-affected areas

Area	SeGSHpx activity[a]	Blood Se (ppm)	Hair Se (ppm)
Non-affected	73.6 ± 3.1[b,d] (20)	0.065[d] (11)	0.237[d] (18)
Affected, healthy children	60.5 ± 0.7[d,e,f] (63)	0.023[d,e,f] (16)	0.058[d,e,f] (22)
Affected, Keshan disease patients	57.1 ± 1.3[e] (22)	0.020[e] (8)	$-$[c]
Affected, healthy children treated with Na_2SeO_3	76.1 ± 1.1[f] (58)	0.050[f] (15)	0.233[f] (21)

[a]Oxidation of GSH determined with 5,5'-dithio-bis (2-nitro-benzoic acid) after
incubation for five minutes with eight µl of whole blood. The nonenxymatic
oxidation has been subtracted.

[b]Mean; sample number is indicated in parentheses.

[c]Not determined.

[df]Means with same superscripts are significantly different (P<0.001).

[e]Means are not significantly different (P>0.05).

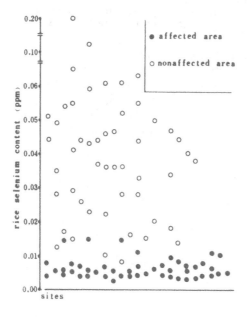

Fig. 10. Concentration of selenium in rice grown in Keshan disease-affected and non-affected areas.

without Keshan disease, as estimated by the intake of Se from cereals, it appears that the minimum Se requirement of adults is about 20 μg per day. This estimate is less than our previous one of 30 μg (Chen *et al.*, 1980). This estimated requirement is substantially lower than the known daily Se intakes in several countries, the lowest of which is 56 μg in New Zealand (Levander, 1976).

Average Se concentrations in maize produced in affected and non-affected areas were 0.005 ± 0.002 ppm and 0.036 ± 0.056 ppm, respectively; those of rice were 0.007 ± 0.003 ppm and 0.024 ± 0.038 ppm, respectively. These differences are highly significant (P<0.01), the Se concentration of rice in affected areas being less than one-fiftieth of that of rice in the U.S.A. as reported by Morris and Levander (1970). Since rural Chinese people live on a rather simple diet composed mainly of locally grown cereals and very small amounts of animal foods, their Se status depends primarily upon the Se content of cereals. These data clearly indicate that the poor Se status of residents in the Keshan disease area, as shown by the low Se levels in their blood, hair, and urine, and their low GSHpx activity is related to the low Se levels in local cereals.

Table II.   Selenium levels (ppm) in Human Hair and Major Staple Foods from Keshan Disease-affected and Non-affected Areas

Areas	Affected By Keshan Disease	Hair	Soy-bean	Sweet Potato	Wheat	Oats	Rice
Hilly Region	+	0.014 (79)[a]	0.028 (9)	0.004 (15)			
Coastal Region	-	0.282 (40)	0.062 (4)	0.009 (4)			
Low Soil Salinity	+	0.056 (20)			0.006 (6)	0.005 (5)	
High Soil Salinity	-	0.170 (20)			0.020 (6)	0.039 (4)	
Surrounding areas	+	0.060 (18)	0.008 (5)				0.006 (4)
"Safety Island"	-	0.144 (20)	0.020 (6)				0.025 (4)

Areas	Affected By Keshan Disease	Hair	Maize	Rice	Soy-bean	Sweet Flour	Potato Pancake
Far from Seleniferous Area	+	0.122 (20)	0.006 (4)	0.017 (4)			
Near Seleniferous Area	-	0.270 (20)	0.019 (4)	0.044 (4)			
Mountainous Region	+	0.084 (20)			0.022 (3)	0.003 (4)	
Plains	-	0.217 (40)			0.061 (4)	0.009 (2)	

[a]Numbers in parentheses are the number of specimens analyzed.

3.3. Epidemiological Characteristics of Keshan Disease and Se

It is well documented that the main epidemiological characteristics of Keshan disease are its regional distribution, seasonal prevalence, and population susceptibility. Therefore, it is of interest to compare the distribution of Se intakes with these major epidemiological characteristics.

Table III.   Differences in Hair Se Concentrations (Mean±SE, ppm) Between
             Children from Farmers' and Non-farming Families in the Same
             Affected Area

Province	Non-farming Families	Farmers' Families
Heilongjiang, Site 1	0.295±0.23[b] (11)[a]	0.151±0.11[b] (20)
Heilongjiang Site 2	0.357±0.021[c] (22)	0.146±0.013[c] (21)
Shandong	0.238±0.011[d] (7)	0.128±0.009[d] (14)
Sichuan	0.161±0.007[e] (16)	0.058±0.003[e] (22)

[a]Numbers in parentheses are the number of samples analyzed

[bcde]Means with same superscripts are significantly different (P<0.01).

3.3.1.   Regional Distribution

In addition to the distribution of this disease in a wide belt across the country, other geographical features have been observed. Numerous Keshan disease-affected areas are located in hilly and mountainous regions, whereas none are located in nearby coastal or plain areas. Keshan disease areas usually have low soil salinization. Within areas generally affected with Keshan disease, there may be an isolated "safety island" with a well-defined boundary line. Se levels in hair and staple food samples collected in these regions agree reasonably well with the distribution of Keshan disease (i.e., all affected areas are uniformly low in Se) (Table II).

3.3.2.   Population Susceptibility

Another outstanding epidemiological characteristic of Keshan disease is its prevalence among the children of farmers' families but not of nearby non-farming families. The results in Table III show that the vulnerable population is lower in Se. This finding may be related to the fact that non-farming families have access to more sources and varieties of food, and have a higher consumption of animal foods.

3.3.3.   Seasonal Prevalence

The peak seasonal prevalence of Keshan disease in northern China is in winter, but in southern China it is in summer. This suggests that seasonal fluctuations in body stores of Se may occur, assuming that Se deficiency alone is responsible for the occurrence of this disease. A total of six affected sites in Heilongjiang, Shandong, Sichuan, and Yunnan Provinces were studied. Hair Se concentrations of local children were monitored in four seasons, from as short a time as one year to as long as three consecutive years (Sun *et al.*, 1980). Although some seasonal variations in hair Se levels were found, no correlations were found between the peak seasonal prevalence of Keshan disease and hair Se concentrations.

4.   The Efficacy of Sodium Selenite in the Prevention of Keshan Disease

Since Keshan disease resembles somewhat the Se-responsive white muscle disease of farm animals with respect to its regional distribution and pathological changes in the heart, we conducted pilot studies on the effect of sodium selenite supplementation on the prevention of this disease in limited populations. After some encouraging results were obtained, from 1974 to 1980 an extensive study among children was carried out in a severely affected area in Sichuan Province to confirm the efficacy of Se intervention in the prevention of Keshan disease.

4.1.   Studies in Mianning County During 1974-1977 (Keshan Research Group, 1979)

In 1974, observations were made on children of susceptible age (1-9 years) in 119 production teams in three communes in Mianning County. In 1975 the study was extended to include 169 teams in four communes. One half of the children were given sodium selenite tablets and the other half a placebo. The assignment of groups was made randomly and was not changed during the two years of investigation. The subjects took sodium selenite once a week, the dosage being 0.5 mg for those aged 1-5 years and 1.0 mg for those 6-9 years old. Because of the convincing results obtained in 1974 and 1975, the control treatment was abandoned and all subjects were given sodium selenite in 1976 and 1977. The tablets were manufactured by the Beijing Fourth Pharmaceutical Factory with each tablet containing 1.0 mg of sodium selenite.

   A Keshan disease hospital was established in the area under investigation. The diagnosis and subtyping of the disease were made according to clinical diagnosis criteria drawn up in 1974 by the National Seminar of

Table IV. Keshan Disease Incidence Rate and Prognosis of Selenium-
Treated and Control (Untreated) Children in Mianning
County During 1974-1977

Group	Year	No. of Subjects	No. of Cases	Cases Alive			Eventual Deaths
				Turned Insidious	Improved	Turned Chronic	
Control	1974	3985	54	16	9	2	27
	1975	5445	52	13	10	3	26
Treated	1974	4510	10	9	0	1	0
	1975	6767	7	6	0	0	1

the Etiology of Keshan Disease. Only the acute and sub-acute cases reported by medical units in the districts or at higher levels were included in the morbidity statistics. Electrocardiograms, heart roentgenograms, and physical examinations were given to patients admitted to the hospital for observation. In some cases, blood Se content, GSHpx activity, serum glutamic oxalacetic transaminase and glutamic pyruvic transaminase were also determined. Patients receiving treatment at other medical units were transferred to the hospital after their general condition improved. Follow-up examinations were carried out each year to study the progress of individual patients.

The incidence rates and progress of the subjects investigated are shown in Table IV. In 1974, of the 3985 children in the control group there were 54 cases of Keshan disease (13.5 per 1000), while only ten of the 4510 Se-treated subjects fell ill (2.2 per 1000). The difference between the incidence rates of the two groups was highly significant (P<0.01). A similar difference was found in 1975, 52 of 5445 children in the control group (9.5 per 1000) and only seven of the 6767 children in the treated group (1.0 per 1000) incurring Keshan disease. In 1976, when all subjects were given Se, four cases occurred out of a total of 12,579 subjects, further lowering the incidence rate to 0.32 per 1000. There was one case of the typical subacute type among 212 children who failed to take Se. In 1977, there were no new cases among the 12,747 treated subjects. Prior to the Se intervention study, the incidence rate of acute and subacute types of Keshan disease in children 1-9 years old in that area had been about 10 per 1000, similar to that of the control groups in 1974 and 1975. These results indicate that sodium selenite intervention had a significant effect in reducing morbidity from Keshan disease.

The course of the disease in patients in the study during the succeeding years is shown in Table IV. Of the 54 cases occurring in the con-

Table V.   Keshan Disease Incidence Rates in Selenium-treated and
Untreated Communes in Mianning County during 1976-1980

Year	Treated Communes (4)[a]			Untreated Communes (3)		
	Total population	No. of cases	Incidence (per 1000)	Total population	No. of cases	Incidence (per 1000)
1976	41,181	4	0.09[b]	5,999	33	5.50[b]
1977	41,758	3	0.07[c]	6,243	12	1.92[c]
1978	41,533	5	0.12[d]	6,310	32	5.07[d]
1979	41,248	8	0.19[e]	6,411	23	3.59[e]
1980	41,384	4	0.10[f]	6,463	18	2.79[f]
Total	207,104	24	0.12[g]	31,426	118	3.75[g]

[a]Numbers in parentheses are the number of communes

[bcdef]Means with same superscripts are significantly different (P<0.01)

[g]Means are significantly different (P<0.00001)

trol group in 1974, 18 patients died that year, the total number of deaths in the subsequent four years being 27. Of the 52 cases occurring in the same group in 1975, 19 patients died that year and the total number of deaths in the subsequent three years was 26. By 1977, 53 deaths had occurred, yielding a case fatality rate of 50%. Among the survivors, six had significant heart failure and 19 had heart enlargement of different degrees, with their heart function never recovering to normal. During the same period, there were 17 cases of Keshan disease in the Se-treated group. Up to 1977, only one of them had died and one remained with pronounced heart failure. The heart function of the remaining 15 cases had all returned to normal, although slight heart enlargement has persisted in these cases. It is obvious that the Se-treated subjects not only had a lower incidence rate, but also a lower mortality and better prognosis.

We found that children were generally willing to take the Se tablets and that there were no untoward side effects. In a few cases nausea occurred after ingestion of the tablet, but this was avoided by taking the pills after meals. In 1976 and 1977, liver function tests (i.e., serum glutamic pyruvic transaminase activity) and general physical examinations were given to 100 subjects who had taken Se tablets weekly for three to four years. The results were not significantly different from those for the untreated children.

4.2. Studies in Mianning County During 1976-1980

Morbidity data for Keshan disease were collected in 1976-1980 by the Prefecture and County Sanitary and Anti-epidemic Station in order to compare the incidence rate in the four Se-treated communes with nearby untreated communes. The diagnostic criteria used previously were employed. The results are shown in Table V.

Although the incidence of Keshan disease was much greater among Se-treated communes than among untreated communes before intervention, the incidence in Se-treated communes during the five-year period from 1976-1980 (0.12 per 1000) was very significantly lower than that in the untreated communes (3.75 per 1000) ($P<0.00001$). Moreover, of the 24 cases which occurred in the Se-treated communes, it was determined that 18 subjects had never taken sodium selenite. Measurement of the hair Se content of six of these latter subjects showed that all values were below 0.005 ppm, the same level found for local untreated children.

4.3. Studies in Five Counties During 1976-1980

Since 1976, observations on the effects of sodium selenite have been extended to include Dechang, Xichang, Yuexi, and Puge Counties in addition to Mianning County. All children (1-12 years of age) in some of the most severely affected communes were treated with Se as described above, while the children in the nearby communes served as untreated

Table VI. Keshan Disease Incidence Rates in Selenium-treated and Untreated Children in Five Counties During 1976-1980

Year	Treated children			Untreated Children		
	No. of Subjects	No. of Cases	Incidence (per 1000)	No. of Subjects	No. of Cases	Incidence (per 1000)
1976	45,515	8	0.17[a]	243,649	488	2.00[a]
1977	67,754	15	0.22[b]	222,944	350	1.57[b]
1978	65,953	10	0.15[c]	220,599	373	1.69[c]
1979	69,910	33	0.47[d]	223,280	300	1.34[d]
1980	74,740	22	0.29[e]	197,096	202	1.07[e]
Total	323,872	88	0.27[f]	1,107,568	1713	1.55[f]

[abcde] Means with same superscripts are significantly different ($P<0.01$)

[f] Means are significantly different ($P<0.00001$).

controls. The results are shown in Table VI. Again, the results show that in each year the incidence rate of Keshan disease among Se-treated children of the five counties was significantly lower than among the untreated children. The total incidence rate in the Se-treated children during the five years (0.2 per 1000) was significantly lower than that in the untreated children (1.55 per 1000) (P<0.00001).

From 1973 on, similar work has been carried out in several provinces (e.g., Heilongjiang, Julin, Shandong, Shanxi, Yunnan, and Sichuan) on millions of people, using sodium selenite tablets manufactured by the same pharmaceutical factory and using the same dosage protocol. All results have shown the same trend as described above. The difference in morbidity between the Se-treated and control subjects has been consistent, especially in more severely affected areas. In all intervention programs, the oral administration of sodium selenite has proved to be effective in reducing the incidence, morbidity, and fatality of Keshan disease.

5.   Discussion

5.1.   Role of Selenium in the Etiology of Keshan Disease

The above findings show that the residents of Keshan disease areas are in poor Se status compared to those in non-affected areas. Among these residents, susceptible groups (e.g., children in rural farming families) have an even lower Se status than non-susceptible groups in the same area. In addition, Se intervention has been proved to be very effective in the prophylaxis of Keshan disease. Thus, it is very likely that Se insufficiency plays an important role in the etiology of this disease. However, some findings indicate that Keshan disease is not due strictly to nutritional Se deficiency. For example, the Se status of residents of affected areas does not change with seasonal fluctuations in the incidence of the disease. Also, sodium selenite has been found not to be effective in treating Keshan disease patients. It has been proposed that molybdenum, magnesium, VE and thiamin insufficiency, or barium and nitrite excess, as well as viruses and mycotoxins may be related to Keshan disease. Present information does not permit ruling out any one of these factors, acting as a co-factor with Se deficiency, in the etiology of this disease.

It is well known that VE and Se interact with each other in their recognized biochemical functions. In order to assess the possible role of VE in Keshan disease, the VE status of the residents of some affected areas was determined (Wang and Yang, unpublished). The results are shown in Table VII. No significant difference was found between patients and normal children, whose plasma VE concentrations were close to the

Table VII. Plasma Vitamin E Level (Mean±SE) of Keshan Disease Patients
and Non-affected Children (1-9 Years Old) in Mianning County

Groups	No. of Subjects	Plasma VE (mg/dl)[a]	Blood SE (ppm)
New Patients	16	0.55±0.09	0.019±0.003
Normal Children	19	0.48±0.06	0.026±0.005

[a]Plasma vitamin E was analyzed by the 4,7-dipenyl-1,10-phenathroline
method of Hashin and Schuttriger (1966).

Table VIII. The Effect of Vitamin E on the Selenium Status of Rats

Dietary Treatments[a]		Urinary Se Excretion	Se Retention	SeGSHpx Activity[b]	
VE	Na$_2$SeO$_3$			Heart	Liver
(IU/Kg)	(ppm)	%	%		
0	0	30.5±4.7 (10)[c]	29.6±1.0 (10)[c]	0.53±0.24 (5)[c]	0.53±0.16 (5)[c]
25	0	25.6±4.5 (10)	33.6±2.4 (10)	1.76±0.44 (9)	3.17±1.48 (9)
0	3	25.7±2.6 (10)	52.2±1.0 (10)	19.74±0.37 (6)	17.50±4.36 (6)
25	3	19.5±2.2 (10)	61.7 3.0 (10)	33.83±3.40 (9)	23.14±1.52 (9)

[a]Vitamin E free, low-selenium basal diet, composed of stripped corn meal and
soybean meal grown in Keshan disease endemic area. The basal diet contained
0.96 mg/Kg of α-tocopherol and 0.008 ppm of Se.

[b]A unit of enzyme activity was defined as a decrease in log[GSH] of 1 per
minute after the decrease in log[GSH] per minute of the non-enzymatic
reaction was subtracted.

[c]Mean±SE; number of samples indicated in parentheses.

deficient range. Chen *et al.* (unpublished) found that VE deficiency
enhanced the pathological and histochemical changes in the heart and
liver of piglets fed a low Se diet composed mainly of cereals grown in the
Keshan disease area. He and Zhu (unpublished) recently found that VE
deficiency increased the urinary excretion of Se and decreased Se retention and GSHpx activities in the heart and liver of rats fed either a low
Se diet or a Se-supplemented diet composed mainly of cereals from the

Keshan disease area (Table VIII). Since low VE intake impairs the bioa-vailability of Se and results in a greater requirement for Se (Reinhold, 1975), it is reasonable to assume that the marginal VE status of residents in Keshan disease areas may act as a co-factor in the occurrence of this disease. If so, VE supplementation might be effective in improving the status of the residents of low Se areas.

In conclusion, a multicause hypothesis is proposed for the etiology of Keshan disease, in which Se deficiency is considered the basic cause because of the endemic occurrence of the disease. However, much further work must be done to elucidate the roles of other factors in the environment which potentiate Keshan disease under conditions of nutritional Se deficiency.

5.2. Possible Mechanisms of the Effects of Sodium Selenite in the Prevention of Keshan Disease

From the known biological functions of Se and from our own work, the following mechanisms are suggested.

5.2.1. Effects on Cell Antioxidant Potential

Selenium has been shown to be an integral part of the Se-dependent glutathione peroxidase (Rotruck *et al.*, 1973) which catalyzes the reduction of a large variety of lipid hydroperoxides and of hydrogen peroxide (Little and O'Brien, 1968). In Se deficiency damage to membrane phospholipids by reactive oxygen metabolites may occur as a result of uncontrolled lipid peroxidation. Hepatic lipid peroxidation *in vitro,* estimated by the thio-barbituric acid (TBA) reaction, was four times as high in rats fed low Se diets composed mainly of cereals grown in Keshan disease areas as it was in rats fed Se-supplemented diets. In the heart this difference was 2.5-fold (He and Zhu, unpublished). Zhu *et al.* (unpublished) also found that the TBA value of GSH-swollen heart mitochondria from rats fed the same low Se diet was greater than that of rats fed a Se adequate diet (Table IX). These results suggest that the mitochondrial membrane peroxidation potential is increased by low Se status. Lipid peroxidation of these organelles might lead to the morphological changes in mitochondria cristae and the increase of acid phosphatase activity observed in the myocardium of Keshan disease patients. Supplementation with Se might protect the membranes from damage caused by lipid peroxidation occurring as a result of a deficiency of Se-dependent glutathione peroxidase.

Table IX.   Effect of Se-supplementation on TBA Value of GSH-swollen
            Heart Mitochondria

| Diet | TBA Value[a] | |
	Before Swelling	After Swelling
Low Se diet[b]	$0.950\pm0.117(9)^{c}$	$3.427\pm0.396(9)^{c,e}$
Low Se diet+Se[d]	$0.750\pm0.077(9)$	$2.120\pm0.197(9)^{e}$

[a] μmoles malonyldialdehyde per mg protein.

[b] Low Se diet was composed of staple cereals grown in Keshan
disease endemic areas.

[c] Mean$\pm$SE; number of samples shown in parentheses.

[d] 0.2 ppm Se added as Na_2SeO_3.

[e] Means with same superscripts are significantly different (P<0.01).

5.2.2.   Effects on Oxygen Metabolism

It has been pointed out (Wang, 1962) that the morphological changes in
the myocardium of Keshan disease patients resemble the changes caused
by ischemia and hypoxia. Rotruck *et al.* (1972) showed that dietary Se
protects hemoglobin against oxidative damage. He and Zhu (unpublished)
found a significant increase in the oxidation of hemoglobin by dihydroxy-
fumaric acid in rats fed a low Se diet composed mainly of cereals from
Keshan disease area. Respiratory decline was evident in the rate of oxygen consumption by liver slices of rats fed the semi-synthetic low Se diet
of Schwarz (1965b), and by heart homogenates of piglets fed a low Se
diet mainly composed of cereals from Keshan disease area (Li *et al.*,
unpublished). Since the respiratory decline in Se deficiency is not associated with an enhanced TBA reaction (Schwarz, 1965b), this effect may
be due to mechanisms other than the accumulation of lipid peroxides. It
is likely that the transport, preservation, and utilization of oxygen in the
myocardium of animals with a low Se intake are impaired, thus rendering
their myocardium more vulnerable to oxidation. Supplementation with Se
may improve the transport and utilization of oxygen and, hence, protect
the myocardium from hypoxic damage.

5.2.3.   The Anti-Infection Effect of Se

Selenium deficiency has been shown to be an endemic factor which,
probably in association with other agents, is responsible for the occurrence

Table X.   The Influence of Supplemental Selenium on Myocardial Necrosis Induced by Viral Infection in Mice

Group	Dietary Se(ppm)	Blood Se (ppm) Adult	Blood Se (ppm) Suckling	Total N	Heart Lesions N	Heart Lesions %
EXPERIMENT A						
Sichuan[a]	0.010	0.053	0.032	56	20	36[x]
Sichuan + Se[b]	0.140[f]	0.183	0.138	30	4	13[y]
Stock[c]	0.518	0.402	0.306	52	8	15[y]
EXPERIMENT B						
Sichuan[a]	0.017	0.054	0.033	44	17	38[x]
Sichuan + Se[d]	0.147[f]	0.150	0.070	58	8	14[y]
Semisynth.[e]	0.015	0.049	0.024	74	23	31[x]
Semisynth. + Se[d]	0.145[f]	0.111	0.088	70	16	23
Stock[c]	0.334	0.440	0.321	72	8	11[y]
EXPERIMENT C						
Sichuan[a]	0.027	0.075	0.059	182	73	40[x]
Sichuan + Se[d]	0.062[f]	0.210	0.146	97	19	20[y]
Sichuan + Se[d]	0.462[f]	0.582	0.316	128	13	10[y]
Semisynth.[e]	0.027	0.130	0.070	68	22	32[x]
Semisynth. + Se[d]	0.062[f]	0.255	0.146	71	14	20
Semisynth.+ Se[d]	0.462[f]	0.599	0.328	74	9	12[y]

[a] A natural low Se diet consisting mainly of cereals produced in the Keshan disease endemic area of Sichuan province.

[b] Se provided as sodium selenite by stomach tube.

[c] A stock laboratory diet used routinely in our Institute.

[d] Se added as sodium selenite to drinking water.

[e] A semisynthetic diet based mainly on starch and low Se yeast.

[f] Estimate based on calculation of total Se intake from Se content of ingredients.

[x,y] Indicating a statistical difference between Se treated and untreated groups by Chi-square analysis.

of Keshan disease. Experiments have been conducted in our laboratory to determine the combined effects of dietary Se and viral infection on the myocardium of mice. Weanling Kunming mice were fed a natural or semi-synthetic low Se diet with or without selenium supplementation. Sodium selenite was administered per os in one experiment and in the drinking water in two other experiments. After six weeks on their experimental diets, the animals were bred within treatment groups. At seven days of age, the offspring were injected intraperitoneally with the virus Coxsackie B4, isolated from the blood of a child who suffered from subacute Keshan disease. By 14 days of age, the blood Se concentrations of suckling mice paralleled that of their parents and were consistent with their respective rates of Se intake (Table X). At the same time, the incidence of heart lesions, revealed by pathological examination, was much greater among Se-deficient mice (31-40%) than in the Se-supplemented group (10-23%) (X^2=54.4, $P<0.00001$). An inverse correlation between blood Se concentrations and incidence of myocardial lesions was observed (r=-0.789). Thus Se treatment protected Se-deficient mice against the heart lesions induced by Coxsackie B4 virus.

Evidence has accumulated that Se is required for normal immune function in many species. As summarized by Spallholz (1980), many measurements of humoral and cellular immunities have been shown to be impaired in Se-deficient animals. Selenium administered at various levels has been shown to enhance immune responses to viral, bacterial, mycotic, and cellular antigens in several species. One way that Se appears to act is as a nonspecific stimulant of immune mechanisms contributing to anti-infection. It is reasonable to assume that if a cardiophilic virus contributes to the prevalence of Keshan disease, Se supplementation might enhance the resistance of the Se-deficient subjects to viral infection of their damaged hearts.

References

Andrews, J.W., Hames, C.G., and Metts, J.C. Jr., 1981, Selenium and Cadmium status in blood of residents from low selenium-high cardiovascular disease area of southeastern Georgia, in: *Selenium in Biology and Medicine* (J.E. Spallholz, J.L. Martin, and H.E. Ganther, eds.), pp. 348-353, AVI Publ. Co. Westport, Connecticut.

Awasthi, Y.C., and Dao, D.D., 1978, Purification and properties of glutathione peroxidase from human placenta, *Fed. Proc.* 37:1340, (abstract).

Awasthi, Y.C., Beutler, E., and Srivastova, S.K., 1975, Purification and properties of human erythrocyte glutathione peroxidase, *J. Biol. Chem.* 250:5144.

Burk, R. Jr., Pearson, W.N., Wood, R.P. II, and Viteri, F., 1967, Blood selenium levels and *in vitro* red blood cell uptake of ^{75}Se in kwashiorkor, *Am. J. Clin. Nutr.* 20:723.

Chen, X., Yang, G., Chen, J., Chen, X., Wen, Z., and Ge, K., 1980, Studies on the relation of selenium and Keshan disease, *Biol. Trace Element Res.* 2:91.

Frost, D.V., and Lish, P.M., 1975, Selenium in biology, *Ann. Rev. Pharmacol.* 15:259.

Godwin, K.O., 1965, Abnormal electrocardiograms in rat fed a low selenium diet, *Quart. J. Exptl. Physiol.* 50:282.

Godwin, K.O., and Frash, J.F., 1966, Abnormal electrocardiograms, blood pressure changes and some aspects of histopathology of selenium deficiency in lambs, *Quart. J. Exptl. Physiol.* 51:94.

Hafeman, D.G., Sunde, R.A., and Hoekstra, W.G., 1974, Effect of dietary selenium on erythrocyte and liver glutathione peroxidase in the rat, *J. Nutr.* 104:580.

Hartley, W.J., and Grant, A.B., 1961, A review of selenium responsive diseases of New Zealand livestock, *Fed. Proc.* 20:679.

Holt, S.J., 1959, Factors governing the validity of staining methods for enzymes, and their bearing upon the Gomori acid phosphatase technique, *Exptl. Cell Res.* suppl. 7:1.

Keshan Disease Research Group of the Chinese Academy of Medical Sciences, 1979, Observation on the effect of sodium selenite in prevention of Keshan disease, *Chinese Med. J.* 92:471.

Levander, O.A., 1976, Selenium in food, Proceedings of the Symposium on Selenium and Tellurium in the Environment, Industrial Health Foundation, Pittsburgh.

Little, C., and O'Brien, P.J., 1968, An intracellular GSH-peroxidase with a lipid peroxide substrate, *Biochem. Biophys. Res. Comm.* 31:145.

McCoy, K.E.M., and Weswig, P.H., 1969, Some selenium responses in the rat not related to vitamin E, *J. Nutr.* 98:383.

McKeehan, W.L., Hamilton, W.G., and Ham, R.G., 1976, Selenium is an essential trace nutrient for growth of WI-38 diploid human fibroblasts, *Proc. Natl. Acad. Sci.* 73:2023.

Morris, V.C., and Levander, O.A., 1970, Selenium content of foods, *J. Nutr.* 100:1383.

Muth, O.H., Oldfield, J.E., Remmert, L.F., and Schubert, J.R., 1958, Effects of selenium and vitamin E on white muscle disease, *Science* 128:1090.

Nachlas, M.M., Tsou, K.C., Souea, E.D., Cheng, C.S., and Seligman, A.M., 1957, Cytochemical demonstration of succinic dehydrogenase by the use of a new p-nitrophenyl substituted ditetrazole, *J. Histochem. Cytochem.* 5:420.

Patterson, E.L., Milstrey, R., and Stokstad, E.L.R., 1957, Effect of selenium in preventing exudative diathesis in chicks, *Proc. Soc. Exp. Biol. Med.* 95:617.

Reinhold, J.G., 1975, Trace elements - A selective survey, *Clin. Chem.* 21:476.

Rotruck, J.T., Pope, A.L., Ganther, H.E., Swanson, A.B., Hafeman, D.G., and Hoekstra, W.G., 1973, Selenium: biochemical role as a component of glutathione peroxidase, *Science* 179:588.

Rotruck, J.T., Pope, A.L., Ganther, H.E., and Hoekstra, W.G., 1972, Prevention of oxidative damage to rat erythrocytes by dietary selenium, *J. Nutr.* 102:689.

Schwarz, K., 1965a, Selenium and kwashiorkor, *Lancet* 1:1335.

Schwarz, K., 1965b, Role of vitamin E, selenium, and related factors in experimental nutritional liver diseases, *Fed. Proc.* 24:58.

Schwarz, K., and Foltz, C.M., 1957, Selenium as an integral part of factor 3 against dietary necrotic liver degeneration, *J. Am. Chem. Soc.* 79:3292, (letter).

Shamberger, R.J., and Willis, C.E., 1976, Epidemiological studies on selenium and heart disease, *Fed. Proc.* 35:578, (abstract).

Spallholz, J.E., 1980, Anti-inflammatory, immunologic and carcinostatic attributes of selenium in experimental animals, in: *Diet and Resistance to Disease* (M. Philips and A. Baetz, eds.) pp. 43-62, Plenum Press, New York.

Su, C., Gong, C., Li, J., Cheng, C., Zhou, D., and Jin, Q., 1979, Preliminary results of viral etiological study of Keshan disease, *Chinese Med. J.* 59:466 (in Chinese).

Sun, S., Yin, T., Wang, H., You, D., and Yang, G., 1980, The relationship between seasonal prevalence of Keshan disease and hair selenium of inhabitants, *Chinese J. Prev. Med.* 14:17 (in Chinese).

Thompson, J.N., and Scott, M.L., 1970, Impaired lipid and vitamin E absorption related to atrophy of the pancreas in selenium-deficient chicks, *J. Nutr.* 100:797.

Thomson, C.D., and Robinson, M.F., 1980, Selenium in human health and disease with emphasis on those aspects peculiar to New Zealand, *Am. J. Clin. Nutr.* 33:303.

Wang, F., 1962, Discussion of the etiology of Keshan disease based on the pathological findings, *Chinese Med. J.* 48:17.

Wang, G., Zhou, R., Sun, S., Yin, T., and Yang, G., 1979, Difference between blood selenium concentrations of residents of Keshan disease-affected and non-affected areas - correlation between the selenium content of blood and hair, *Chinese J. Prev. Med.* 13:204 (in Chinese).

Watkinson, S.H., 1966, Fluorometric determination of selenium in biological material with 2,3-diaminonaphthalene, *Anal. Chem.* **38**:92.

WHO, 1973, Trace elements in human nutrition, WHO Technical Reports Series, no. 532, Geneva, 1973.

Wu, A.S.H., Oldfield, J.E., Shull, L.R., and Cheeke, R.R., 1979, Specific effect of selenium deficiency on rat sperm, *Biology of Reproduction* **20**:793.

Yin, T., Sun, S., Wang, H., You, D., and Yang, G., 1979, Difference in the amount of selenium excreted in urine between children in Keshan disease-affected and non-affected areas, *Chinese J. Prev. Med.* **13**:207.

Chapter 9

Sucrose-Isomaltose Malabsorption

E. Gudmand-Høyer, P.A. Krasilnikoff, and H. Skovbjerg

1. Introduction

In the beginning of this century diarrhea occurring in some children and adults was shown to be provoked, apparently, by carbohydrates in the food (Schmidt and Strasburger, 1901; Jacobi, 1901; Howland, 1921; Hurst and Knott, 1931). Fermentative dyspepsia, fermentative diarrhea, starch intolerance, intestinal carbohydrate dyspepsia, and intolerance of carbohydrate are terms which have been used to describe these conditions.

Iversen (1942) showed that sucrose in the food was responsible for a chronic diarrhea in an 18-year-old man. Administration of sucrase caused the diarrhea to disappear. Only when disaccharidases were found to be localized in the small intestinal mucosa, and when their significance for the digestion of carbohydrates was realized (Borgström *et al.*, 1957; Dahlqvist and Borgström, 1961; Miller and Crane, 1961) did a real understanding of the pathogenesis of disaccharide malabsorption emerge; only then were clinicians able to categorize the various types of sugar intolerance.

Weijers *et al.* (1960, 1961) employing disaccharide tolerance tests, found sucrose malabsorption in three children, while Anderson *et al.* (1962) were the first to demonstrate the absence of sucrase activity in small-intestinal biopsies from patients suffering from this disorder.

E. Gudman-Høyer and H. Skovbjerg • Medical Department F (Gastroenterology), Copenhagen University Hospital in Gentofte, 2900 Hellerup, Denmark. P. A. Krasilnikoff • Paediatric Department L, Copenhagen University Hospital in Gentofte, 2900 Hellerup, Denmark.

During the next few years several cases of diarrhea caused by ingestion of sucrose were reported in Western Europe, the U.S.A., and Australia (Prader et al., 1961; Bach et al., 1962; Chaptal et al., 1962; Grenet et al., 1962; Clement, 1962; Corouben, 1963; Rey et al., 1963; Francois et al., 1963; Nordio et al., 1961; Anderson et al., 1963; Jensen, 1962; Launiala et al., 1964). By 1965, 63 cases had been described (Prader and Auricchio), by 1973 approximately 100 cases (Ament et al.), and by 1982 almost 200 cases.

Thus sucrose malabsorption is a relatively rare disease. The incidence of sucrose malabsorption among Eskimos is much higher, however, than among other populations. In 190 Greenlandic Eskimos McNair et al. (1972) found sucrose malabsorption in 20, or 10.5%, of the subjects. It was soon realized that sucrose malabsorption occurred familially (Auricchio et al., 1961; Jensen, 1962). Kerry and Townley (1965) showed that the mode of genetic transmission of sucrase deficiency is autosomal and recessive.

Isomaltase deficiency always accompanies sucrase deficiency. This fact was first cited by Prader et al. (1961), who based this conclusion on the results of disaccharide tolerance tests. It was later confirmed by Anderson et al. (1962), Auricchio et al. (1963a) and Anderson et al. (1963), among others. The small intestinal mucosa is histologically normal in cases with sucrose-isomaltose malabsorption (Anderson et al., 1963).

This review focuses primarily on specific (primary) sucrose-isomaltose malabsorption caused by a deficiency of sucrase and isomaltase. The role of the brush border hydrolases in carbohydrate digestion and their localization in the normal intestine are discussed, together with diagnostic procedures, clinical symptoms, and theories concerning the molecular background of sucrase-isomaltase deficiency. Finally, non-specific (secondary) sucrose-isomaltose malabsorption, as seen in cases with diffuse lesions of the small intestine, is discussed.

2.   The Structure of the Brush Border Membrane

2.1.   The Morphology of the Enterocyte

The small intestinal enterocytes are columnar cells covering the villi in a single layer. They are formed at the bottom of the crypts and move upward along the villi until they are shed at the top. The life span of an enterocyte is 2-5 days. The brush border is constituted by the part of the enterocyte facing the intestinal lumen. It is formed by a dense layer of parallel microvilli, each surrounded by a highly specialized membrane (the microvilli membrane). Inside each microvillus there is a cytoskeleton

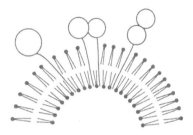

Fig. 1. Section of the micro-villus membrane showing the integral stalked enzymes.

reaching into the terminal web which is localized just beneath the brush border membrane. Laterally the microvilli membrane is connected with the remaining plasma membrane of the cell, known as the baso-lateral membrane.

2.2. The Structure of the Membrane and Location of the Enzymes (Fig. 1)

The microvillus membrane, like other animal cell membranes, is structured as a double lipid membrane with polar (hydrophilic) groups at the two surfaces and non-polar (hydrophobic) groups towards the centre (Singer and Nicholson, 1972). The proteins in the membrane act as receptors, as carriers and as enzymes. The so-called peripheral proteins are linked superficially to the membrane by means of electrostatic linkages and hydrogen bonds, while the integral proteins are linked primarily to, and more or less buried in, the inside of the membrane by means of hydrophobic linkages. Stalked, integral membrane proteins (Brunner *et al.,* 1979) are a subgroup of the integral proteins; the greater part of the protein molecule is localized outside the membrane, but it is anchored to the inside by means of a hydrophobic, relatively small part of the molecule (the anchor). Cytochrome b_5 (Spatz and Strittmatter, 1971) and Semliki Forest Virus spike proteins (Utermann and Simons, 1974) are examples of such proteins. From studies of aminopeptidase in pig intestine (Louvard *et al.,* 1975; Maroux and Louvard, 1976) and sucrase-isomaltase in rabbit intestine (Sigrist *et al.,* 1975) a model has been constructed which seems to be common to the microvillus enzymes. These enzymes also belong to the group of stalked, integral membrane proteins. The hydrophilic, enzymatically active part is located outside the membrane, turned toward the intestinal lumen and anchored in the membrane by means of one or two relatively low molecular weight, hydrophobic segments of the peptide chain. The microvillus enzymes are often dimers.

Table I.   Intestinal Brush Border Enzymes

Name and EC number[a]	Species Demonstrated in	Species Purified from[b]	References[c]
Sucrase Isomaltase 3.2.1.48-3.2.1.10	human pig rat rabbit	human pig rat rabbit	Conklin et al. (1975) Sjostrom et al. (1980) Kolinska & Kraml (1972) Sigrist et al. (1975)
Lactase-Phlorizin Hydrolyase 3.2.1.23-3.2.1.62	human pig rat monkey hamster calf	human pig rat monkey	Skovbjerg et al. (1981) Skovbjerg et al. (1982) Schlegel-Haueter et al. (1972) Ramaswamy & Radhakrishnan (1975) Colombo et al. (1973) Wallenfels & Fischer (1960)
Maltase (Glucoamylase)[d] 3.2.1.20	human pig rat rabbit monkey	human pig rat rabbit	Kelly & Alpers (1973) Hedeager-Sorensen et al. (1982) Lee et al. (1980) Sivakami & Radhakrishnan (1973) Seetharam et al. (1970)
Trehalase 3.2.1.28	human pig rat	rat	Maestracci et al. (1975) Dahlqvist (1960) Sasajima et al. (1975)
Aminopeptidase N 3.4.11.2	human pig rat rabbit	pig rat rabbit	Sterchi & Woodley (1978). McClellan & Garner (1980). Skovbjerg et al. (1978) Maroux & Louvard (1976). Sjostrom et al. (1978) Gray & Santiago (1977) Takesue (1975)
Aspartate Aminopeptidase 3.4.11.7	human pig rabbit	pig	Sterchi & Woodley (1978). Skovbjerg (1981b) Benajiba & Maroux (1980) Andria et al. (1976)
Dipeptidyl Peptidase IV 3.4.14	human pig rat rabbit	pig	Sterchi & Woodley (1978). Skovbjerg et al. (1978). Svensson et al. (1978). Hopsu-Havu & Ekfors (1969). Auricchio et al. (1978)
Carboxy-Peptidase 3.4.12	human rabbit		Andria et al. (1980). Skovbjerg (1981b) Auricchio et al. (1978)
Entero-Peptidase 3.4.21.9	human pig cow	human pig cow	Grant & Hermon-Taylor (1976) Baratti et al. (1973) Anderson et al. (1977)
Neutral Endopeptidase 3.4.24	human pig rat		Welsh et al. (1972) Danielsen et al. (1980) Kocna et al. (1980)
γ-Glutamyl Transferase 2.3.2.2	human pig rabbit		Sterchi & Woodley (1978) Noren et al. (1979) Ross et al. (1973)
Alkaline Phosphatase 3.1.3.1	human pig rat calf	pig rat calf	Komoda & Sakagishi (1976) Colbeau & Maroux (1978) Malik & Butterworth (1976) Fosset et al. (1974)

[a] Systematic number recommended by Enzyme Commission (1978).

[b] Purified to homogeneity without any detectable contamination with other proteins as shown by enzymatic analysis, immunoelectrophoresis or polyacrylamide gel electrophoresis.

3.   The Digestive Functions of the Brush Border Enzymes

A survey of the brush border enzymes appears in Table I. The majority of the substrate specificities of the enzymes have been studied completely, or partially, in purified preparations from various species. The assumed function of the glycosidases in the complete digestion of carbohydrates is outlined briefly below.

Food carbohydrates are hydrolyzed to monosaccharides before being transported across the microvillus membrane. Sixty per cent of the ingested carbohydrates are polysaccharides composed of glucose molecules: α-amylase (unbranched α-1,4 glucoside linkage), amylopectin and glycogen (α-1,4 glucoside linkage branched with α-1,6 glucoside linkages). Amylase from the salivary glands and the pancreas acts upon α-1,4 linkages in the intestinal lumen and hydrolyzes the polysaccharides to maltose, maltotriose (α-1,4 glucoside linkages) and limit dextrins containing 5-9 glucose units (α-1,4 glucoside branched linkages with one or more α-1,6 glucoside bonds) (Gray, 1975). Further hydrolysis takes place as the result of the interaction involving the brush border glucosidases. The brush border contains α-1,4 as well as α-1,6 glucosidase activities; maltase hydrolyzes α-1,4 linkages in maltose (composed of two glucose molecules), as well as in the largest malto-oligosaccharides in which the enzyme splits the glucose molecules from the end groups (glucoamylase activity). Isomaltase (from the sucrase-isomaltase complex) splits α-1,6 linkages in border dextrins and also acts, as does sucrase, upon α-1,4 linkages in maltose and maltotriose. Sucrose (cane sugar) constitutes approximately 30% of the carbohydrates ingested daily in Western countries. Sucrose is a disaccharide composed of fructose and glucose (α-1,2 glucoside linkages). It is hydrolyzed exclusively by the brush border enzyme sucrase. Another α-glucosidase is found in the brush border, *viz.* trehalase. It acts upon trehalose (two glucose molecules linked by an α,α-1,1 glucoside linkage) and is found in mushrooms and insect hemolymph.

About 10% of food carbohydrates are composed of lactose (milk sugar) and small amounts of monosaccharides (glucose and fructose). Lactose is a disaccharide composed of galactose and glucose. In contrast

---

[c]For each enzyme one or two central references are given priority as follows:
(1) The enzyme has been purified to homogeneity. If possible the enzyme was purified in its amphiphilic form. (2) The enzyme has been partly purified.
(3) The activity has been shown to be attached to a specific brush border protein.
(4) Activity has been demonstrated in a purified brush border membrane preparation.

[d]This enzyme has both maltase and glucoamylase activity and is designated differently in the literature (maltase, glucoamylase, maltase/glucoamylase). In this article the enzyme is named maltase.

to the carbohydrates mentioned above, glucose and galactose are linked by β-glycoside linkages (β-1,4) which are split by lactase. The brush border contains yet another β-glycosidase (phlorizin hydrolase) which splits the β-glycoside linkage between glucose and the aglycone phlorizin. Its natural substrate is thought to be complex lipids, however, in which a ceramide (N-acyl sphingosine) is linked with a β-glycoside linkage to a glucose or galactose remnant (Leese and Semenza, 1973). Such lipids are found in the fatty globular membrane in milk. Lactase and phlorizin hydrolase activities are closely linked and the enzyme is also designated lactase-phlorizin hydrolase.

A theory postulating the significance of the glycosidases for glucose transport as such ("hydrolase-related transport") has been put forward (Malathi *et al.* 1973) but has not been finally established. It has been shown, however, that glucose is transported across the microvillus membrane by means of an active Na^+ - dependent process which probably takes place by means of an unidentified carrier molecule (Gray, 1975). Galactose apparently follows the same procedure, while fructose is transported by means of a Na^+ - independent mechanism (Crane, 1977).

4.   Distribution of Brush-Border Enzymes

4.1.   The Distribution of Enzymes Along the Small Intestine

The location of carbohydrate-digesting enzymes in the small intestine has been investigated by Asp *et al.* (1975b), Newcomer and McGill (1966), Skovbjerg *et al.* (1979a) and Skovbjorg (1981b).

The specific activity of the individual brush border enzymes changes gradually from the proximal jejunum to the distal ileum. Sucrase-isomaltase activity is distributed along the whole length of the small intestine, the highest activity occurring in the middle with 20-30% less activity at Treitz' ligament and distally in the ileum. In contrast, lactase shows a distinct maximum 50-200 cm distal to Treitz' ligament; activity is 25% lower at Treitz' ligament and is almost absent in the distal part of the ileum. Maltase and all the peptidases (except carboxypeptidase which is distributed evenly) show increasing activity distally.

Amino acids were thought to be absorbed mainly in the proximal part of the jejunum, but recent studies indicate that they are absorbed mainly in the distal part of the jejunum and in the ileum (Chung *et al.,* 1979; Curtis *et al.,* 1978). This observation is in keeping with the demonstration of increased specific activity of the brush border peptidases in the distal part of the small intestine (Skovbjerg *et al.,* 1979a; Skovbjerg, 1981b).

4.2. The Distribution of Enzymes Along the Villus-Crypt Axis

The enterocytes are formed in the bottom of the crypts and move toward the top of the villi where they are shed. The amount of brush border enzymes changes along the villus-crypt axis (Nordström et al., 1967). In the undifferentiated crypt cells enzyme activity is very low. It increases gradually and reaches maximal values in the villi. Earlier immunofluorescence studies of rabbit intestine led to the suggestion that catalytically inactive sucrase molecules were synthesized in the plasma membrane of the crypt cells, and were incorporated into the membrane to be activated as the enterocytes moved upward along the villi (Dubs et al., 1975; Silverblatt et al., 1974). However, studies using radioactive amino acids have shown the synthesis of brush border enzymes to take place continuously even in mature villus enterocytes (Alpers, 1977; James et al., 1971). Immunoelectrophoretic results (Skovbjerg, 1981a) indicate that there is a constant relation between enzyme activity and the amount of immunoreactive protein, and hence that the synthesis of immunoreactive, inactive enzyme precursors takes place in the crypt cells is not significantly greater than that in the villus cells.

5.   Biochemical Background of Sucrase-Isomaltase Deficiency

In all normal mammal intestines studied so far (with the exception of the sea lion) sucrase-isomaltase is composed of two polypeptide chains (subunits) of almost identical size (M_r 120,000 and 130,000) linked by non-covalent bonds (Semenza, 1976). The sucrase and isomaltase activities are localized in their respective subunits, but apart from this the two polypeptide chains show a high degree of overlapping substrate specificity and resemble each other very closely. Therefore it has been suggested that the enzymes are synthesized on the basis of duplicate genes (Semenza, 1976). Recent investigations on pig and rat intestine (Hauri et al., 1979; Sjöström et al., 1980) have shown sucrase-isomaltase to be synthesized as one polypeptide chain which later, by means of pancreatic proteases, is split into the two subunits. Sucrase-isomaltase is anchored to the membrane by the hydrophobic N-terminal end of isomaltase, while sucrase apparently is anchored by its linkage to isomaltase (Brunner et al., 1979; Frank et al., 1978) (Fig. 2).

Small amounts of free isomaltase (i.e., not associated with sucrase) have been demonstrated in normal jejunal biopsies and in purified jejunal-microvillus membranes by means of crossed immunoelectrophoresis (Skovbjerg et al., 1979b). This finding may be due to the physiological decomposition of the sucrase-isomaltase complex. The fact that free isomaltase, and not sucrase, is found in the brush border membrane corre-

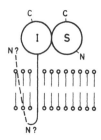

Fig. 2. Suggested psoition of sucrase-isomaltase in the small intestinal brush border membrane. I - Isomaltase, S - Sucrase.

sponds with the fact that isomaltose acts as the anchor while sucrase is linked to the membrane via isomaltase.

Sucrase-isomaltase is already well developed in the human fetus. Antonowicz and Lebenthal (1977) reported that sucrase activity after 10-26 weeks of gestation was already 60% of that found in full term infants and 70% of that at 26-30 weeks of gestation. It is noteworthy that sucrase-isomaltase in fetuses of 16 weeks gestational age occurs as one polypeptide (pro-sucrase-isomaltase) (Skovbjerg, 1982) with a molecular weight of 260,000. This is probably due to lack of activity of pancreatic proteases in the fetal intestine. Treatment with elastase converts the pro-sucrase-isomaltase into two polypeptides of the apparent molecular weight of sucrase and isomaltase from normal adult intestine (Skovbjerg, 1982).

Our present knowledge of the biosynthesis of sucrase-isomaltase as a one polypeptide chain may explain the double enzyme defect in most sucrase-isomaltase deficient patients. The question whether the molecular defect in sucrase-isomaltase deficiency is due to a complete switching off of biosynthesis or to biosynthesis of an inactive protein has been investigated by various methods. Preiser *et al.* (1974) and Schmitz *et al* (1974) noted the absence of protein bands in the sucrase-isomaltase position in polyacrylamide gels of four sucrose intolerant patients. This finding was supported by that of Gray *et al.* (1976) using radioimmunoassays on biopsies from seven sucrose intolerant patients. In contrast, Dubs *et al.* (1973) and Freiburghaus *et al.* (1977) demonstrated, by immunohisto-chemical means using specific antibodies against the sucrase-isomaltase complex, cross reacting protein in sucrose intolerant patients, suggesting the presence of an inactive enzyme variant.

By rocket immunoelectrophoresis against a specific antibody raised against pure human sucrase-isomaltase, we recently studied biopsies from six sucrose intolerant patients (Skovbjerg and Krasilnikoff, 1981; Skovbjerg *et al.*, 1982). Five of these patients lacked both sucrase and isomaltase activities, and no precipitates were evident in the immunoelectrophoretic plate. However, the sixth patient had a low but significant isomaltase activity (7.8 units/g protein) (normal range 36-93). These patients also exhibited a precipitate on the immunoelectrophoretic plate. This precipitate had isomaltase activity but no sucrase activity. On the basis of this finding of an isomaltase polypeptide without sucrase activity, we suggested that sucrase-isomaltase deficient patients are divisible into at least two different groups, one lacking both sucrase and isomaltase activities and another with high residual isomaltase activity.

A review of the literature (Auricchio *et al.*, 1965; Auricchio *et al.*, 1972; Dubs *et al.*, 1973; Eggermont and Hers, 1969; Freiburghaus *et al.*, 1977; Gray *et al.*, 1976; Greene *et al.*, 1972; Preiser *et al.*, 1974; Schmitz *et al.*, 1974) showed that approximately 20% of the sucrose intolerant patients described had considerable isomaltase activity (10-25% of the normal value) similar to the one case in our study. However, the activity seemed always to be depressed compared with that of normal biopsies, showing that the synthesis of isomaltase is affected and that isomaltase is more easily degraded when not associated with sucrase. By two-dimensional crossed immunoelectrophoresis of small intestinal biopsies from sucrase-isomaltase deficient patients, it was also shown that the isomaltase of the patient with the high residual isomaltase activity had the same electrophoretic mobility as free isomaltase in normal intestinal biopsies (Skovbjerg *et al.*, 1979b). The molecular weight of the isomaltase in these patients has not been investigated, and the possibility exists that the isomaltase polypeptide has a molecular weight different from that of normal isomaltase.

6.   Pathophysiology

The pathophysiology in the various forms of disaccharide malabsorption syndromes is identical. In children, tolerance tests with the disaccharide in question are followed by an increased fecal excretion of the disaccharide, the corresponding monosaccharides and low molecular weight organic acids, 90% of which are lactic acid and acetic acid produced by bacterial fermentation. The pH of the feces, normally in the neutral range, drops to values between 5 and 6 (Weijers *et al.*, 1961; Prader *et al.*, 1961; Auricchio *et al.*, 1963a; Kerry and Anderson, 1964; Anderson *et al.*, 1966). These changes may also be found in adults (Auricchio *et al.*, 1963a; Dunphy *et al.*, 1965; Haemmerli *et al.*, 1965; Cuatrecasas *et al.*, 1965), but not constantly enough to be used as a screening test for disaccharide malabsorption (Kern *et al.*, 1964; McMichael *et al.*, 1965; Newcomer and McGill, 1967; Dahlqvist *et al.*, 1968).

The unabsorbed disaccharide and its hydrolysates exert a pronounced

Table II. Clinical Data on 7 Children Suffering from Sucrose Malabsorption[a]

Case No.	Sex	Age at onset of symptoms	Age at diagnosti-cation	Previous diagnoses	Symptoms prior to dietetic treatment	Symptoms after institution of treatment
1	F	3.5 m	11 m	Chronic dyspepsia	Watery diarrhea 8–10 times daily	Pasty stools 3 times daily
2	F	2 m	4 y	Allergic dyspepsia	Watery diarrhea 7–10 times daily	None
3	F	1.5 m	2 y	Chronic dyspepsia observed for milk allergy	Watery diarrhea 2–4 times daily Crying spells	None
4	M	4 d	7 d	None	Watery diarrhea	None
5	F	14 d	20 m	None	Pasty-watery diarrhea	None
6	M	1 m	4 m	Chronic dyspepsia	Watery diarrhea meteorism	None
7	M	1.5 m	20 m	Chronic dyspepsia, lactose intolerance	Watery diarrhea 8–16 times daily Meteorism	Occasionally pasty stools 3 times daily

[a]From Gudmand-Hoyer, E. and Krasilnikoff (1977).

osmotic effect, drawing water into the intestinal canal (Kern and Struthers, 1966; Launiala, 1968). The result is an accelerated intestinal passage of the intestinal contents and diarrhea. By adding disaccharide to a barium sulphate suspension it has been possible radiologically to demonstrate these changes. A rapid intestinal passage and a dilution of the contrast medium with fluid can be detected (Laws and Neale, 1966; Preger and Amberg, 1967: McNeish and Sweet, 1968; Bowdler and Walker-Smith, 1969; Gudmand-Høyer and Folke, 1970).

7.   Symptoms of Sucrose-Isomaltose Malabsorption (S-I Malabsorption)

The symptoms of S-I malabsorption become manifest only when sucrose is introduced into the diet. Many milk formulas contain sucrose; thus bottle-fed babies may develop the symptoms immediately after birth. In breast-fed babies the symptoms appear when more solid foods containing sucrose are introduced into the diet. The main symptom experienced by these children is watery diarrhea, often accompanied by excoriation of the buttocks. Other symptoms may be crying spells, meteorism and sometimes vomiting. The attacks of diarrhea may become so intensive that they lead to dehydration. An example of this is seen in Table II, which lists the symptoms in seven of our patients (Gudmand-Høyer and Krasilnikoff, 1977). The first symptoms in all the patients were closely related to the time when sucrose was introduced into the diet. The diarrhea frequently led to excoriation of the buttocks which was often difficult to treat. In case No. 2, treatment with parenteral nutrition and rehydration for a longer period was necessary. Only one mother had noticed the relation between the attacks of diarrhea and the intake of sucrose, and had acted in accordance with this observation. Five of the patients had been hospitalized at least once earlier, in one case (No. 7) as many as six times as the symptoms became exacerbated. Five of the patients had at some earlier stage been labelled with a symptomatic diagnosis.

In the first 43 cases of S-I malabsorption described in the literature (Kühni, 1967) the degree of severity of the clinical picture was categorized. Eight patients belonged to the group with slight symptoms, the most pronounced being watery diarrhea. Out of the 43 patients, 22 belonged to the group with fairly severe symptoms. Growth rate with regard to length as well as weight was distinctly retarded in these patients. Nine patients showed severe symptoms and had been hospitalized in a dehydrated condition necessitating intravenous rehydration. Other studies have also shown that patients with S-I malabsorption, especially during the first year of life, may show a decreased weight gain (Ament et al., 1973; Burgess et al., 1964). If S-I malabsorption is diagnosed at a later stage,

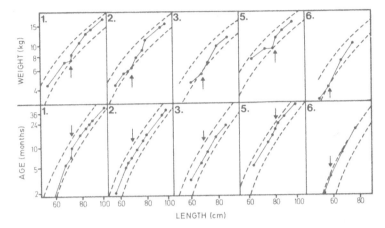

Fig. 3. Growth charts for 5 children suffering from sucrose malabsorption. The dotted lines indicate the normal range (± 2 SD).
The arrows indicate the time of institution of a diet poor in
sucrose. Gudmand-Høyer and Krasilnikoff (1977).

the patient's height and general rate of growth are usually found to be
normal (Ament *et al.,* 1973). In five of our patients (Gudmand-Høyer and
Krasilnikoff, 1977) we succeeded in collecting information on earlier
records of height and weight (Fig. 3). It can be seen that, before treatment with a sucrose deficient diet was instituted, all five patients suffered
from a relative loss of weight. In three patients (Nos. 1, 5, and 6) this
coincided with the time of diagnosis. In two cases (Nos. 2 and 3) the
mother had noticed sucrose intolerance in the children even before the
diagnosis had been established and had instituted a sucrose deficient diet
on her own initiative. Just prior to the institution of dietetic treatment
weight in relation to height was at the lower normal limit, and in cases 2
and 3 was below this limit.

Before the institution of treatment, cases 1 and 5 also showed a relative decrease in growth rate in relation to length. The other three patients,
who maintained the growth rate they had established with regard to
height, were the ones who had had dietetic treatment instituted the earliest in life. At the time of institution of dietetic treatment, length in relation to age was definitely within the normal range in the four Danish
children (Nos. 1, 2, 3 and 5), while the Greenlandic child (No. 6) was at
the lower limit of the normal range for Scandinavian children. When fed
a sucrose-free diet all the patients immediately underwent an increase in
absolute weight as well as in weight relative to length. Subsequently two
patients (Nos. 1 and 5) further increased their age-adjusted growth rate
relative to length. The increase in length lagged behind the increase in
weight, and normal length was reached only after 1.5 to 2 years of treat-

ment. The remaining patients continued along the established curve for increasing length. Our results thus may be interpreted as showing that S-I malabsorption may be significant for the long term development of children.

An example of how great the variation in severity of symptoms may be from one patient to the next appears in the work of McNair et al. (1972) on S-I malabsorption in Greenland. Out of 20 Greenlandic Eskimos in whom they found sucrose malabsorption, eight denied having any gastrointestinal symptoms. The remaining 12 all had chronic or intermittent diarrhea. One of these was a 65-year-old woman, while the remainder were children between 15 months and 16 years. Thus only one of eight adults had diarrhea. One possible explanation may be that in the food eaten in Greenland the sugar content has reached European levels only within the last few decades (Helms, 1981). It may be assumed that older Greenlanders eat more Greenlandic food with its large content of protein and fat and low carbohydrate content than do the younger generations. The fact that 8 out of 20 cases in this study had no abdominal symptoms is not unique. Similar observations have been made by Anderson et al. (1963). Five of the Greenlanders suffering from diarrhea had themselves noticed that ingestion of sugar and other sweet foods resulted in diarrhea and abdominal symptoms. Three of the children suffered from severe malnutrition during their first year of life when they were given cow's milk with sucrose. In Greenland the mortality rate for "enteritis" among infants under one year of age is several times higher than in Denmark (McNair et al., 1972). Undiagnosed S-I malabsorption may be a contributory cause. Isomaltose as such is not found in foods, but isomaltose and various maltotrioses result from the hydrolysis of polysaccharides such as starch. Usually these patients tolerate the presence of starch in food, but some may have slight symptoms. A persistent slight tendency to diarrhea despite treatment with a sucrose-free diet may be ascribed to isomaltase deficiency and a resulting starch intolerance (Gudmand-Høyer and Krasilnikoff, 1977). The severity of the symptoms in cases with disaccharidase deficiencies relates to the amount of non-tolerable carbohydrates (Gudmand-Høyer and Simony, 1977). In many patients a slight steatorrhea has been recorded which is undoubtedly caused by rapid intestinal passage (Nordio et al., 1961; Rey et al., 1963; Anderson et al., 1963; Francois et al., 1963; Corouben et al., 1963; Burgess et al., 1964; Lifshitz and Holman, 1964).

8.   Diagnosis

In principle, one distinguishes between the concepts of S-I malabsorption and S-I intolerance. The latter is characterized by clinical symptoms in the form of diarrhea and vomiting provoked by the ingestion of sucrose, while S-I malabsorption is characterized only by the demonstration in the laboratory of an incomplete absorption of sucrose-isomaltose from the small intestine. According to these definitions S-I malabsorption is not necessarily accompanied by S-I intolerance, while a recorded S-I intolerance is an indication also of S-I malabsorption (Walker-Smith, 1979; Gardiner *et al.*, 1981b). Usually the term S-I malabsorption covers both concepts.

8.1. Clinical Features

The first indication of a S-I malabsorption is based on the clinical symptoms. These arise typically a few hours after the intake of sucrose and disappear within 24 hours when a sucrose-free diet is instituted. The symptoms appear for the first time in the infant when the diet changes from mother's milk to a diet containing cane sugar.

A careful history of eating habits, including a record of the type and amount of various foods ingested at each meal and an evaluation of the time of appearance of symptoms, as well as an exact knowledge of the sucrose content of various foods, thus are important for all age groups. Lists of the sugar content of fresh foods are available (Hardinge *et al.*, 1965). The same is true for manufactured foods, including milk powders and tinned foods. Shannon (1978) has recently investigated the concentration of sugars in commercial baby foods in the U.S.A. but only from three different factories. The significance of such lists of manufactured foods is limited by the fact that the producers often change the composition of the products. It is therefore recommended that the list of contents, however incomplete, found on many wrappings be studied carefully.

When clinical suspicion has been aroused, the diagnosis is confirmed by the use of various diagnostic tests. The relatively simple examinations which may be carried out by ordinary means and equipment will be mentioned first, while investigations demanding greater resources will be mentioned last.

8.2. Investigations of Stools

In children the stools are typically watery and acid and contain reducing substances. Investigation of the stools should be carried out on the fluid portion. A careful collection of stools is necessary – either by means of a plastic insertion in the diaper (to stop watery stools from being absorbed by the diaper) or by nursing the baby on a metabolic frame. In adults, in whom the stools often are less watery and contain less disaccharide, tests for pH and reducing substances are unreliable (McMichael *et al.*, 1965).

8.2.1. Clinitest

An excess of reducing substances (i.e., 0.5% or more) can be demonstrated by the clinitest method introduced by Kerry and Anderson (1964). The method depends on the stools being examined (or placed in a freezer) to stop bacterial decomposition of any carbohydrates immediately after they are passed. This method has a drawback in the diagnosis of S-I malabsorption, in that sucrose is not a reducing carbohydrate and thus must be hydrolyzed by boiling the feces with 0.1 N HCl. When the test is carried out before and after this hydrolysis, an increase in reducing substances may indicate the presence of sucrose. In practice, this test often affords many false negative results (Soeparto et al., 1972).

8.2.2. Chromatography

Paper chromatography seems to be a more reliable method for demonstrating an increased excretion of sucrose in the feces (Soeparto et al., 1972). The method is sensitive and hence a positive chromatography test, when accompanied by a negative clinitest, need not be a sign of disaccharide intolerance (Anderson and Burke, 1975).

8.2.3. pH

In S-I malabsorption the pH of the stools will typically lie between 5 and 6 as a result of the formation of organic acids. As a screening test for S-I malabsorption this test is unreliable, however, as many results are either false positive or negative. A clinitest carried out together with the pH determination may increase its diagnostic value (Soeparto et al., 1972; Walker-Smith, 1973; Lindquist and Wranne, 1976).

8.2.4. Lactic Acid

Weijers et al. (1961) found that children suffering from S-I malabsorption excreted more than 50 mg lactic acid/100 ml feces as a result of an increased fermentation. Due to the inaccuracy of the method, it is no longer used as a routine test.

8.3. Barium-Sucrose Meal

Simultaneous ingestion of disaccharide and barium sulphate may demonstrate radiologically the pathophysiological changes taking place in the form of a dilation of the small intestine, a dilution of contrast and a reduction in transit time (Laws and Neale, 1966; Preger and Amberg, 1967; McNeish and Sweet, 1968; Bowdler and Walker-Smith, 1969; Gud-

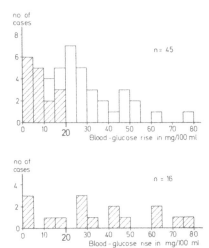

Fig. 4. The results of oral sucrose tolerance tests in 45 children.
In 16 cases with the lowest rise in blood glucose (hatched area)
the test was repeated with intraduodenal instillation of sucrose.
Krasilnikoff *et al.* (1975).

mand-Høyer and Folke, 1970). This method represents a radiological
means of demonstrating S-I malabsorption. Connolly *et al.* (1978) recommended this method as a means of excluding S-I malabsorption in cases
of chronic diarrhea, but we do not recommend it because of the radiation
risk.

8.4. Oral Sucrose Tolerance Test

A tolerance test involving oral administration of sucrose followed by a
determination of the rise in blood glucose constitutes the most common
screening test for S-I malabsorption. In children a rise in blood glucose of
less than 20 mg per 100 ml is usually considered an indication of S-I
malabsorption.

    However, we found a high incidence of false positive tests in a group
of 61 children when the disaccharide was administered orally (Krasilnikoff
et al., 1975). One hundred and five oral disaccharide tolerance tests were
carried out with either lactose or sucrose (dose 2 g per kg body weight).
In the cases in which the increase in blood sugar was <20 mg per 100
ml, the test was repeated with direct instillation of the disaccharide in the
duodenum by means of a tube. Following the administration of sucrose
the increase in blood glucose was found to be ≤20 mg/100 ml in 20 out
of 45 patients (Fig. 4). In 16 patients with a rise in blood glucose of ≤20
gm/100 ml this test was repeated with intraduodenal instillation of suc-

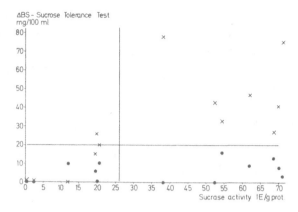

Fig. 5. Rise in blood glucose during oral (•) and intraduodenal
(×) sucrose tolerance test correlated with the sucrase activity in
a peroral small-intestine biopsy. Krasilnikoff *et al.* (1975).

rose. In all but five cases the rise in blood glucose was >20 mg/100 ml,
the exceptions comprising three patients with coeliac disease and nonspecific sucrose malabsorption and two patients with specific S-I malabsorption. In 12 children the rise in blood glucose following the two forms of
administration of sucrose was correlated with sucrase activity in a peroral
small-intestinal biopsy at the ligament of Treitz. In children a good correlation was found between low sucrase activity in the intestinal mucosa
(<26 I U per g protein) and rise in blood glucose for both forms of
administration (Fig. 5). In all seven patients showing a normal sucrase
activity, the rise in blood glucose was <20 mg/100 ml following oral
administration. After intraduodenal administration the figures became
normal and in all cases above 25 mg/100 ml. The incidence of false positive oral sucrose tests was between 24 and 33 per cent.

The conclusion from this study was that a flat disaccharide tolerance
curve following an oral test can be caused by a slow gastric emptying rate
and must be verified by a tolerance test with direct intra-intestinal instillation of the disaccharide. The investigation also showed that the figure
<20 mg/100 ml rise in blood glucose following an intraduodenal disac-
charide tolerance test distinguishes in a very satisfactory manner between
patients with normal and abnormal absorption of disaccharides. Furthermore, blood glucose determinations more than one hour after ingestion of
the disaccharide are without diagnostic significance and may be omitted.

These tolerance tests may be supplemented with simultaneous
administration of the products of hydrolysis of sucrose, i.e, fructose and
glucose, in doses of 1 g/kg each, with blood glucose being recorded as
described above. If the diagnosis of specific S-I malabsorption is correct,
blood glucose will increase more than 20 mg/100 ml.

Apart from these laboratory tests, it is very important to register any symptoms simultaneously, especially such symptoms as abdominal pains, meteorism and diarrhea. Without such additional symptoms, it can be concluded that the complaints which led to the tests were not caused by S-I malabsorption.

8.5. Sucrose Hydrogen Breath Test

Another test has been introduced for the diagnosis of disaccharide malabsorption (Levitt, 1969; Bond and Levitt, 1972; Newcomer *et al.*, 1975). The theoretical basis for the test is as follows. Carbohydrates which are not absorbed in the small intestine will, when reaching the colon, be metabolized by bacteria to gaseous hydrogen, carbon dioxide and short-chain fatty acids. The hydrogen diffuses across the colonic wall and is carried in solution to the lungs, where it is excreted in the breath and can be estimated. In normal persons the production of hydrogen is localized mainly in the colon (Levitt, 1969), and there is a direct correlation between the total excretion of hydrogen and the amount of carbohydrate ingested (Bond and Levitt, 1972). As mentioned by Gardiner *et al.* (1981a) the method is based on a number of assumptions: 1) hydrogen-producing bacteria are not present in the small intestine; 2) the colonic flora will metabolize the sugar, releasing hydrogen; 3) hydrogen produced in the colon is not prevented from appearing in expired air by diseases of the colonic mucosa, colonic microcirculation, or the lungs; and 4) the hydrogen detected in expired air does not originate from an extra-intestinal source such as oral bacteria.

In adults the method is claimed to be reliable for the demonstration of lactose malabsorption (Newcomer *et al.*, 1975), although Gilat *et al.* (1978) found only 20% non-hydrogen producers in a group of normal adults. In children the value of the method has not yet been determined fully. While the first reports were promising (Maffei *et al.*, 1977; Douwes *et al.*, 1978; Hyams *et al.*, 1980; Nose *et al.*, 1979; Perman *et al.*, 1978), later reports have been more critical and variable with respect to both lactose malabsorption (Barr *et al.*, 1981; Gardiner *et al.*, 1981a) and S-I malabsorption (Douwes *et al.*, 1980; Gardiner *et al.*, 1981b). The last authors even conclude that the hydrogen breath test is of no value in diagnosing primary sugar malabsorption in children, probably because of altered activity of the bacterial flora in these patients.

Further investigations leading to an increased understanding of factors influencing the decomposition of carbohydrates in the intestinal canal are necessary before this method can become a routine test.

8.6. Disaccharidase Assay

Final proof of the absence of sucrase-isomaltase activity in the small intestine is obtained by direct determination of enzyme activity in a small-intestinal biopsy. This is usually carried out according to Dahlqvist's method (1968), but may also be determined by means of an immunoelectrophoretic method developed by Skovbjerg *et al.* (1978, 1979a). At the same time a routine determination of the other disaccharidases, and of alkaline phosphatase as a reference enzyme, is carried out.

Specific S-I malabsorption will result in sucrase-isomaltase values of zero or practically zero, while the figures for lactase, trehalase and alkaline phosphatase will be normal. Maltase activity will be reduced; this is due to the fact that isomaltase activity constitutes 70% of the total "maltase activity" when determined by Dahlqvist's method. In the case of non-specific S-I malabsorption the other disaccharidases will show low activity or none.

8.7. Conclusion

In conclusion, it should be mentioned that most authors (Anderson and Burke, 1975; Krasilnikoff *et al.*, 1975; Gudmand-Høyer and Krasilnikoff, 1978; Walker-Smith, 1979) prefer to confirm a clinically suspected S-I malabsorption primarily by carrying out a sucrose tolerance test while at the same time recording any accompanying clinical symptoms. The diagnosis is then verified by a small-intestinal biopsy with determination of the disaccharide-splitting enzymes, while a stereomicroscopic and/or a histological examination is carried out to distinguish between specific and non-specific S-I malabsorption. The diagnosis is finally confirmed by the clinical response to the institution of a sucrose-free or sucrose-restricted diet.

9.   Incidence and Genetic Conditions

S-I malabsorption is considered a rare disease. Only about 200 cases have been reported so far, but undoubtedly the number of cases diagnosed is considerably higher. It is also likely that the problems attributable to S-I malabsorption are still underestimated in most populations, and that numerous patients suffering from diarrhea remain undiagnosed (McNair *et al.*, 1972; Ament *et al.*, 1973; Gudmand-Høyer and Krasilnikoff, 1977). In one study on six children the delay in diagnosis varied from one to eight years (Ament *et al.*, 1973), and in our investigation (Gudmand-

Høyer and Krasilnikoff, 1977) a delay of four years, including six inconclusive hospitalizations, was encountered in one patient. It also seems obvious that centers especially interested in this disease record more cases. Thus, Anderson and Burke (1975) estimated that a large pediatric gastroenterologic department should diagnose three to four families with S-I malabsorption annually. Out of 13 adult Caucasians diagnosed by means of a small-intestinal biopsy, four were found in one center (Ringrose *et al.*, 1980). From the number of heterozygote carriers of the abnormal gene, determined by sucrase tests in small-intestinal biopsies, Peterson and Herber (1967) estimated the incidence of S-I malabsorption to be about 1:500. This is not true of Eskimos, however. In 1972 McNair *et al.* showed that the incidence of S-I malabsorption in Greenlandic Eskimos must be much higher. Studies by Bell *et al.* (1973) indicate that there is also a high incidence in Alaskan Eskimos. They described five adult Alaskan Eskimos with histories of intolerance to sugar-containing foods and a flat glucose curve after a sucrose load. Ellestad-Sayed and Haworth (1977) found that two of 55 Canadian Indians tested had S-I malabsorption, and Ellestad *et al.* (1978) reported that four of 56 (7%) of Canadian Eskimos had S-I malabsorption. About 50% of the Canadian Eskimos experienced abdominal symptoms when sweet foods were consumed.

In a later study (Gudmand-Høyer *et al.*, unpublished) in which 54 Greenlanders, chosen at random, had a small-intestinal biopsy removed surgically during an abdominal operation, we found only one patient with S-I malabsorption. This seems to indicate an incidence of approximately 2%, an incidence which is from 10 to 1000 times that in Western countries.

The reason for the high incidence of S-I malabsorption in the Arctic regions may be the fact that for thousands of years the population has eaten mainly animal foods. This has led to a loss of sucrase and isomaltase as a negative adaptive response, as sucrose and isomaltose are of plant origin. A better explanation may be that patients with S-I malabsorption eating mostly animal foods will have no symptoms and their chances of survival will be just as great as if they had a normal sucrase-isomaltase activity. Thus there would be no positive selection of individuals with a normal sucrase-isomaltase activity, and the abnormal gene would remain in the population.

It seems reasonable to assume that the condition S-I malabsorption is of an autosomal recessive nature. Numerous cases among siblings and parents have been described (for instance, Prader and Auricchio, 1965; Kerry and Townley, 1965; McNair *et al.*, 1972; Ament *et al.*, 1973; Gudmand-Høyer and Krasilnikoff, 1977).

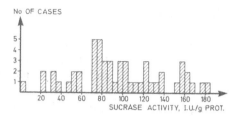

Fig. 6. Intestinal sucrase activity in 54 Greenland Eskimos.
Gudmand-Høyer *et al.* (1982).

One of the problems involved in proving the recessive nature of the condition is the determination of the heterozygotes. Kerry and Townley (1965) used the determination of disaccharidases in small intestinal biopsies to establish the carriers of the abnormal genes for sucrase-isomaltase deficiency. In four families, each having one child with S-I malabsorption, peroral small intestinal biopsies were carried out. The parents of the affected children were found to have sucrase and isomaltase activities equal to, or less than, the lower limits in a control group. From this it was concluded that S-I malabsorption is recessively inherited in a simple Mendelian fashion. Ament *et al.* (1973), in their familial investigation, reached a similar conclusion. Sucrase activities of 54 Greenlanders are shown in Fig. 6 (Gudmand-Høyer *et al.*, unpublished). It would seem that these 54 subjects can be divided into three distinct groups. One group consists of one Greenlander with S-I malabsorption; another group consists of 10 with an activity between 20 and 60 I U /g protein, and a third (normal) group has an activity above 70 I U /g protein. According to Kerry and Townley (1965) the heterogene carriers of the abnormal gene have a sucrase activity which is below the lower limit for the normal section of the population; this criterion is fulfilled by the middle group of Greenlanders. Thus, they are probably heterogene carriers of the abnormal gene. Furthermore, sucrase activity in this group is half, or less than half, the average activity in the third, largest group. This may be a useful new criterion of heterogeneity.

According to the Hardy-Weinberg equation (Hardy, 1908; Weinberg, 1908), the number of heterozygotes when a group of 54 includes one patient with S-I malabsorption should be 11. The middle group includes 10 Greenlanders. This supports findings that the incidence of S-I malabsorption among Greenlanders is higher than it is in other populations.

Disaccharidase activity in surgically removed biopsies from Greenlanders is higher than that in peroral small-intestinal biopsies from Danes. Direct comparisons of the normal range thus have not been possible.

10.   S-I Malabsorption Diagnosis in Adults

Even though the published number of patients with primary S-I malabsorption approximates 200, only a little more than a tenth are adult patients. Enzyme determination in a small-intestinal biopsy has been carried out in 16 of these adult patients (Sontag *et al.*, 1964; Jansen *et al.*, 1965; Neale *et al.*, 1965; Welsh and Brown, 1966; Laws and Neale, 1966; Pink, 1967; Peterson and Herber, 1967; Starnes and Welsh, 1970; Asp *et al.*, 1975b; Ringrose *et al.*, 1980). Three were Greenlandic Eskimos (Asp *et al.*, 1975a) among whom the incidence of S-I malabsorption is much higher than in other populations (McNair *et al.*, 1972), and who also had a primary lactase deficiency. For these two reasons the Greenlandic patients are not directly comparable to the other patients described. The reason for the small number of documented adult cases is undoubtedly that pediatricians are more aware of S-I malabsorption as a cause of chronic diarrhea and retarded development, as the first cases described were in children. Also, the symptoms are less pronounced in adults, and a small-intestinal biopsy on the indication of chronic or intermittent diarrhea is carried out much less frequently in adults than in children. The latter reason is corroborated by the fact that four of the adult cases described had a small-intestinal biopsy carried out at the same centre (Ringrose *et al.*, 1980) where a special interest is taken in disaccharide malabsorption syndromes. Another reason for the relatively small number of published adult cases may be that some of the children died within the first months/years of life at a time when knowledge of the disease was more limited (McNair *et al.*, 1972).

Case stories of 13 adult patients who were not Greenlanders and in whom the diagnosis was established by means of a small-intestinal biopsy have been summarized by Ringrose *et al.* (1980). Eight were women and five were men. Six were Caucasians; race was not recorded in the remainder. Only in two cases did the deficiency occur familially. Five, or half of the patients on whom data on the age of onset were available, had had intermittent or persistent symptoms since childhood. Two of the others had had symptoms in early childhood but then had been without symptoms until they reappeared when the patients were 20 and 40 years of age, respectively. In the last three patients the symptoms appeared for the first time in the first and second decades of life. Only a few patients had detected the relationship between the intake of sugar and the symptoms. The most common symptoms were abdominal pains, distension, flatulence and diarrhea; a few had intermittent diarrhea and constipation.

Individuals with a sucrase:lactase ratio of 0.9 or lower are usually considered heterogene carriers of the gene for sucrase-isomaltase deficiency. Among 339 Caucasians with a normal small-intestinal histology, Welsh *et al.* (1978) found, by this criterion, six, or 2%, heterozygotes for

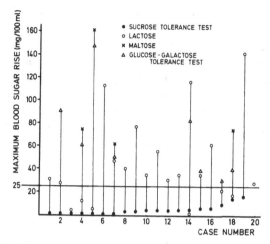

Fig. 7. Sugar tolerance tests performed in 20 Greenland Eskimos with S-I malabsorption. McNair *et al.* (1972).

sucrase-isomaltase deficiency. In another sample comprising 101 individuals, Peterson and Herber (1967) found an incidence of 8.9% heterozygotes, equivalent to 500,000 individuals with sucrase-isomaltase deficiency in the U.S.A. This figure is undoubtedly too high, but on the other hand the incidence of sucrase-isomaltase deficiency as the cause of diarrhea in adults is no doubt higher than the number of published cases suggests.

11. S-I Malabsorption Combined With Other Specific Disaccharide Deficiencies

In certain populations, such as the Greenlandic, in which not only sucrase-isomaltase deficiency occurs frequently but also lactase deficiency (Gudmand-Høyer and Jarnum, 1969; Gudmand-Høyer *et al.*, 1973; McNair *et al.*, 1972), and no doubt trehalase deficiency (Asp *et al.*, 1975a), a random sample will include some individuals with multiple specific disaccharidase deficiencies. Thus, McNair *et al.* (1972), using sugar tolerance tests, found that six out of the 20 Greenlanders with S-I malabsorption also suffered from lactose malabsorption (Fig. 7). Out of 19 patients in whom disaccharidase activity was determined by means of small-intestinal biopsies, three suffered from sucrase-isomaltase as well as lactase deficiency (Asp *et al.*, 1975a). The sample included two Greenlanders with trehalase deficiency in which the activity of the other disaccharidases was normal.

One of the patients with S-I malabsorption as well as lactose malabsorption (shown by sugar tolerance tests) but with normal absorption of maltose and monosaccharides was an 18-month-old boy. He represented the most severe case, having spent almost 1.5 years in hospital from his 10th week of life because of malnutrition, diarrhea and abdominal pains. As a result of the investigation, he was treated with a diet low in sucrose and lactose, and this led to rapid improvement, diarrhea disappearing and an increase in weight of two kilos occurring in the course of one month.

One may wonder what the introduction of Westernized foods will mean to such populations, and to what extent such foods will lead to malnutrition. Not all of the so-called benefits of Western civilization can be transferred directly to other peoples.

12. Unspecific (Secondary) S-I Malabsorption

Unspecific disaccharide malabsorption occurs in diseases which involve all, or most of the small intestinal mucosa, e.g., tropical sprue (Sheehy and Anderson, 1965; Desai *et al.*, 1967; Gray *et al.*, 1968), coeliac disease (Shmerling *et al.*, 1964; Plotkin and Isselbacher, 1964; Weser and Sleisenger, 1965; de Larrechea *et al.*, 1965; Lifshitz *et al.*, 1965; Arthur *et al.*, 1966; Jos *et al.*, 1967; Dahlqvist *et al.*, 1970; Gudmand-Høyer and Krasilnikoff, 1978; Sjöström *et al.*, 1981) protein-calorie malnutrition (Bowie *et al.*, 1965; Cook and Lee, 1966; Wharton *et al.*, 1968; Chandra *et al.*, 1968; Prinsloo *et al.*, 1971), gastroenteritis (Sunshine and Kretchmer, 1964; Burke *et al.*, 1965; Lifschitz *et al.*, 1971a, 1971b; Welsh and Porter, 1967; Barnes and Townley, 1973; Gudmand-Høyer and Söeberg, 1974; Davidson and Barnes, 1979), Crohn's disease of the small intestine (Pena and Truelove, 1970) as well as diseases not primarily affecting the small intestine, such as acute viral hepatitis (Gudmand-Høyer and Söeberg, 1973), ulcerative colitis (Gudmand-Høyer *et al.*, 1975), and iron deficiency anemia (Lanzkowsky *et al.*, 1981).

Lactase activity is not as great as that of the other disaccharidases such as sucrase (Dahlqvist, 1962) and is therefore reduced more readily to values below the critical level where symptoms occur. Lactase is also the enzyme which is affected most frequently and most severely when diffuse lesions of the small intestine occur. Further, it is the enzyme which is restituted most slowly once the original disease has been cured (Shmerling *et al.*, 1964; Plotkin and Isselbacher, 1964; Cook and Lee, 1966; Gray *et al.*, 1968; Gudmand-Høyer and Söeberg, 1974). It is therefore not surprising that intolerance to lactose associated with diffuse intestinal lesions is described far more frequently in the literature than intolerance to sucrose, or that secondary sucrose intolerance is always found in combination with secondary lactose intolerance.

For the same reason sucrase activity is reduced only in conditions accompanied by severe mucosal damage such as coeliac disease. As shown

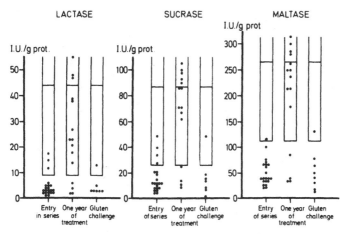

Fig. 8. Disaccharidase activities in small intestinal biopsies from Treitz ligament before treatment was instituted, after one year on a gluten-free diet and after 3 months gluten challenge on an ordinary diet. Columns show mean value and lower normal range for the disaccharidase activities. Gudmand-Høyer and Krasilnikoff (1978).

previously by other authors, Gudmand-Høyer and Krasilnikoff (1978) found a diffusely reduced disaccharidase activity in untreated coeliac disease (Fig. 8). The condition became normal after one year's treatment with a gluten-free diet except in four cases, three of whom did not maintain the diet. Before treatment was instituted, almost half the patients showed an increase in blood glucose of less than 20 mg/100 ml following lactose as well as sucrose tolerance tests (Fig. 9), and thus according to this criterion suffered from lactose as well as sucrose malabsorption. The clinical significance of secondary disaccharide malabsorption cannot have been great in this sample. All the children responded rapidly to a gluten-free diet with an increase in weight and cessation of diarrhea. It is generally accepted that secondary disaccharide malabsorption is of little clinical significance in children suffering from coeliac disease (Anderson *et al.,* 1966; McNeish and Sweet, 1968), but some investigators recommend an initial treatment with a gluten-free as well as a lactose- and sucrose-free diet (Arthur *et al.,* 1966). In our experience, however, adult patients with untreated coeliac disease often have clinically significant secondary lactose malabsorption (Gudmand-Høyer, unpublished). This may be explained by the fact that coeliac disease in adults has been present for a much longer period of time and therefore has led to more severe changes in the small intestinal mucosa.

Clinically significant unspecific disaccharide malabsorption has been demonstrated in children with acute gastroenteritis (Sunshine and Kretch-

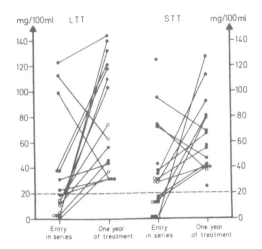

Fig. 9. Maximal blood glucose rise during lactose tolerance test (LTT) and sucrose tolerance test (STT), respectively, before treatment was instituted and after one year on a gluten-free diet. *Gudmand-Høyer and Krasilnikoff (1978).

mer, 1964; Burke *et al.*, 1965; Lifshitz *et al.*, 1971a, 1971b; Barnes and Townley, 1973). Also, in this case the secondary lactose malabsorption is quantitatively most significant. Unspecific sucrose malabsorption probably occurs only in cases in which the mucosal lesion is as pronounced as it is in coeliac disease (Barnes and Townley, 1973; Gudmand-Høyer and Söeberg, 1974; Antonowicz *et al.*, 1972). Consequently, it is recommended that children with acute gastroenteritis be treated initially with a lactose-free diet. In children with prolonged attacks of diarrhea it may be necessary to institute treatment with a completely disaccharide-free diet (Lifshitz *et al.*, 1971a). When present, disaccharide malabsorption is transitory in patients with gastroenteritis and disappears in the course of a few months. The fact that unspecific S-I malabsorption is of little clinical significance in cases of acute gastroenteritis has been demonstrated in studies on rotavirus infection in children (Rahilly *et al.*, 1976; Sack *et al.*, 1980; Black *et al.*, 1981). In controlled investigations the children were treated with a peroral electrolyte mixture to which either glucose or sucrose had been added. In none of the cases were the results worse when sucrose was added than when glucose was added.

In kwashiorkor the situation is the same as it is in gastroenteritis. Secondary transitory disaccharide malabsorption is present. Lactose malabsorption occurs most frequently, but many patients suffer from sucrose malabsorption as well. In these cases it is recommended that primary treatment with a lactose-free diet be instituted, and that a completely disaccharide-free diet be introduced if this does not prove effective (Chandra *et al.*, 1968; Wharton *et al.*, 1968).

13. Treatment

The treatment of primary S-I malabsorption consists of the elimination of sucrose from the diet. The number of 1,6-linkages in amylopectin or starch (which are split by isomaltase) is usually small, and thus it is seldom necessary to make the diet starch-free as well. Institution of a sucrose-free diet usually leads to the disappearance of symptoms within 24 hours.

If a sucrose-free diet does not lead to disappearance of the symptoms (which is rarely the case) the starch content of the diet must be reduced, and special attention should be paid to foods having a high amylopectin content. This includes such common foods as wheat and potatoes (Lindquist and Meeuwisse, 1966).

A thorough knowledge of the sucrose and starch content of various foods is necessary when planning the diet (Hardinge *et al.*, 1965; Somogyi and Trautner, 1974). Instructions for a sucrose-free diet, as well as a combined sucrose- and lactose-free diet, for children have been compiled by Gracey and Anderson (1975). As the child with S-I malabsorption grows, the tolerance to sucrose will increase, and the diet then may be made less restrictive. This change in tolerance is probably due to the fact that the adult has a larger total absorption capacity for sucrose in the intestine. It is not due to development of the deficient enzymes; this is demonstrated by the finding that in older children and adults who, despite their S-I malabsorption, follow a normal diet there is no significant sucrase or isomaltase activity (Ringrose *et al.*, 1980), nor any increase in the activity of these enzymes (Kilby *et al.*, 1978).

In normal adults the ingestion of sucrose has been shown to be followed by a distinct increase in the activity of sucrase and maltase — the so-called enzyme adaptation (Rosensweig and Herman, 1968, 1969, 1970). However, lactase activity was not affected, not even when lactose was administered (Rosensweig, 1971). This investigation also showed that fructose, a product of sucrose hydrolysis, was as effective as sucrose as a stimulator of sucrase and maltase activity. This indicates that fructose is the active principle in the sucrose molecule and that the specific substrate is not necessary for the adaptive response.

On the basis of these results, Greene *et al.* (1972) examined the possible value of fructose in the treatment of a 7-year-old girl with S-I malabsorption. With fructose feedings the activity of sucrase and isomaltase increased fourfold and the sucrose tolerance test became normal. Continued administration of a sucrose-deficient diet containing 20% fructose led to complete disappearance of the symptoms. The child's weight increased and gradually she was able to tolerate small amounts of sucrose.

The significance of these findings awaits final evaluation. As far as we have been able to ascertain, no similar results have been published, and it seems possible that the patient did not have a specific, but rather an unspecific, S-I malabsorption. If fructose actually has as positive an effect as described, one would expect an increase in sucrase-isomaltase activity in older children and adults with S-I malabsorption who, despite their illness, do not keep up their diet, but this has not been the case (Kilby *et al.,* 1981).

Symptomatic treatment of S-I malabsorption might in principle be effected by adding the missing enzyme to the food. This has been attempted in individual patients and it has led to a decrease in their symptoms (Iversen, 1942; Weijers *et al.,* 1960). A sucrase supplement to the food could be an alternative to a sucrose-free diet, especially in patients who find it difficult to follow such a diet on a permanent basis or even occasionally.

References

Alpers, D.H., 1977, Protein turnover in intestinal mucosal villus and crypt brush border membranes, *Biochem. Biophys. Res. Commun.* **75**:130.

Ament, M.E., Perera, D.R., and Esther, L.J., 1973, Sucrase-isomaltase deficiency — a frequently misdiagnosed disease, *J. Pediatr.* **83**:721.

Anderson, C.M., and Burke, V., 1975, Disorders of carbohydrate digestion and absorption, in: *Paediatric Gastroenterology* (C.M. Anderson and V. Burke, eds.), pp. 199-217, Blackwell Scientific Publications, Oxford.

Anderson, C.M., Messer, M., Townley, R.R.W., Freeman, M., and Robinson, M.J., 1962, Intestinal isomaltase deficiency in patients with hereditary sucrose and starch intolerance, *Lancet* **2**:556 (letter).

Anderson, C.M., Messer, M., Townley, R.R.W., and Freeman, M., 1963, Intestinal sucrase and isomaltase deficiency in two siblings, *Pediatrics* **31**:1003.

Anderson, C.M., Burke, V., Messer, M., and Kerry, K.R., 1966, Sugar intolerance and coeliac disease, *Lancet* **1**:1322.

Anderson, L.E., Walsh, K.A., and Neurath, H., 1977, Bovine enterokinase: purification, specificity, and some molecular properties, *Biochemistry* **16**:3354.

Andria, G., Marzi, A., and Auricchio, S., 1976, α-glutamyl-β-naphthylamide hydrolase of rabbit small intestine. Localization in the brush border and separation from other brush border peptidases, *Biochim. Biophys. Acta* **419**:42.

Andria, G., Cucchiara, S., de Vizia, B., de Ritis, G., Mazzacca, G., and Auricchio, S., 1980, Brush border and cytosol peptidase activities of human small intestine in normal subjects and celiac patients, *Pediatr. Res.* **14**:812.

Antonowicz, I., and Lebenthal, E., 1977, Developmental pattern of small intestinal enterokinase and disaccharidase activities in the human fetus, *Gastroenterology* **72**:1299.

Antonowicz, I., Lloyd-Still, J.D., Khaw, K.T., and Shwachman, H., 1972, Congenital sucrase-isomaltase deficiency, *Pediatrics* **49**:847.

Arthur, A.B., Clayton, B.E., Cottom, D.G., Seakins, J.W.T., and Platt, J.W., 1966, Importance of disaccharide intolerance in the treatment of coeliac disease, *Lancet* **I**:172.

Asp, N.G., Berg, N.O., Dahlqvist, A., Gudmand-Høyer, E., Jarnum, S., and McNair, A., 1975a, Intestinal disaccharidases in Greenland Eskimos, *Scand. J. Gastroenterol.* **10**:513.

Asp, N.G., Gudmand-Høyer, E., and Andersen, B., 1975b, Distribution of disaccharidases, alkaline phosphatase, and some intracellular enzymes along the human small intestine, *Scand. J. Gastroenterol.* **10**:647.

Auricchio, S., Prader, A., Mürset, G., and Witt, G., 1961, Saccharose intoleranz. Durchfall infolge hereditären Mangels an intestinalen saccharaseaktivität, *Helv. Paediatr. Acta* 16:483.

Auricchio, S., Dahlqvist, A., Mürset, G., and Prader, A., 1963a, Isomaltose intolerance causing decreased ability to utilize dietary starch, *J. Pediatr.* 62:165.

Auricchio, S., Rubino, A., Landall, M., Semenza, G., and Prader, A., 1963b, Isolated intestinal lactase deficiency in the adult, *Lancet* 2:324.

Auricchio, S., Rubino, A., Prader, A., Rey, J., Jos, J., Frézal, J., and Davidson, M., 1965, Intestinal glycosidase activities in congenital malabsorption of disaccharides, *J. Pediatr.* 66:555.

Auricchio, S., Ciccimarra, F., Moauro, L., Rey, F., Jos, J., and Rey, J., 1972, Intraluminal and mucosal starch digestion in congenital deficiency of intestinal sucrase and isomaltase activities, *Pediatr. Res.* 6:832.

Auricchio, S., Greco, L., de Vizia, B., and Buonocore, V., 1978, Dipeptidylaminopeptidase and carboxypeptidase activities of the brush border of rabbit small intestine, *Gastroenterology* 75:1073.

Bach, Chr., Theriez, H., Schaefer, Ph., and Cayrocke, P., 1962, Intolerance au saccharose chez un nourrisson, *Ann. Fr. Pediatr.* 19:1138.

Baratti, J., Maroux, S., Louvard, D., and Desnuelle, P., 1973, On porcine enterokinase. Further purification and some molecular properties, *Biochim. Biophys. Acta* 315:147.

Barnes, G.L., and Townley, R.R.W., 1973, Duodenal mucosal damage in 31 infants with gastroenteritis, *Arch. Dis. Child.* 48:343.

Barr, R.G., Watkins, J.B., and Perman, J.A., 1981, Mucosal function and breath hydrogen excretion: comparative studies in the clinical evaluation of children with nonspecific abdominal complaints, *Pediatrics* 68:526.

Bell, R.R., Draper, H.H., and Bergan, J.G., 1973, Sucrose, lactose, and glucose tolerance in northern Alaskan Eskimos, *Am. J. Clin. Nutr.* 26:1185.

Benajiba, A., and Maroux, S., 1980, Purification and characterization of an amino-peptidase A from hog intestinal brush-border membrane, *Eur. J. Biochem.* 107:381.

Black, R.E., Merson, M.H., Taylor, P.R., Yolken, R.H., Yunus, M., Alim, A.R.M.A., and Sack, D.A., 1981, Glucose vs sucrose in oral rehydration solutions for infants and young children with rotavirus-associated diarrhea, *Pediatrics* 67:79.

Bond, J.H., and Levitt, M.D., 1972, Use of pulmonary hydrogen (H_2) measurements to quantitate carbohydrate absorption. Study of partially gastrectomized patients, *J. Clin. Invest.* 51:1219.

Borgström, B., Dahlqvist, A., Lundh, G., and Sjövall, J., 1957, Studies of intestinal digestion and absorption in the human, *J. Clin. Invest.* 36:1521.

Bowdler, J.D., and Walker-Smith, J.A., 1969, Le role de la radiologie dans le diagnostic de l'intolerance au lactose chez l'enfant, *Ann. Radiol. (Paris)* 12:467.

Bowie, D., Brinkman, G.L., and Hansen, D.L., 1965, Acquired disaccharide intolerance in malnutrition, *J. Pediatr.* 66:1083.

Brunner, J., Hauser, H., Braun, H., Wilson, K.J., Wacker, H., O'Neill, B., and Semenza, G., 1979, The mode of association of the enzyme complex sucrase isomaltase with the intestinal brush border membrane, *J. Biol. Chem.* 6:1821.

Burgess, E.A., Levin, B., Mahalabis, D., and Tonge, R.E., 1964, Hereditary sucrose intolerance: levels of sucrase activity in jejunal mucosa, *Arch. Dis. Child.* 39:431.

Burke, V., Kerry, K.R., and Anderson, C.M., 1965, The relationships of dietary lactose to refractory diarrhoea in infancy, *Aust. Paediatr. J.* 1:147.

Chandra, R.K., Pawa, R.R., and Ghai, O.P., 1968, Sugar intolerance in malnourished infants and children, *Br. Med. J.* 4:611.

Chaptal, J. Dossa, J.R., Meylan, D., Guillaumat, P., Morel, R., Vernier, G., and Kesch, M., 1962, Diarrhees chroniques ni infectieuses, ni parasitaires du nourrisson, *Arch. Fr. Pediatr.* **19**:463.

Chung, Y.C., Kim, Y.S., Shadchehr, A., Garrido, A., MacGregor, I.L., and Sleisenger, M.H., 1979, Protein digestion and absorption in human small intestine, *Gastroenterology* **76**:1415.

Clément, M., 1962, Diarrhées chroniques par deficit de saccharase, *Thèse Méd.* no. 680, Paris.

Colbeau, A., and Maroux, S., 1978, Integration of alkaline phosphatase in the intestinal brush border membrane, *Biochim. Biophys. Acta* **511**:39.

Colombo, V., Lorenz-Meyer, H., and Semenza, G., 1973, Small intestinal phlorizin hydrolase: The "β-gycosidase complex," *Biochim. Biophys. Acta* **327**:412.

Conklin, K.A., Yamashiro, K.M., and Gray, G.M., 1975, Human intestinal sucrase-isomaltase. Identification of free sucrase and isomaltase and cleavage of the hybrid into active distinct subunits, *J. Biol. Chem.* **250**:5735.

Connolly, K.D., Kearney, P.J., and Gordon, I.R.S., 1978, The barium sucrose meal, *J. Irish Med. Ass.* **71**:212.

Cook, G.C., and Lee, F.D., 1966, The jejunum after kwashiorkor, *Lancet* **2**:1263.

Corouben, J.C., Bedu, J., Le balle, J.C., Crumback, R., Yongo, J., Weill, J., and Kaplan, M., 1963, Intolerance au saccharose. Etude clinique et biologique de cinq cas, *Arch. Franc. Pédiat.* **20**:253.

Crane, R.K., 1977, Digestion and absorption: water-soluble organics, in: *International Review of Physiology, Gastrointestinal Physiology II,* vol. 12 (R.K. Crane, ed.), p. 325, University Park Press, Baltimore.

Cuatrecasas, P., Lockwood, D.H., and Caldwell, J.R., 1965, Lactase deficiency in the adult: a common occurrence, *Lancet* **1**:14.

Curtis, K.J., Kim, Y.S., Perdomo, J.M., Silk, D.B.A., and Whitehead, J.S., 1978, Protein digestion and absorption in the rat, *J. Physiol.* **274**:409.

Dahlqvist, A., 1960, Characterization of hog intestinal trehalase, *Acta Chem. Scand.* **14**:9.

Dahlqvist, A., 1962, Specificity of the human intestinal disaccharidases and implications for hereditary disaccharide intolerance, *J. Clin. Invest.* **41**:463.

Dahlqvist, A., 1968, Assay of intestinal disaccharidases, *Anal. Biochem.* **22**:99.

Dahlqvist, A., and Borgström, B., 1961, Digestion and absorption of disaccharides in man, *Biochem. J.* **81**:411.

Dahlqvist, A., Lindquist, B., and Meeuwisse, G., 1968, Disturbances of the digestion and absorption of carbohydrates, in: *Carbohydrate Metabolism and Its Disorders,* Vol. 1, (F. Dickens, P.J. Randle, and W.J. Whelan, eds.), pp. 199-222, Academic Press, London and N.Y.

Dahlqvist, A., Lindberg, T., Meeuwisse, G., Akerman, M., 1970, Intestinal dipeptidases and disaccharidases in children with malabsorption, *Acta Paediatr. Scand.* **59**:621.

Danielsen, E.M., Vyas, J.P., and Kenny, A.J., 1980, A neutral endopeptidase in the microvillar membrane of pig intestine. Partial purification and properties, *Biochem. J.* **191**:645.

Davidson, G.P., and Barnes, G.L., 1979, Structural and functional abnormalities of the small intestine in infants and young children with rotavirus enteritis, *Acta Paediatr. Scand.* **68**:181.

Desai, H.G., Chitre, A.V., Parekh, D.V., and Jeejeebhoy, K.N., 1967, Intestinal discchari-dases in tropical sprue, *Gastroenterology* **53**:375.

Douwes, A.C., Fernandes, J., and Rietveld, W., 1978, Hydrogen breath test in infants and children: sampling and storing expired air, *Clin. Chim. Acta* **82**:293.

Douwes, A.C., Fernandes, J., and Jongbloed, A.A., 1980, Diagnostic value of sucrose tolerance test in children evaluated by breath hydrogen measurement, *Acta Paediatr. Scand.* **69**:79.

Dubs, R., Steinmann, B., and Gitzelmann, R., 1973, Demonstration of an inactive enzyme antigen in sucrase-isomaltase deficiency, *Helv. Paediatr. Acta* **28**:187.

Dubs, R., Gitzelmann, R., Steinmann, B., and Lindenmann, J., 1975, Catalytically inactive sucrase antigen of rabbit small intestine: the enzyme precursor, *Helv. Paediatr. Acta* **30**:89.

Dunphy, J.V., Littman, A., Hammond, J.B., Forstner, G., Dahlqvist, A., and Crane, R.K., 1965, Intestinal lactase deficit in adults, *Gastroenterology* **49**:12.

Eggermont, E., and Hers, H.G., 1969, The sedimentation properties of the intestinal α-glucosidases of normal human subjects and of patients with sucrose intolerance, *Eur. J. Biochem.* **9**:488.

Ellestad-Sayed, J.J., and Haworth, J.C., 1977, Disaccharide consumption and malabsorption in Canadian Indians, *Am. J. Clin. Nutr.* **30**:1977.

Ellestad-Sayed, J.J., Haworth, J.C., and Hildes, J.A., 1978, Disaccharide malabsorption and dietary patterns in two Canadian Eskimo communities, *Am. J. Clin. Nutr.* **31**:1473.

Fosset, M., Chappelet-Tordo, D., and Lazdunski, M., 1974, Intestinal alkaline phosphatase. Physical properties and quaternary structure, *Biochemistry* **13**:1783.

Francois, R., Frederich, A., Vicens-Calvet, E., Bertrand, M., and Ruitton-Ugliengo, M., 1963, Intolerance isolée au saccharase, *Pédiatrie* **18**:563.

Frank, G., Brunner, J., Hauser, H., Wacker, H., Semenza, G., and Zuber, H., 1978, The hydrophobic anchor of small-intestinal sucrase-isomaltase. N-terminal sequence of the isomaltase subunit, *FEBS Lett.* **96**:183.

Freiburghaus, A.U., Dubs, R., Hadorn, B., Gaze, H., Hauri, H.P., and Gitzelmann, R., 1977, The brush border membrane in hereditary sucrase-isomaltase deficiency: abnormal protein pattern and presence of immunoreactive enzyme, *Eur. J. Clin. Invest.* **7**:455.

Gardiner, A.J., Tarlow, M.J., Sutherland, I.T., and Sammons, H.G., 1981a, Lactose malabsorption during gastroenteritis, assessed by the hydrogen breath test, *Arch. Dis. Child.* **56**:364.

Gardiner, A.J., Tarlow, M.J., Symonds, J., Hutchison, J.G.P., and Sutherland, I.T., 1981b, Failure of the hydrogen breath test to detect primary sugar malabsorption, *Arch. Dis. Child.* **56**:368.

Gilat, T., Ben Hur, H., Gelman-Malachi, E., Terdiman, R., and Peled, Y., 1978, Alterations of the colonic flora and their effect on the hydrogen breath test, *Gut* **19**:602.

Gracey, M., and Anderson, C.M., 1975, Nutrition: normal requirements and dietary therapy, in: *Paediatric Gastroenterology* (C.M. Anderson and V. Burke, eds.), pp. 611-632, Blackwell Scientific Publications, Oxford.

Grant, D.A.W., and Hermon-Taylor, J., 1976, The purification of human enterokinase by affinity chromatography and immunoadsorption. Some observations on its molecular characteristics and comparisons with the pig enzyme, *Biochem J.* **155**:243.

Gray, G.M., 1975, Carbohydrate digestion and absorption. Role of the small intestine, *N. Engl. J. Med.* **292**:1225.

Gray, G.M., and Santiago, N.A., 1977, Intestinal surface amino-oligopeptidases. I. Isolation of two weight isomers and their subunits from rat brush border, *J. Biol. Chem.* **252**:4922.

Gray, G.M., Walter, W.M., and Colver, E.H., 1968, Persistent deficiency of intestinal lactase in apparently cured tropical sprue, *Gastroenterology* **54**:552.

Gray, G.M., Conklin, K.A., and Townley, R.R.W., 1976, Sucrase-isomaltase deficiency. Absence of an inactive enzyme variant, *N. Engl. J. Med.* **294**:750.

Greene, H.L., Stifel, F.B., and Herman, R.H., 1972, Dietary stimulation of sucrase in a patient with sucrase-isomaltase deficiency, *Biochem. Med.* **6**:409.

Grenet, P., Lestradet, H., Dugas, M., Iniguez, M., and Gourgon, R., 1962, Absence of saccharase, a cause of chronic diarrhea in an infant, *Arch. Franc. Pediatr.* **19**:1131.

Gudmand-Høyer, E., and Folke, K., 1970, Radiological detection of lactose malabsorption, *Scand. J. Gastroenterol.* **5**:565.

Gudmand-Høyer, E., and Jarnum, S., 1969, Lactose malabsorption in Greenland Eskimos, *Acta Med. Scand.* **186**:235.

Gudmand-Høyer, E., and Krasilnikoff, P.A., 1977, The effect of sucrose malabsorption on the growth pattern in children, *Scand. J. Gastroenterol.* **12**:103.

Gudmand-Høyer, E., and Krasilnikoff, P.A., 1978, Disaccharide malabsorption in coeliac disease in children, *Ugeskr. Laeg.* (Copenhagen) **140**:653.

Gudmand-Høyer, E., and Simony, K., 1977, Individual sensitivity to lactose in lactose malabsorption, *Am. J. Dig. Dis.* **22**:177.

Gudmand-Høyer, E., and Söeberg, B., 1973, Jejunal brush-border disaccharidase and alkaline phosphatase activity in acute viral hepatitis, *Scand. J. Gastroenterol.* **8**:377.

Gudmand-Høyer, E., and Söeberg, B., 1974, Disaccharidase activity in the small intestinal mucosa in cases with acute enteritis, *Scand. J. Gastroenterol.* **9**:405.

Gudmand-Høyer, E., McNair, A., and Jarnum, S., 1973, Lactose malabsorption in Western Greenland, *Ugeskr. Laeg.* (Copenhagen) **135**:169.

Gudmand-Høyer, E., Binder, V., and Söltoft, J., 1975, The small intestinal disaccharidase activity in ulcerative colitis, *Scand. J. Gastroenterol.* **10**:209.

Haemmerli, U.P., Kistler, H.J., Ammann, R., Auricchio, S., and Prader, A., 1965, Acquired milk intolerance in the adult caused by lactose malabsorption due to a selective deficiency of intestinal lactase activity, *Am. J. Med.* **38**:7.

Hardinge, M.G., Swarner, J.B., and Crooks, H., 1965, Carbohydrates in foods, *J. Am. Diet. Assoc.* **48**:307.

Hardy, G.H., 1908, Mendelian proportions in a mixed population, *Science* **28**:48.

Hauri, H.-P., Quaroni, A., and Isselbacher, K.J., 1979, Biogenesis of the intestinal plasma membrane: post-translational route and cleavage of sucrase-isomaltase, *Proc. Natl. Acad. Sci. USA.* **76**:5183.

Hedeager-Sørensen, S., Norén, O., Sjöström, H., and Danielsen, M.D., 1982, Amphiphilic pig intestinal microvillus maltase-glucoamylase: structure and specificity, *Eur. J. Biochem.* **126**:555.

Helms, P., 1981, Changes in disease and food patterns in Angmagssalik, 1949-1979, in: *Circumpolar Health 81*, Proceedings of 5th International Symposium on Circumpolar Health (B. Harvald and J.P. Hart Hansen, eds.), pp. 243-261, Stougaard Jensen, Copenhagen.

Hopsu-Havu, V.K., and Ekfors, T.O., 1969, Distribution of a dipeptide naphthylamidase in rat tissues and its localization by using diazo coupling and labeled antibody techniques, *Histochemie* **17**:30.

Howland, J., 1921, Prolonged intolerance to carbohydrates, *Trans. Am. Pediatr. Soc.* (N.Y.) **38**:393.

Hurst, A.F., and Knott, F.A., 1931, Intestinal carbohydrate dyspepsia, *Q. J. Med.* **24**:171.

Hyams, J.S., Stafford, R.J., Grand, R.J., and Watkins, J.B., 1980, Correlation of lactose breath hydrogen test, intestinal morphology, and lactase activity in young children, *J. Pediatr.* **97**:609.

Iversen, P., 1942, Et tilfaelde af kulhydratdyspepsi, *Nord. Med.* (Copenhagen) **16**:2860.

Jacobi, A., 1901, Milk-sugar infant feeding, *Trans. Am. Pediatr. Soc.* **13**:150.

James, W.P.T., Alpers, D.H., Gerber, J.E., and Isselbacher, K.J., 1971, The turnover of disaccharidases and brush border proteins in rat intestine, *Biochim. Biophys. Acta* **230**:194.

Jansen, W., Que, C.S., Weeger, W., 1965, Primary combined saccharase and isomaltase deficiency, *Arch. Intern. Med.* **116**:879.

Jensen, P.E., 1962, Intolerance of cane sugar as sequel of enzyme deficiency, *Acta Paediatr. Uppsala* **51**:227.

Jos, J., Frézal, J., Rey, J., and Lamy, M., 1967, Histochemical localization of intestinal disaccharidases: application to peroral biopsy specimens, *Nature* **213**:516.

Kelly, J.J., and Alpers, D.H., 1973, Properties of human intestinal glucoamylase, *Biochim. Biophys. Acta* **315**:113.

Kern, F. Jr., and Struthers, J.E. Jr., 1966, Intestinal lactase deficiency and lactose intolerance in adults, *J. Am. Med. Ass.* **195**:927.

Kern, F. Jr., Singleton, J.W., and Struthers, J.E. Jr., 1964, Fecal lactic acid in carbohydrate malabsorption, *J. Lab. Clin. Med.* **64**:874.

Kerry, K.R., and Anderson, C.M., 1964, A ward test for sugar in faeces, *Lancet* **1**:981.

Kerry, K.R., and Townley, R.R.W., 1965, Genetic aspects of intestinal sucrase-isomaltase deficiency, *Aust. Paediatr. J.* **1**:223.

Kilby, A., Burgess, E.A., Wigglesworth, S., and Walker-Smith, J., 1978, Sucrase-isomaltase deficiency: a follow-up report, *Arch. Dis. Child.* **53**:677.

Kocna, P., Fric, P., Slaby, J., and Kasaffirek, E., 1980, Endopeptidase of the brush border membrane of rat enterocyte: separation from aminopeptidase and partial characterization, *Z. Physiol. Chem.* **361**:1401.

Kolinská, J., and Kraml, J., 1972, Separation and characterization of sucrase-isomaltase and of glucoamylase of rat intestine, *Biochim. Biophys. Acta* **284**:235.

Komoda, T., and Sakagishi, Y., 1976, Partial purification of human intestinal alkaline phosphatase with affinity chromatography: some properties and interaction of Concanavalin A with alkaline phosphatase, *Biochim. Biophys. Acta* **445**:645.

Krasilnikoff, P.A., Gudmand-Høyer, E., and Moltke, H.H., 1975, Diagnostic value of disaccharide tolerance tests in children, *Acta Paediatr. Scand.* **64**:693.

Kühni, M., 1967, Kongenitaler primärer Saccharase- und Isomaltase-mangel des Dünndarms, *Praxis* **4**:130.

Lanzkowsky, P., Karayalcin, G., Miller, R., and Lane, B.P., 1981, Disaccharidase values in iron-deficient infants, *J. Pediatr.* **99**:605.

Launiala, K., 1968, The effect of unabsorbed sucrose and mannitol on the small intestinal flow rate and mean transit time, *Scand. J. Gastroenterol.* **3**:665.

Launiala, K., Perheentupa, J., Visakorpi, J., and Hallmann, N.P., 1964, Disaccharidases of intestinal mucosa in a patient with sucrose intolerance, *Pediatrics* **34**:615.

Larrechea, I. de, Sampayo, R.R.L., and Miatello, C.S., 1965, Disaccharidase deficiency in non-tropical sprue, Whipple's disease and congenital intestinal lymphangiectasia, *Gastroenterology* **48**:829.

Laws, J.W., and Neale, G., 1966, Radiological diagnosis of disaccharidase deficiency, *Lancet* **2**:139.

Lee, L.M.Y., Salvatore, A.K., Flanagan, P.R., and Forstner, G.G., 1980, Isolation of a detergent-solubilized maltase/glucoamylase from rat intestine and its comparison with a maltase/glucoamylase solubilized by papain, *Biochem. J.* **187**:437.

Leese, H.J., and Semenza, G., 1973, On the identity between the small intestinal enzymes phlorizin hydrolase and glycosylceramidase, *J. Biol. Chem.* **248**:8170.

Levitt, M.D., 1969, Production and excretion of hydrogen gas in man, *N. Engl. J. Med.* **281**:122.

Lifshitz, F., and Holman, G.H., 1964, Disaccharidase deficiencies with steatorrhea, *J. Pediatr.* **64**:34.

Lifshitz, F., Klotz, A.P., and Holman, G.H., 1965, Intestinal disaccharidase deficiencies in gluten-sensitive enteropathy, *Am. J. Dig. Dis.* **10**:47.

Lifshitz, F., Coello-Ramirez, P., and Contreras-Gutierrez, M.L., 1971a, The response of infants to carbohydrate oral loads after recovery from diarrhea, *J. Pediatr.* **79**:612.

Lifshitz, F., Coello-Ramirez, P., Gutierrez-Topete, G., and Cornado-Cornet, M.C., 1971b, Carbohydrate intolerance in infants with diarrhea, *J. Pediatr.* **79**:760.

Lindquist, B., and Meeuwisse, G., 1966, Diets in disaccharidase deficiency and defective monosaccharide absorption, *J. Am. Diet. Assoc.* **48**:307.

Lindquist, B.L., and Wranne, L., 1976, Problems in analysis of faecal sugar, *Arch. Dis. Child.* 31:189.

Louvard, D., Maroux, S., Vannier, C., and Desnuelle, P., 1975, Topological studies on the hydrolases bound to the intestinal brush border membrane. I. Solubilization by papain and Triton X-100, *Biochim. Biophys. Acta* 375:236.

Maestracci, D., Preiser, H. Hedges, T., Schmitz, J., and Crane, R.K., 1975, Enzymes of the human intestinal brush border membrane: identification after gel electrophoretic separation, *Biochim. Biophys. Acta* 382:147.

Maffei, H.V.L., Metz, G., Bampoe, V., Shiner, M., Herman, S, and Brook, C.G.D., 1977, Lactose intolerance detected by the hydrogen breath test in infants and children with chronic diarrhea, *Arch. Dis. Child.* 52:766.

Malathi, P., Ramaswamy, K., Caspary, W.F., and Crane, R.K., 1973, Studies on the transport of glucose from disaccharides by hamster small intestine *in vitro*. I. Evidence for a disaccharidase-related transport system, *Biochim. Biophys. Acta* 307:613.

Malik, N., and Butterworth, P.J., 1976, Molecular properties of rat intestinal alkaline phosphatase, *Biochim. Biophys. Acta* 446:105.

Maroux, S., and Louvard, D., 1976, On the hydrophobic part of aminopeptidase and maltases which bind the enzyme to the intestinal brush border membrane, *Biochim. Biophys. Acta* 419:189.

McClellan, J.B. Jr., and Garner, C.W., 1980, Purification and properties of human intestine alanine aminopeptidase, *Biochim. Biophys. Acta* 613:160.

McMichael, H.B., Webb, J., and Dawson, A.M., 1965, Lactase deficiency in adults: a cause of "functional" diarrhoea, *Lancet* 1:717.

McNair, A., Gudmand-Høyer, E., Jarnum, S., and Orrild, L., 1972, Sucrose malabsorption in Greenland, *Br. Med. J.* 2:19.

McNeish, A.S., and Sweet, E.M., 1968, Lactose intolerance in childhood coeliac disease, *Arch. Dis. Child.* 43:433.

Miller, D., and Crane, R.K., 1961, The digestive function of the epithelium of the small intestine. II. Localization of disaccharide hydrolysis in the isolated brush border portion of intestinal epithelial cells, *Biochim. Biophys. Acta* 52:293.

Neale, G., Clark, M., and Levin, B., 1965, Intestinal sucrase deficiency presenting as sucrose intolerance in adult life, *Br. Med. J.* 2:1223.

Newcomer, A.D., and McGill, B., 1966, Distribution of disaccharidase activity in the small bowel of normal and lactase deficient subjects, *Gastroenterology* 51:481.

Newcomer, A.D., and McGill, D.B., 1967, Disaccharidase activity in the small intestine: prevalence of lactase deficiency in 100 healthy subjects, *Gastroenterology* 53:881.

Newcomer, A.D., McGill, D.B., Thomas, P.J., and Hofmann, A.F., 1975, Prospective comparison of indirect methods for detecting lactase deficiency, *N. Engl. J. Med.* 293:1232.

Nordio, S., La Medica, G.M., and Vignolo, L.G., 1961, Un caso di diarrea cronica, connatale di intoleranza al saccarosio edalle destrine, *Minerva Pediatr.* 13:1766.

Nordström, C., Dahlqvist, A., and Josefsson, L., 1967, Quantitative determination of enzymes in different parts of the villi and crypts of rat small intestine. Comparison of alkaline phosphatase, disaccharidases and dipeptidases, *J. Histochem. Cytochem.* 15:713.

Norén, O., Sjöström, H., Danielsen, E.M., Staun, M., Jeppesen, L., and Svensson, B., 1979, Comparison of two pig intestinal brush border peptidases with the corresponding renal enzymes, *Z. Physiol. Chem.* 360:151.

Nose, O., Ilda, Y., Kai, H., Harada, T., Ogawa, M., and Yabuuchi, H., 1979, Breath hydrogen test for detecting lactose malabsorption in infants and children, *Arch. Dis. Child.* 54:436.

Pena, S., and Truelove, S.C., 1970, Lactase deficiency in ulcerative colitis and Crohn's disease, in: Advance Abstracts, 4th World Congress of Gastroenterology, Copenhagen, (P. Riis, P. Anthonisen, and H. Baden, eds.), p. 587, The Danish Gastroenterological Association, Copenhagen.

Perman, J.A., Barr, R.G., and Watkins, J.B., 1978, Sucrose malabsorption in children: Non-invasive diagnosis by interval breath hydrogen determination, *J. Pediatr.* **93**:17.

Peterson, M.L., and Herber, R., 1967, Intestinal sucrase deficiency, *Trans. Assoc. Am. Physicians* **80**:275.

Pink, I.J., 1967, Diarrhoea due to sucrase and isomaltase deficiency, *Gut* **8**:373.

Plotkin, G.R., and Isselbacher, K.J., 1964, Secondary disaccharidase deficiency in adult celiac disease (non-tropical sprue) and other malabsorption states, *N. Engl. J. Med.* **271**:1033.

Prader, A., and Auricchio, S., 1965, Defects of intestinal disaccharide absorption, *Ann. Rev. Med.* **16**:345.

Prader, A., Auricchio, S., and Mürset, G., 1961, Durchfall infolge hereditären Mangel an intestinaler Saccharaseaktivität, *Schweiz. Med. Wschr.* **91**:465.

Preger, L., and Amberg, J.R., 1967, Sweet diarrhea. Roentgen diagnosis of disaccharidase deficiency, *Am. J. Roentgenol.* **101**:287.

Preiser, H., Menard, D., Crane, R.K., and Cerda, J.J., 1974, Deletion of enzyme protein from the brush border membrane in sucrase-isomaltase deficiency, *Biochim. Biophys. Acta* **363**:279.

Prinsloo, J.G., Wittmann, W., Kruger, H., and Freier, E., 1971, Lactose absorption and mucosal disaccharidases in convalescent pellagra and kwashiorkor children, *Arch. Dis. Child.* **46**:474.

Rahilly, P.M., Shepherd, R., Challis, D., Walker-Smith, J., and Manly, J., 1976, Clinical comparison between glucose and sucrose additions to a basic electrolyte mixture in the outpatient management of acute gastroenteritis in children, *Arch. Dis. Child.* **51**:152.

Ramaswamy, S., and Radhakrishnan, A.N., 1975, Lactase-phlorizin hydrolase complex from monkey small intestine: purification, properties and evidence for two catalytic sites, *Biochim. Biophys. Acta* **403**:446.

Rey, J., Frezal, J., Jos, V., Bauche, P., and Lamy, M., 1963, Diarrhoe par toruble de l'hydrolyse du saccharoses, du maltose et de l'isomaltose, *Arch. Franc. Pédiat.* **20**:381.

Ringrose, R.E., Preiser, H., and Welsh, J.D., 1980, Sucrase-isomaltase (palatinase) deficiency diagnosed during adulthood, *Dig. Dis. Sci.* **25**:384.

Rosensweig, N.S., 1971, Adult lactase deficiency: genetic control or adaptive response? *Gastroenterology* **60**:464.

Rosensweig, N.S., and Herman, R.H., 1968, Control of jejunal sucrase and maltase activity by dietary sucrose or fructose in man: a model for the study of enzyme regulation in man, *J. Clin. Invest.* **47**:2253.

Rosensweig, N.S., and Herman, R.H., 1969, Time response of jejunal sucrase and maltase activity to a high sucrose diet in normal man, *Gastroenterology* **56**:500.

Rosensweig, N.S., and Herman, R.H., 1970, Dose response of jejunal sucrase and maltase activities to isocaloric high and low carbohydrate diets in man, *Am. J. Clin. Nutr.* **23**:1373.

Ross, L.L., Barber, L., Tate, S.S., and Meister, A., 1973, Enzymes of the γ-glutamyl cycle in the ciliary body and lens, *Proc. Natl. Acad. Sci. USA.* **70**:2211.

Sack, D.A., Islam, S., Brown, K.H., Islam, A., and Kabir, I., 1980, Oral therapy in children with cholera: a comparison of sucrose and glucose electrolyte solutions, *J. Pediatr.* **96**:20.

Sasajima, K., Kawachi, T., Sato, S., and Sugimura, T., 1975, Purification and properties of α, α-trehalase from the mucosa of rat small intestine, *Biochim. Biophys. Acta* **403**:139.

Schlegel-Haueter, S., Hore, P., Kerry, K.R., and Semenza, G., 1972, The preparation of lactase and glucoamylase of rat small intestine, *Biochim. Biophys. Acta* **258**:506.

Schmidt, A., and Strasburger, J., 1901, Ueber die intestinale Gährüngsdyspepsie der Erwachsenen (Insufficienz der Stärkeverdaurung), *Dtsch. Arch. Klin. Med.* **69**:570.

Schmitz, J., Commegrain, C., Maestracci, D., and Rey, J., 1974, Absence of brush border sucrase-isomaltase complex in congenital sucrose intolerance, *Biomedicine* (Paris) **21**:440.

Seetharam, B., Swaminathan, N., and Radhakrishnan, A.N., 1970, Studies on mammalian glucoamylases with special reference to monkey intestinal glucoamylase, *Biochem. J.* 117:939.

Semenza, G., 1976, Glycosidases of small intestinal brush borders, in: *Membranes and Disease* (L. Bolis, J.F. Hoffman, and A. Leaf, eds.), p. 243, Raven Press, New York.

Shannon, I.L., 1978, Concentration of sugars in commercial baby foods, *J. Dent. Child.* 19:451.

Sheehy, T.W., and Anderson, P.R., 1965, Disaccharidase activity in normal and diseased small bowel, *Lancet* 2:1.

Shmerling, D.H., Auricchio, S., Rubino, A., Hadorn, B., and Prader, A., 1964, Der sekundäre mangel an intestinaler disaccharidase-aktivität bei der Cöliakie. Quantitative bestimmung der enzymaktivität und klinische beurteiling, *Helv. Paediat. Acta* 19:507.

Sigrist, H., Ronner, P., and Semenza, G., 1975, A hydrophobic form of the small-intestinal sucrase-isomaltase complex, *Biochim. Biophys. Acta* 406:433.

Silverblatt, E.R., Conklin, K., and Gray, G.M., 1974, Sucrase precursor in human jejunal crypts, *J. Clin. Invest.* 53:76a.

Singer, S.J., and Nicolson, G.L., 1972, The fluid mosaic model of the structure of cell membranes. Cell membranes are viewed as two-dimensional solutions of oriented globular proteins and lipids, *Science* 175:720.

Sivakami, S., and Radhakrishnan, A.N., 1973, Purification of rabbit intestinal glucoamylase by affinity chromatography on Sephadex G-200, *Indian J. Biochem. Biophys.* 10:283.

Sjöström, H., Norén, O., Jeppesen, L., Staun, M., Svensson, B., and Christiansen, L., 1978, Purification of different amphiphilic forms of a microvillus aminopeptidase from pig small intestine using immunoadsorbent chromatography, *Eur. J. Biochem.* 88:503.

Sjöström, H., Norén, O., Christiansen, L., Wacker, H., and Semenza, G., 1980, A fully active, two-active-site, single-chain sucrase isomaltase from pig small intestine: implications for the biosynthesis of a mammalian integral stalked membrane protein, *J. Biol. Chem.* 255:11332.

Sjöström, H., Norén, O., Krasilnikoff, P.A., and Gudmand-Høyer, E., 1981, Intestinal peptidases and sucrase in coeliac disease, *Clin. Chim. Acta* 109:53.

Skovbjerg, H., 1981a, Immunoelectrophoretic studies on human small-intestinal brush border proteins. Relation between enzyme activity and immunoreactive enzyme along the villuscrypt axis, *Biochem. J.* 193:887.

Skovbjerg, H., 1981b, Immunoelectrophoretic studies on human small intestinal brush border proteins. The longitudinal distribution of peptidases and disaccharidases, *Clin. Chim. Acta* 112:205.

Skovbjerg, H., 1982, High molecular weight pro-sucrase-isomaltase in human fetal intestine, *Pediatr. Res.* 16:948.

Skovbjerg, H., and Krasilnikoff, P.A., 1981, Immunoelectrophoretic studies on human small intestinal brush border proteins. The residual isomaltase in sucrose intolerant patients, *Pediatr. Res.* 15:214.

Skovbjerg, H., Norén, O., and Sjöström, H., 1978, Immunoelectrophoretic studies on human small intestinal brush border proteins. A qualitative study of the protein composition, *Scand. J. Clin. Lab. Invest.* 38:723.

Skovbjerg, H., Sjöström, H., Norén, O., and Gudmand-Høyer, E., 1979a, Immunoelectrophoretic studies on human small intestinal brush border proteins. A quantitative study of brush border enzymes from single small intestinal biopsies, *Clin. Chim. Acta* 92:315.

Skovbjerg, H., Sjöström, H., and Norén, O., 1979b, Does sucrase-isomaltase always exist as a complex in human intestine?, *FEBS Lett.* 108:399.

Skovbjerg, H., Sjöström, H., and Norén, O., 1981, Purification and characterization of amphiphilic lactase-phlorizin hydrolase from human small intestine, *Eur. J. Biochem.* 114:653.

Skovbjerg, H., Norén, O., Sjöström, H., Danielsen, E.M., and Enevoldsen, B., 1982, Further characterization of intestinal lactase-phlorizin hydrolase, *Biochim. Biophys. Acta* **707**:89.

Soeparto, P., Stobo, E.A., and Walker-Smith, J.A., 1972, Role of chemical examination of the stool in diagnosis of sugar malabsorption in children, *Arch. Dis. Child.* **47**:56.

Somogyi, J.C., and Trautner, K., 1974, Der glukose-, fruktose-, und saccharosegehalt vershiedener Gemüsearten, *Schweiz. Med. Wschr.* **104**:177.

Sontag, W.M., Brill, M.L., Troyer, W.C., Welsh, J.D., Semenza, G., and Prader, A., 1964, Sucrose-isomaltose malabsorption in an adult woman, *Gastroenterology* **47**:18.

Spatz, L., and Strittmatter, P.A., 1971, A form of cytochrome b_5 that contains an additional hydrophobic sequence of 40 amino acid residues, *Proc. Natl. Acad. Sci. USA.* **68**:1042.

Starnes, C.W., and Welsh, J.D., 1970, Intestinal sucrase-isomaltase deficiency and renal calculi, *N. Engl. J. Med.* **202**:1023.

Sterchi, E.E., and Woodley, J.F., 1978, Peptidases of the human intestinal brush border membrane, in: *Perspectives in Coeliac Disease* (B. McNicholl, C.F. McCarthy, and P.F. Fottrell, eds.), p. 437, MTP Press, Lancaster, England.

Sunshine, P., and Kretchmer, N., 1964, Studies of small intestine during development. III. Infantile diarrhea associated with intolerance to disaccharides, *Pediatrics* **34**:38.

Svensson, B., Danielsen, M., Staun, M., Jeppesen, L., Norén, O., and Sjöström, H., 1978, An amphiphilic form of dipeptidyl peptidase IV from pig small-intestinal brush-border membrane. Purification by immunoadsorbent chromatography and some properties, *Eur. J. Biochem.* **90**:489.

Takesue, Y., 1975, Purification and properties of leucine β-naphthylamidase from rabbit small-intestinal mucosal cells, *J. Biochem.* **77**:103.

Utermann, G., and Simons, K., 1974, Studies on the nature of the membrane proteins in Semliki Forest virus, *J. Mol. Biol.* **85**:569.

Walker-Smith, J., 1973, Screening tests for sugar malabsorption, *J. Pediatr.* **82**:893.

Walker-Smith, J., 1979, Sugar malabsorption, in: *Diseases of the Small Intestine in Childhood* (J. Walker-Smith, ed.), pp. 250-278, Pitman Medical Publishing, London.

Wallenfels, K., and Fischer, J., 1960, Untersuchungen über milchzuckerspatende Enzüme, X. Die Lactase des Kälberdarms, *Z. Physiol. Chem.* **321**:223.

Weijers, H.A., van de Kamer, J.H., Mossel, D.A., and Dicke, W.K., 1960, Diarrhea caused by deficiency of sugar splitting enzymes, *Lancet* **2**:296.

Weijers, H.A., van de Kamer, J.H., Dicke, W.K., and Ijsseling, J., 1961, Diarrhoea caused by deficiency of sugar-splitting enzymes, *Acta Paediatr. Scand.* **50**:55.

Weinberg, W., 1908, Uber den Nachweis der Vererbung beim Menschen, *Jahr. Verein. Vatest. Naturk.* (Württ) **64**:368.

Welsh, J.D., and Brown, R.C., 1966, Sucrase-palatinase deficiency, *Lancet* **2**:139.

Welsh, J.D., and Porter, M.G., 1967, Reversible secondary disaccharidase deficiency, *Am. J. Dis. Child.* **113**:716.

Welsh, J.D., Preiser, H., Woodley, J.F., and Crane, R.K., 1972, An enriched microvillus membrane preparation from frozen specimens of human small intestine, *Gastroenterology* **62**:572.

Welsh, J.D., Poley, J.R., Bhatia, M., and Stevenson, D.E., 1978, Intestinal disaccharidase activities in relation to age, race, and mucosal damage, *Gastroenterology* **75**:847.

Weser, E., and Sleisenger, M.H., 1965, Lactosuria and lactase deficiency in adult celiac disease, *Gastroenterology* **48**:571.

Wharton, B., Howells, G., and Phillips, I., 1968, Diarrhoea in kwashiorkor, *Br. Med. J.* **4**:608.

Chapter 10

Nutrient Absorption in Gnotobiotic Animals

Géza Bruckner and Jozsef Szabó

1. Introduction

Animals, whatever the species, are constantly exposed to a fluctuating external population of microorganisms. These microorganisms constitute part of the host's "external environment," and if the delicate balance regulating their numbers is disrupted, or if they gain access to the host's "internal environment," they may be detrimental to the host's homeostatic mechanisms. The interactions between the gastrointestinal (GI) microflora and the host animal were first recognized by Pasteur (1885) nearly a century ago. Pasteur postulated that the microflora were indispensable to the host and assumed that in the course of phylogenetic evolution, microbial associates developed that assisted in the "survival of the fittest" (Schottelius, 1902). Nencki (1886) and later Metchnikoff (1903) perceived the microflora as antagonistic to the host's well-being. Some sixty years elapsed before definite proof of "reproducing germfree existence" was obtained by Gustafsson (1948) and Reyniers and co-workers (1946, 1949). With the establishment of the gnotobiotic (known flora) animal as a tool, it became possible to assess the total effect of the microflora on the host's nutritional, physiological, biochemical, and morphological status, to study the relationship between individual species of microorganisms in mono-, di-, or polyassociation with the host, to study the homeostatic effects

Géza G. Bruckner • Department of Clinical Nutrition, University of Kentucky, Medical Center Annex Two, Lexington, Kentucky 40536-0080. Jozsef Szabó • Department of Animal Hygiene, University of Veterinary Science, Budapest, Hungary.

which microorganisms exert on each other directly or indirectly, and to determine the effect of the host on its microflora. With knowledge of these interrelationships the nutritional needs of the microbes as well as the hosts could be studied separately.

Gnotobiotic is a word derived from the Greek "gnotos" and "biota" meaning known flora and fauna. Gnotobiote means one of an animal stock or strain derived by aseptic Caesarean section (or sterile hatching of eggs) which are reared and continuously maintained with germfree techniques under isolator conditions and in which the composition of any associated fauna and flora, if present, are fully defined by accepted current methodology. Germfree animal (GF) (axenic) is a gnotobiote which is free from all demonstrable associated forms of life including bacteria, viruses, fungi, protozoa, and other saprophytic or parasitic forms. Conventional animal (CV) is usually the same species as the gnotobiotic animal and fed the same sterilized diet but which lives in an open environment (Gordon and Pesti, 1971).

2.   The Normal Intestinal Microflora

The absorptive and secretory functions and the immunological status of the host are decisive factors in the development and maintenance of a "normal" intestinal microflora. Changes in diet composition bring about shifts in microbial populations which may have a harmful or beneficial effect on the host.

Bacteria have been associated with animals for as long as we are able to trace phylogenetic evolution. This indigenous (autochthonous) microflora affects the host animal's physiological, biochemical, and morphological characteristics and can be considered as an "organ system." This "external organ" is greater in its total number of cells than the host itself (prokaryotic vs. eukaryotic cells) (Savage, 1977).

The fetus *in utero* is sterile. At birth it comes in contact with various vaginal microorganisms, the first of which is generally of the Lactobacillus type (Speck, 1976). How these microbes become established and what influences their succession to a relatively stable climax population is not clearly understood, but they are subject to various regulatory factors: microbe-microbe interactions, both synergistic and antagonistic; the type of gastrointestinal substrates (nutrients) and the availability of these substrates to the various microbes; microorganisms encountered by the host (microbial load); and the gastrointestinal milieu, i.e., pH, redox potential,

Table 1. Microorganisms Isolated from the Gastrointestinal Tract of Nonruminant Animals[a]

Microorganisms	Pig			Rat			Mouse		
	Stomach[b]	Small Intestine Lower[b]	Large Intestine[c]	Stomach[b]	Small Intestine Lower[b]	Large Intestine[c]	Stomach[b]	Small Intestine Lower[b]	Large Intestine[c]
Lactobacilli	+	+	+	+	+	+	+	+	+
Bifidobacteria			+			+			+
Clostridia	+		+	+		+	+		+
Veillonella	+		+	+		+			+
Coliforms	+	+	+	+	+	+			+
Bacteroides			+			+			+
Eubacterium			+			+			+
Propionibacterium						+			+
Fusobacterium						+			+
Other bacterial types	+	+	+	+	+	+			+
Candida	+			+			+		
Torulopsis				+			+		
Other yeasts		+	+		+	+		+	+
Streptococci	+	+	+	+	+	+	+	+	+

[a] Savage (1977)   [b] To be indicated positive a microbial type had to be isolated at an estimated population exceeding 10³ microbes/g of wet contents   [c] To be indicated positive a microbial type had to be isolated at an estimated population exceeding 10⁶ microbes/g wet contents

bile acids, enzymes, volatile fatty acids, etc. These factors generally favor the proliferation of facultative anaerobic microorganisms, such as Lactobacilli or *E. coli* shortly after birth. These organisms decline in number as the animal begins consuming solid food and is weaned from its milk diet. This weaning process starts the succession of a new group of organisms which predominate in the climax population, *viz.,* the strict anaerobes. The strict anaerobes outnumber the facultative organisms by as much as 1000 to 1 (Gordon and Dubos, 1970; Lee *et al.,* 1971; Savage *et al.,* 1968; Savage, 1979). A general overview of the microorganisms found in stable climax populations from various segments of the GI tract is presented in Table I. Some of the factors controlling and influencing microbial succession recently have been discussed (Berg, 1980; Sakai *et al.,* 1980).

3.   Intestinal Morphology of Gnotobiotic Animals

The intimate association which the gastrointestinal microflora have with structural, anatomical and secretory functions of the intestinal epithelium of the host is extremely important. The presence of the microflora and its byproducts brings about profound changes in the intestinal morphology of the host animal.

3.1.   The Effects of Germfree Status on Various Intestinal Parameters

3.1.1.   Intestinal Length and Weight

One consistent effect noted in the absence of the microflora is a lower small intestine weight (Reyniers *et al.,* 1960; Gordon *et al.,* 1966; Bruckner-Kardoss and Gordon, 1958, unpublished) (Tables II and III). The weight of the cecal wall, however, is increased in germfree rodents (Bruckner-Kardoss and Gordon, 1958, unpublished; Gordon *et al.,* 1960) and the obvious cecal distention has been the subject of numerous studies (Gordon, 1968; Wostmann and Knight, 1961). Cecal weight is reduced in GF chickens (Reyniers *et al.,* 1960) and pigs (Miniats and Valli, 1973).

3.1.2.   Intestinal Surface Area

The surface area of the small intestinal mucosa in germfree rats was found by Gordon and Bruckner-Kardoss (1961a) to be decreased by about 30% compared to its CV counterpart. Stevens and Heneghan (1976) found a 16% decrease in the upper small intestine of GF rats and this difference increased to 28% in the lower ileum. Meslin (1971) and Meslin *et al.* (1973) observed similar trends in GF rodents. The mucosal surface area of the GF dog small intestine also showed an overall decrease (26%) but the changes were unlike those observed in the rat (Stevens and Heneghan, 1976). The upper segments in the dog intestine showed the greatest decrease in surface area, whereas the lower segments did not show any difference. In GF piglets results similar to those reported for the canine have been found, i.e., no apparent differences from CV animals were noted in the mid small intestinal surface area (Heneghan *et al.*, 1979).

3.1.3.   Intestinal Histology and Cell Constituents

The small intestinal mucosa of GF animals appears generally to be of the "fetal" type, in that the epithelial layer remains more uniform and the *lamina propria* is less developed than they are in CV animals. In the GF animal there are smaller Peyer's patches with relatively fewer reactive centers and a decreased number of plasmocytes. The numbers of granulocytes, lymphocytes, and histocytes are also decreased in the *lamina propria* of GF animals (Gordon and Bruckner-Kardoss, 1958-59; Knight and Wostmann, 1964; Miniats and Valli, 1973; Bekkum, 1966). These observations have led to the commonly used term "sterile physiological inflammatory status" to describe the influence of the microflora on the mucosal morphology of the animal.

In the GF rat the increase in cecal wall weight arises from a proportionate increase in the mucosa, submucosa, and muscular tissue. The bulk of these increments originates in the cytoplasmic components. The individual smooth muscle cells are hypertrophied and elongated (Gordon, 1968). While the villi in the upper small intestine of the GF rat, mouse, and pig are longer and more uniform than in their CV counterparts (Galjaard *et al.*, 1972; Miniats and Valli, 1973; Lesher *et al.*, 1964), this difference was not observed in the lower small intestine of the rat (Meslin *et al.*, 1973). GF guinea pigs showed no change in villi length; however, a poorly developed *lamina propria* and much thinner villi were noted (Sprinz, 1962; Sprinz *et al.*, 1961). Atypical mitochondria associated with the absorptive epithelial cells of GF rats were noted by Nakao and Levenson (1967).

The distended GF cecum is characterized by narrower and higher epithelial cells, larger nuclei, and longer microvilli. In the cecum the arrangement of the mucosa is more irregular, with the *lamina propria* containing more labrocytes than it does in the CV animal. The *tunica muscularis* appears to be hypertrophied (Gustafsson and Maunsbach, 1971). Dupont *et al.* (1965) observed "absolutely monstrous" myenteric neurons with reduced levels of diphosphopyridine nucleotide diaphorase activity in the GF rat's cecal wall.

Other intestinal epithelial constituents in which differences have been

noted in the absence of the microflora are: (a) a decreased deoxyribonucleic acid in the rat cecal wall (Combe *et al.*, 1965); (b) increased levels
of 5-hydroxytryptamine in the small intestine of GF mice (Phillips *et al.*,
1961); (c) decreased 5-hydroxytryptamine levels in rats and mice (Beaver
and Wostmann, 1962); and (d) increased levels of alkaline phosphatase
and decreased acid phosphatase in the small intestine of GF rats (Jervis
and Biggers, 1964).

3.1.4. Cell Renewal and Mitotic Index

It is generally accepted that there is a decrease in mitotic activity in the
crypts of the germfree mouse and rat and a concomitant slowing down of
epithelial cell migration along the villi (Abrams *et al.*, 1963; Lesher *et al.*,
1964; Meslin *et al.*, 1974; Guenet *et al.*, 1970; Sugiyama and Stewart,
1975; Khoury *et al.*, 1969). Epithelial cell migration in the GF mouse is
reduced by approximately 50% (Abrams *et al.*, 1963; Lesher *et al.*, 1964;
Matsuzawa and Wilson, 1965). In the GF rat this decrease is on the order
of 30% (Guenet *et al.*, 1970; Galjaard *et al.*, 1972). It has been specu-
lated that the increased transit time is due to the increased villi length in
the GF rat (Galjaard *et al.*, 1972). However, Meslin *et al.* (1974) have
shown that this is not the case, since the lower small intestine villi height
is the same in GF and CV rats whereas epithelial cell migration in the
GF rat is still decreased. The literature relative to enterocyte kinetics in
GF animals recently has been reviewed by Heneghan (1979).

Conventionalization (Khoury *et al.*, 1969), supplementation of the
diet with cholic acid (Ranken *et al.*, 1971), and cecectomy (Sacquet *et
al.*, 1972) seem to accelerate the migration rate of epithelial cells in GF
mice and rats. Meslin *et al.* (1974) showed that gnotobiotic rats having a
defined microflora which did not deconjugate bile salts or increase the
level of free cholic acid still exhibited increased migration of intestinal
epithelial cells. These observations indicate that there is more than one
mechanism involved in the regulation of the crypt mitotic index and the
subsequent epithelial cell migration rate in GF and CV rodents.

3.2.1. Intestinal Histology and Cell Renewal Rate

Whereas the microflora has specific effects on the host, the magnitude of
these effects is not the same along the entire gastrointestinal tract. Some
microbial monoassociations produce no noticeable effects on intestinal
morphology, while others bring about drastic epithelial and subepithelial
alterations (Table II). Lactobacillus and Clostridia (type A *Clostridium
perfringens)* did not produce any effect on the intestinal mucosa of
monoassociated mice (Fry *et al.*, 1966). Monoassociation of mice with *E.
coli,* however, brought about enlargement of the villi and an increase in

lamina propria cellularity (Hershovic *et al.,* 1967). Furthermore, association of germfree animals with *E. coli* has been found to result in deeper crypt gland development and a decrease in mucus content, the total mucosal thickness remaining unchanged (Sprinz, 1962; Sprinz *et al.,* 1961). Contrary to these observations, other workers have found no differences in the intestinal morphology of GN mice monoassociated with *E. coli* (Syed *et al.,* 1970). Strain differences in *E. coli* may explain these conflicting observations.

Kenworthy and Allen (1966) found that piglets associated with *E. coli* or *Staphylococcus albus* exhibited an increase in mitotic index, deeper crypts, and smaller villi. Piglets monoassociated with *Staphylococcus* had better developed microvilli than do polyassociated animals. Staley *et al.* (1970a, 1970b) observed a smoother and more regularly developed mucosal surface in the large intestine of piglets associated with *E. coli* than in GF counterparts. Meslin *et al.* (1974) implanted 14 bacterial genera (none of which metabolizes bile salts) in GF rats and observed morphological changes in the small intestine, in addition to an increased epithelial cell migration rate, which were comparable to those seen in CV animals. Recently, Wells and Babish (1980) noted that rats monoassociated with *Fusobacterium necrophorum* VPI 6054A, *Bacteroides fragilis* or *Propionibacterium acnes* RC248-2 exhibited thickening of the *lamina propria* similar to that seen in CV rats. While most of the histological characteristics seen by these investigators were "normalized," there were still some apparent abnormalities.

Gordon and Bruckner-Kardoss (1958-1959) found that GN chickens associated with *Streptococcus faecalis* or *Clostridium perfringens* had elevated levels of ileal subepithelial lymphocytes, plasma cells and globule leukocytes similar to those of CV chicken (Table III). Knight and Wostmann (1964) found a different response in GN rats exposed to *Salmonella typhimurium*. During the period of monoassociation the ileal subepithelial lymphocyte counts and globule leukocytes decreased, and the subepithelial plasma cells showed a slight elevation but did not reach CV values. In addition to the intestinal microflora, dietary fiber and/or lactose have been shown to alter intestinal epithelial cell renewal (Komai and Kimura, 1980; Meslin *et al.,* 1981). The alterations in intestinal morphology which organisms elicit may affect absorptive capacity.

3.2.2. Cecal Reduction

The distended cecum of the GF rodent can be reduced to CV levels by introducing a CV microflora (Syed *et al.,* 1970) (Table IV). Apparently several microorganisms in synergistic association are required to eliminate the GF cecal anomalies (Sacquet *et al.,* 1972; Celesk *et al.,* 1976). Some investigators believe that only anaerobic bacterial populations can main-

Table II.   Morphological Characteristics of the Intestine Associated with
            Gnotobiotic and Conventional Animals (Mean ± S.D.)

	Chicken		Rat	
	GF	CV	GF	CV
Small Intestine				
Wt, g/100 g body wt.	15 ± 4	27 ± 9[a]	12 ± 2	13 ± .9[b]
Surface Area:				
Total, cm^2			466 ± 37	671 ± 35[j]
Villi length, microns				
No. epithelial cells				
Crypt length, no. epithelial cells				
Mucosal thickness, no. epithelial cells				
Lamina propria, microns				
Muscularis (cir), microns				
Muscularis (lon), microns				
Mitotic Index:				
Mitotic divisions/no.			12 ±.2	18 ± .8
epithelial cells counted			2.6	4.1
				4.1[j]
Thymidine labelled nuclei/ crypt/25 crypt sections				
Thymidine labelled cells/ villus/24 hr				
Transit time, % Y-91 left 24 hr after feeding				
Cecum				
Wt, g/100 g body wt.	2.0 ± .5	3.2 ± .8[a]	4.3 ± .6	2.1 ± .2[b]
Crypt length, microns				
Lamina propria, microns				
Muscularis (cir), microns				
Muscularis (lon), microns				
Mitotic index			4.2 ± .5	9.9 ± .7[m]
Transit time, % Y-91 left 24 hr after feeding				
Colon				
Wt, g/100 g body wt.	91	67[a]	3.3 ± .5	2.7 ± .2[b]
Transit time, % Y-91 left 24 hr after feeding				

[a]Reyniers et al. (11960)   [b]Gordon et al. (1966)   [c]Bruckner-Kardoss & Gordon (1958)
[g]Lesher et al. (1964)   [h]Khoury et al. (1969)   [i]Guenet et al. (1970)
[m]Sugiyama & Stewart (1975)

Mouse		Pig		
GF	CV	GF	CV	GN
26 ± .6	39 ± 6[c]	39 ± .7	35 ± .9[d]	
		32 ± .7	57 ± 13[e]	
36 ±.8	40 ± .6[l]	640 ± 118	473 ± 162[e]	132[f]
				148[f]
				149[f]
				128[f]
13 ± .1	17 ± .3[l]			43[f]
				42[f]
				36[f]
				34[f]
49 ± .9	58 ± .8[l]	81 ± 7	113 ± 58[e]	
		85 ± 24	168 ± 63[e]	
		62 ± 26	109 ± 31[e]	
		3.0-4.3		2.1[f]
				1.5[f]
				1.2[f]
				1.6[f]
8	13[g]			
9	31[h]			
6-11	0[k]			
7.8 ± .7	3.6 ± .8	2.2 ± .2	4.6 ± 1.0[e]	
		266 ± 29	437 ± 137[e]	
		27 ± 7	54 ± 28[e]	
		91 ± 10	118 ± 62[e]	
		54 ± 14	86 ± 26[e]	
17-31	.1[k]			
		14 ± 1	23 ± 5[e]	
4-8	0[k]			

[d]Waxler & Drees (1972) [e]Miniats & Valli (1973) [f]Kenworthy & Allen (1966)
[j]Gordon & Bruckner-Kardoss (1961) [k]Abrams & Bishop (1967) [l]Abrams et al. (1963)

Table III. Morphological Characteristics of the Ileum of Gnotobiotic and Conventional Animals

	GF	CV	GN-Salmonella typhimurium 3 days exposure	11 days exposure
			RAT[a]	
Epithelium				
Villus epithelium	288 ± 17	199 ± 22	103 ± 40	160 ± 25
Lieberkuhn epithelium	97 ± 14	124 ± 10	245 ± 23	230 ± 8
Goblet	34 ±	25 ± 4.7	2 ± 0.4	8 ± 1.0
Lymphocytes	2 ± 0.1	6 ± 1.1	0 ± 0.4	1
Schollen leucocytes	1 ± 0.1	1 ± 0.2	0	1
Lamina propria and submucosa				
Connective elements	151 ± 22	137 ± 7	206 ± 37	180 ± 7
Lymphocytes	13 ± 4.6	52 ± 7	8 ± 0.4	9 ± 0.1
Ring	3 ± 1.0	4 ± 0.2	9 ± 2.6	5 ± 2.5
Eosinophils	4 ± 1.3	10 ± 2.1	2 ± 0.4	4 ± 1.0
Mast	0	1 ± 0.2	0 ± 0.4	0
Plasma	1 ± 0.1	16 ± 7	3 ± 0.6	6 ± 1.0
Muscle	5 ± 2.3	6 ± 1.8	20 ± 5	14 ± 1
Muscle				
Circular	263 ± 19	251 ± 15	251 ± 14	230 ± 6
Longitudinal	142 ± 11	159 ± 17	152 ± 32	148 ± 7

	GF	CV	GN-Cl. perfringens	GN-Strep. faecalis
			CHICKEN[b]	
Epithelial tissue, total	360 ± 93	1250 ± 370	530 ± 22	380 ± 210
Lymphocytes	218 ± 81	641 ± 370	313 ± 23	178 ± 98
Schollen leucocytes	116 ± 40	535 ± 207	210 ± 13	183 ± 112
Monocytes	1(0-3)	3(0-8)	4(0-9)	3(0-14)
Macrophages	1(0-12)	2(0-14)	0	1(0-3)
Heterophils	10(0-42)	42(0-147)	0	1(0-2)
Basophils	9(0-31)	24(0-62)	9(4-16)	11(8-20)
Eosinophils	1(0-4)	1(0-3)	0	1(0-3)
Subepithelial tissue, total	570 ± 160	1610 ± 520	1470 ± 300	1080 ± 260
Lymphocytes	498 ± 143	1180 ± 489	1240 ± 217	826 ± 181
Plasma cells	26 ± 23	268 ± 98	165 ± 217	159 ± 65
Schollen leukocytes	14(0-40)	24(0-64)	24(0-42)	22(0-51)
Monocytes	6(0-18)	19(0-48)	20(8-34)	26(19-36)
Macrophages	9(0-44)	71(13-270)	3(0-9)	32(12-50)
Heterophils	11(0-40)	29(0-49)	4(0-9)	11(0-23)
Basophils	6(0-35)	7(0-23)	0	0
Eosinophils	4(0-15)	14(0-41)	8(0-25)	2(0-5)

[a]Distribution of cells (counts/10³) in the villi, submucosa and muscularis of 120-day-old rats (Mean ± S.D.). Knight and Wostmann (1964). [b]Total counts and differential counts of scattered reticulo-endothelial cells in mucosa and submucosa of lower ileum in GN and CV White Leghorn chickens 30-68 days. Cells/mm² tissue in 4 μ sections (Mean ± S.D. or range). Gordon and Bruckner-Kardoss (1958-59).

tain cecal size (Savage and McAllister, 1971), and that a direct correlation exists between cecal redox potential (Eh) and cecal size in mice (Celesk *et al.,* 1976). Skelly *et al.* (1962) found cecal reduction in GF rodents comparable to that observed in CV animals after monoassociation with *Clostridium difficile* or di-association with two Bacteroides strains. Attempts to reproduce these observations have been unsuccessful (Raibaud *et al.,* 1966). Clostridia and fusiform bacteria apparently play a role in eliminating the GF cecal anomalies (Itoh and Mitsuoka, 1980) but not

Table IV. Effect of Various Microbial Species on Cecal Weight in Gnotobiotic and Conventional Rats and Mice

GN Rats[a,b,d,g]			GN Mouse[c,d,e,f]		
Cecum, % body wt.	Microbial Species	Association weeks	Cecum, % body wt.	Microbial Species	Association weeks
10.6 ± 1.4	Germfree		11.0 ± 1.6	Germfree	
2.7 ± 1.1	Conventional		16.4 ± 3.3	Conventional	
10.4 ± 1.6	Staph. epidermidis	2-9	4.0	Cl. difficile	
7.4 ± 0.4	E. coli	3-5	10.1	Group K	
10.7 ± 1.5	Bact. fragilis	6		Streptococcus	
			10.5	Bact. fragilis	
11.3 ± 1.0	E. coli	6			
	Bact. fragilis				
11.9 ± 1.4	Bacillus macerans	4-12	4.2	Peptococcus	
	aerobic diptheroid		5.5	Clostridium and	
				Peptococcus	
10.9 ± 0.5	Germfree	1-2			
2.0 ± 0.1	Conventional	1-2	7.1 ± 2.2	Clostridium	
9.1	Strep. (strict	1-2	4.3 ± .9	Strep. faecalis +	
	anerobe)			L. brevis + Staph.	
14.4	Strep. (strict	1-2		epidermidis + Ent.	
	anerobe)			aerogenes + Bact.	
8.2 ± 0.5	Sphaerophrous	1-2		fragilis var. vulgatus	
10.0 ± 0.4	Butyr. bacterium	1-2		and Torulopris	
13.5	Catana-bacterium	1-2			
11.8	Catana-bacterium	1-2	2.0-2.5	Cl. difficile	0.5
16.1	Catana-bacterium	1-2	2.0-2.5	Bacteroides	1.3
12.3	Rami-bacterium	1-2	6.2-6.5	Bacteroides	0.5
8.9	Rami-bacterium	1-2	5.0-5.5	Bacteroides	0.5
12.9	Fusiformis	1-2			
7.5 ± 0.2	Cl. bifermentans	1-2			
11.0	Plectridium	1-2			
4.0 ± 0.4	Cl. difficule	1-2			
9.0 -11.0	Streptococcus	1-14			

[a] Bornside et al. (1976)   [b] Sacquet et al. (1973)   [c] Skelly et al. (1962)   [d] Asano, (1967)
[e] Loesche, (1969)   [f] Celesk et al. (1976)   [g] Hudson and Luckey, (1964)

as effectively as an association of several strict anaerobic species (Sacquet et al., 1972; Wells and Babish, 1980). Pleasants et al. (1981) reported that the cecal size of mice fed a chemically defined diet supplemented with cellulose (paper bedding) was reduced by approximately 40 percent. This implies that dietary factors as well as the presence of the microflora contribute to "normal" cecal function.

Hence, the impact of the microflora on the morphological characteristics of the intestinal tract is complex. Microbial species specificity is important in maintaining proper intestinal function and needs to be considered in referring to the host animal. The intestinal tract of GF animals reflects a lack of "physiological inflammation". The small intestine generally shows a decreased wet weight in the absence of the microflora and a reduced mitotic index with a concomitant decrease in epithelial cell transit time. These various conditions can be altered by the introduction of a gnotoflora. Complete remission of GF intestinal anomalies (including the enlarged cecum) requires the presence of several synergistic microorganisms, among which strict anaerobes are of primary importance.

Table V. Physical Characteristics of the Intestinal Contents of Germfree Animals (Mean ± S.D.)

	Chicken		Rat		Mouse
	GF	CV	GF	CV	GF
Small Intestinal Contents					
Wt, g/100 g body wt	1.79	1.79[a]			2.55 ± .39
Dry wt, %			24 ± 0.8	26 ± 0.7[o]	27 ± 2.0
Rel. viscosity					
Osmolality, mOsm/L			310 ± 10	369 ± 35[g]	
pH	6.3	6.5[i]			6.6-6.7
Reduction potential, mV					
Cecal Contents					
Wt, g/100 g body wt	0.34	.347[a]	3.6 ± 1.5	0.7 ± 0.2[c]	4.89 ± 1.11
			11.8 ± 1.9	0.6 ± 0.2[d]	10.4 ± 0.40
Dry wt, %	13 ± 0.8	26 ± 2.7[i]	21.9 ± 0.6	27 ± 0.3[a]	22 ± 1.8
	13 ± 2.2	19 ± 4.4[i]	15.3	21.9	
			16.1 ± 0.4	26.3 ± 0.5[o]	
Rel. viscosity (supernate)			2.5 ± 0.2	1.18 ± .05[o]	
			6.4 ± 1.0	3.3 ± 1.3	
Colloid Osm. pressure, mm Hg			107 ± 9	40 ± 8[g]	
Osmolality, mOsm/L			313 ± 1	475 ± 509[f]	310 ± 3
pH	6.4	6.5[i]	7.5 ± 0.1	6.6 ± 0.1[t]	7.5
			7.6 ± 0.1	6.9 ± 0.1[v]	7.0-7.1
Reduction potential, mV			+35 ± 22	-246 ± 24	+90
			+46 ± 8	-275 ± 12	+210
Colon Contents					
Wt, g/100 g body wt			.36 ± .17	.23 ± .14[c]	1.01 ± 0.24
Dry wt, %	16 ± 3.0	19 ± 2.9[i]	19.6 ± 0.8	39.2 ± 2.3[d]	22.8 ± 2.7
Rel. viscosity					
Colloid Osm pressure, mm Hg					
Osmolality, mOsm/L			279 ± 6	304 ± 27[g]	
pH	6.3	6.4	7.8 ± 0.2	7.0 ± 0.1[u]	6.8-6.9
Reduction potential, mV			+57 ± 6	-217 ± 15[u]	
Intracolonic pO_2, mm Hg			12.8 ± 0.8	11.1 ± 1.0[b]	
pCO_2, mm Hg			54.6 ± 0.9	83.4 ± 11.5[b]	

[a]Gordon and Bruckner-Kardoss (1961) [b]Bornside et al. (1976) [c]Gordon et al. (1966)
[g]Bruckner-Kardoss and Gordon (unpublished 1958) [h]Asano (1969b) [i]Reyniers et al. (1960)
[m]Bruckner (unpublished 1976) [n]Celesk et al. (1976) [o]Wostmann and Bruckner-Kardoss (1959)
[t]Wostmann and Bruckner-Kardoss (1966) [u]Heneghan and Gordon (unpublished 1974)

4. Physical Characteristics of the Intestinal Contents of Gnotobiotic Animals

The presence of microorganisms in the GI tract alters not only the morphological and biochemical characteristics of the intestinal milieu, but also the physical properties of the gut contents. This alteration can be detrimental or beneficial to the host animal.

Mouse	Pig		Dog		Lamb	
CV	GF	CV	GF	CV	GF	CV
.91 ± .30[r]			3.4	1.7[u]	6.7	6.7[m]
27 ± 1.4[r]					1.7	1.7[m]
			164	189[u]	288	396[m]
6.4-6.7[s]	7.3	7.4[e]				
			+156	−177[u]		
.53 ± 0.12[g]	.15	.40				
1.59 ± 2.26[h]						
32[l] ± 2.7[g]	12 ± 1.3	13 ± 4.7[p]	22 ± 3.4	31.2[j]	19.4	17.1[m]
	3.48 ± .022[b]	2.50 ± 0.02[l]	7.46	1.8[j]		
					3.1[c]	1.7[m]
			99	10[j]		
392-470[h]	234 ± 13	505 ± 50[l]	296 ± 52	418[j]		
	282 ± 19	335 ± 54[p]			251	301[m]
7.2[n]	7.4 ± 0.3	7.0 ± 0.2[p]	7.5	6.8[u]		
6.4-6.5[s]	8.1	6.7[e]				
−290[k]						
−200[m]						
.54 ± 0.24[s]						
30.8 ± 3	9.0 ± 1.3	25.3 ± 6.2[e]	25.0	31.5[j]		
	12.4 ± 1.1	25.4 ± 6.6[p]				
	5.94 ± 0.84	2.98 ± 0.04[l]	8.0 ± 1.0	4.0 ± 1.1[j]	20.2	43.1[m]
	85 ± 10	39 ± 8[l]	118 ± 10	26 ± 16[j]		
	301 ± 10	519 ± 37[l]	249 ± 58	482 ± 58[j]		
	282 ± 31	355 ± 233	274	402[u]	515	993[m]
6.4-6.7	6.4 ± 0.4	7.2 ± 0.2	7.2	5.5[u]		
	7.1	6.5	+63	−104[u]		

---

[d]Bruckner and Gordon (unpublished 1977)   [e]Miniats and Valli (1973)   [f]Asano (1969)
[j]Heneghan and Gordon (unpublished 1973)   [k]Yoshida (1972)   [l]Miniats and Gordon (unpublished 1974)
[p]Cherian et al. (1979)   [q]Gordon and Wostmann (1973)   [r]Combe et al. (1965)   [s]Gordon (1968)
[v]Eyssen et al. (1972)

4.1. Dry Weight Percent

The presence of the microflora alters, through direct and indirect mechanisms, the amount of water present in the intestinal contents. The microflora do not seem to affect the movement of water in the small intestine. The dry weight percent of the small intestinal contents has been found to be similar in the rat, mouse, and lamb (Wostmann and Bruckner-Kardoss, 1959; Combe et al., 1965; Bruckner et al., 1984). The percent dry matter in the cecal contents of the CF rat, mouse, dog, and chicken is significantly decreased compared to their CV counterparts (Asano, 1969b; Bruckner-Kardoss and Gordon, 1958, unpublished; Heneghan and Gordon, 1973, unpublished; Reyniers *et al.*, 1949; Wostmann and Bruckner-Kardoss, 1959). No difference was observed in the pig or the lamb (Cherian *et al.*, 1979; Bruckner *et al.*, 1984). In all species studied (chicken, rat, mouse, pig, dog, and lamb) the colon contents showed a decreased percent dry matter in the GF state (Bruckner *et al.*, 1984; Bruckner-Kardoss and Gordon, 1958, unpublished; Miniats and Valli, 1973; Bruckner *et al.*, 1984; Heneghan and Gordon, 1973, unpublished; Reyniers *et al.*, 1960). A general correlation exists between the regions of the intestinal tract in CV animals where microbial proliferation is the greatest and the sites of anomalies in the GF condition (Tables V, VI, and VII).

4.2. Relative Viscosity (RV)

The relative viscosity (RV) of the supernates of the cecal and colon contents of GF animals is generally increased (Bruckner *et al.*, 1984; Heneghan and Gordon, 1973, unpublished; Miniats and Gordon, 1974, unpublished). Gordon and Wostmann (1973) showed that the increase in RV is correlated with elevated levels of mucopolysaccharides. This acidic undegraded mucus is implicated in the binding of Cl⁻ and in the increased colloid osmotic pressure of the intestinal contents of CF animals.

4.3. Colloid Osmotic Pressure (COP) and Total Osmolality

Undegraded mucinous material contributes to the COP (water retention force) of the intestinal contents. Other substances originating from dietary as well as endogenous sources can add to total COP. The microflora is directly responsible for the decreased COP of CV lower bowel contents. This difference between GF and CV animals, which in some cases may be fourfold, has been observed in the rat, dog, and pig (Gordon and Wostmann, 1973; Heneghan and Gordon, 1973, unpublished; Gordon and Miniats, 1974).

Supernates from the lower bowel contents of the CV rat, mouse, pig, dog, and lamb have osmolality values significantly greater than their GF counterparts (Asano, 1969a, 1969b; Cherian *et al.*, 1979; Heneghan and Gordon, 1973, unpublished; Bruckner *et al.*, 1984). The lack of mucinases in the GF lower bowel contents (Hoskins, 1968) may result in the accumulation of large mucopolysaccharides and thereby an increased COP. The lower levels of certain electrolytes (Cl⁻, HCO₃⁻) in the GF cecal contents also may contribute to a change in osmolality.

4.4. pH

The pH of the small intestinal contents is unchanged in several species of GF animals (Reyniers *et al.*, 1960; Gordon, 1968). However, the lower bowel contents of GF animals are slightly more alkaline (Wostmann and Bruckner-Kardoss, 1959; Celesk *et al.*, 1976; Cherian *et al.*, 1979; Heneghan and Gordon, 1974). The chicken seems to be an exception (Reyniers *et al.*, 1960). The lower pH value noted for CV animals may be due to the presence of lactic acid-producing bacteria in the lower gastrointestinal tract.

4.5. Redox Potential (Eh)

The presence of a gastrointestinal microflora comprised mainly of strict anaerobes brings about a change from a positive Eh in GF animals to the negative Eh observed in CV animals (Wostmann and Bruckner-Kardoss, 1959; Yoshida, 1972; Celesk *et al.*, 1976; Heneghan and Gordon, 1974). Yoshida (1972) has shown that *E. coli* or Pseudomonas in monoassociation with a GN rodent can induce a change from positive to negative Eh values in the lower bowel contents. Celesk *et al.* (1976) found similar results after introducing a pentaflora to GF rats (Table VII). Eh values have been positively correlated to a rodent's cecal size, i.e., the more negative the value the smaller the size of the cecum, expressed as percent body weight (Celesk *et al.*, 1976).

Hence, the differences between the physiological characteristics of the intestinal contents of GF and CV animals seem to be related to the net water loss of the host in the absence of the microflora. The microflora are responsible directly or indirectly for the increase in colloid osmotic pressure and decrease in osmolality observed in GF intestinal contents. The correlation between Eh and cecal size in rodents emphasizes again the importance of the anaerobic microflora in maintaining normal gastrointestinal absorptive function.

5.   Biochemical Characteristics of the Intestinal Contents of Gnotobiotic Animals

The biochemical characteristics of the intestinal contents reflect the "absorptive status" of the host. By comparing the gut contents of GF and CV animals, it is possible to identify the end products associated with bacterial degradation of endogenous materials (enzymes secreted by the host, sloughed epithelial cells, etc.), to determine the extent of bacterial

Table VI. Biochemical Characteristics of the Intestinal Content of Germfree Animals (Mean ± S.D.)

	Rat		Mouse	
	GF	CV	GF	CV
Small Intestine				
Na$^+$, mEq/L				
K$^+$, mEq/L				
Cl$^-$, mEq/L				
Protein, mg/g dry matter	129 ± 3.8	90 ± 4.7[1]		
Hexosamines, mg/g dry matter	8.0 ± 1.2	7.5 ± 1.8[1]		
Glucuronic acid, mg/g dry matter	8.7 ± 0.4	5.7 ± 1.1[1]		
Uric acid, mg/g wet wt/100 g body wt	.73	1.23[j]		
Cecum				
Na$^+$, mEq/L	45 ± 1.8	34-40[c]	57 ± 2	76-145[h]
K$^+$, mEq/L	10 ± .4	11-14[c]	13 ± 0.2	14.7-22.4[h]
Cl$^-$, mEq/L	1.74 ± .2	11. ± 13[c]	3.7 ± 0.2	16.5-27.4[h]
HCO$_3$	8.0	30[n]		
Nitrogen, mg/100 g body wt	77 ± 30	15 ± 6[a]		
Protein, mg/g dry wt	206 ± 12	83 ± 6[1]		
Protein, mg/g wet wt	15 ± 1.3	24 ± 3[1]	37 ± 6.1	25 ± 5.7[b]
Carbohydrates, mg/g wet wt	13 ± 1.3	1.2 ± .2[1]	19 ± 1.0	5.1 ± 0.8[b]
"Mucinous material", mg protein	134 ± 11	11 ± 1[b]	85 ± 7	10 ± 6[b]
Hexosamines, mg/100 g body wt	97 ± 48	2.9 ± 1.0[a]		
Hexosamines, mg/g dry matter	15 ± 1.5	1.4 ± 0.2[1]		
Glucuronic acid, mg/g dry matter	16 ± 1.9	1.6 ± 0.3[1]		
Uric acid, mg/wet wt/100 g body wt	.37	0[j]		
Urea, mg/100 g wet wt	20	0[m,o]		
NH$_3$, mg/100 g wet wt	2.3	22[m,o]		
Colon				
Na$^+$, mEq/L				
K$^+$, mEq/L				
Cl$^-$, mEq/L				
Nitrogen, mg/day	41	32[k]		
Protein, mg/g dry matter	158 ± 15	45 ± 6[k],[1]		
Fat, mg/day	19	37[k]		
Hexosamines, mg/g dry matter	14 ± .7	1.7 ± 0.2[1]		
Glucuronic acid, mg/g dry matter	12 ± 1.1	1.1 ± 0.2[1]		
Uric acid, mg/g wet wt/100 g body wt	.32	0[j]		

[a]Lindstedt et al. (1965)   [b]Loesche (1968)   [c]Asano (1969b)
[f]Gordon (1967)   [g]Heneghan and Gordon (unpublished 1973)   [h]Asano (1969a)
[1]Bruckner and Gordon (unpublished 1977)   [m]Combe et al. (1965)

degradation of exogenous nutrients, and whether the end products of these degradative processes influence the nutritional status of the host by altering absorptive capacity or by affecting nutrient availability. Tables VI and VII contain information on the effects of the intestinal microflora on protein, carbohydrate, fat, and electrolyte absorption.

Pig GF	Pig CV	Dog GF	Dog CV	Lamb GF	Lamb CV
				91(69-128)	77(29-143)
	223 ± 133[l]			22(10-30)	28(23-31)
	9.2 ± 4.8[l]			97(20-112)	61(30-82)
	1.7 ± 0.5[l]				
76 ± 3	71 ± 10[e]	75 ± 30	68[g]	29(25-31)	121(75-140)
46 ± 10	64 ± 10[d]				
11 ± 1	39 ± 13[e]	3.8 ± 0.8	12[g]	21(9-46)	28(27-30)
8.6 ± 2.8	32 ± 8[d]				
14 ± 8	35 ± 17[e]	9.3 ± 2.9	27[g]	36(22-36)	71(58-88)
11 ± 3	31 ± 3[d]				
287 ± 20	131 ± 21[l]				
66 ± 0.1	4.3 ± 0.7[e]				
9.9 ± 0.9	1.2 ± 0.4[e]				
65 ± 5	86 ± 25[l]	58 ± 12	49 ± 15[g]	137(119-125)	193(140-240)
51 ± 8	35 ± 10[d]				
3.5 ± 2.3	70 ± 38[e]	6.1 ± 1.2	9.3 ± 2.3[g]	67(22-52)	67(10-140)
8.2 ± 1.8	40 ± 5[d]				
47 ± 23	40 ± 34[e]	6.3 ± 2.2	14 ± 3.7[g]	152(140-162)	141(130-160)
9.7 ± 0.2	25 ± 5[d]				
225 ± 15	79 ± 22[l]				
63 ± 7	3.6 ± 1.6[e]				
9.8 ± 1.6	1.7 ± 0.7[e]				

[d]Miniats and Gordon (unpublished 1974)   [e]Cherian et al. (1979)
[i]Bruckner (1975)   [j]Cole and Wiseman (1971)   [k]Luckey (1963)
[n]Gordon and Wostmann (1973)   [o]Ducluzeau et al. (1966)

6.   Water and Electrolyte Transport

In the digestive tract an immense surface area is in contact with the food, microflora, etc., and water flow, absorption and elimination are bidirectional. The balance of these processes determines whether there is net

Table VII. Biochemical and Physical Characteristics of the Intestinal Contents of

	Rat	
	GN	Microbial Species
Small Intestinal Contents		
Dry wt, %	24.2	Cl. difficile[d]
Na, mM	124 ± 5	Cl. difficile[d]
K, mM	9 ± .4	Cl. difficile[d]
Cl, mM	4.1 ± .4	Cl. difficile[d]
Uric acid, mg/g wet wt/100 g body wt		
Cecal Contents		
Dry wt, %	13.6 ± .6	L-acidophilus[a]
	14.5	Cl. difficile[d]
	16.9	Peptococcus & Clostridium
Na, mM	88 ± 6	Cl. difficile[d]
K, mM	13.5 ± .4	Cl. difficile[d]
Cl, mM	13.8 ± 2.5	Cl. difficile[d]
Osmolality, mOsmol/L	321 ± 4	Cl. difficile[d]
CO_2, mM	8.6 ± .3	Germfree[a]
	4.9 ± .6	Cl. difficile[d]
Uric acid, mg/g wet wt/100 g body wt	40 ± 6	Salivarius[h]
Urea and ammonia, mg/100 g	0 – 18	L. salivarius[h]
	8 – 20	L. fermenti[h]
	44 ± 8	L. acidophilus & L. salivarius[h]
	40 ± 11	L. acidophilus & L. fermenti[h]
	0 – 25	L. acidophilus & L. fermenti & L. salivarius[h]
	0 – 32	Staph. pyogenes[h]
	0 – 25	Actinobacillus[h]
	33 ± 18	Proteus morganii[h]
	41 ± 7	Bifidobacterium[h]
	41 ± 7	Streptococcus. strict anaerobe
	36 ± 22	Veillonella
	34 ± 23	Bifidobacterium, Streptococcus, Eubacterium, Endosporus, Acuformis, Clostridium, Veillonella[h]
Protein & CHO, mg/100 g body wt		
Reduction potential, mV	-185 ± 14	Eubacterium & Clostridium[j]
pH	6.8 ± 0.1	Eubacterium & Clostridium[j]
Colon Contents		
Uric acid, mg/g wet contents/ 100 g body wt		
pO_2 and pCO_2, mm Hg	13.1 ± 2.0 48.1 ± 2.4	Staph, epidermidis[b]
	14.9 ± 1.3 63.1 ± 3.9	E. coli[b]
	9.3 ± 0.8 55.8 ± 1.4	Bacteroides fragilis[b]
	19.4 ± 0.8 48.7 ± 1.5	E. coli & B. fragilis[b]
	9.5 ± 1.4 47.7 ± 1.1	Aerobic diptheroid & Bacillus macerans[b]
pH	6.8 ± 0.1	Eubacterium & Clostridium[j]
Reduction potential, mV	-178 ± 10	Eubacterium & Clostridium

[a]Luckey (1963) [b]Bornside et al. (1976) [c]Cole and Wiseman (1971) [d]Asano (1967)
[h]Ducluzeau et al. (1966) [i]Asano (1969a) [j]Eyssen et al. (1972)

Gnotobiotic and Conventional Rats and Mice   (Mean ± S.E. or Range)

Mouse		
GN		Microbial Species

.63		Aerobacter[c]
22.0		Cl. difficile[i]
14.0		Cl. difficile[i]
2-7		Germfree[i]
13.0		Cl. difficile[i]
280 ± 46	288 ± 31	Germfree[f]
25 ± 6	10 ± 2	Conventional[f]
345 ± 34	500 ± 15	Bacteroides fragilis[f]
546 ± 72	698 ± 86	Streptococcus[f]
285 ± 62		Peptococcus[f]
274 ± 54	201 ± 49	Clostridium[f]
232 ± 16	184 ± 11	Peptococcus & Clostridium[f]
-270		E. coli[e]
-170		Pseudomonas[e]
0 to +10		Clostridium[g]
-210		Strept. faecalis ± L. brevis + Staph. epidermidis + Enterobacter aerogenes + Bacteroides fragilis[g]
7.8		Torulopsis sp.[g]
7.2		Clostridium sp.[g]
0		Aerobacter[c]

[e]Yoshida (1972) [f]Loesche (1969) [g]Celesk et al. (1976)

water absorption or excretion. Disruption of the normal indigenous micro-flora can give rise to electrolyte and water imbalances and result in diar-rhea. Invasion by nonindigenous (allochthonous) microorganisms can result in damage to the small and large intestinal mucosa and/or the production of enterotoxins. Lesions of the intestinal epithelium can be precipitated by Shigella, Staphylococci, Clostridia, certain *E. coli* strains and Salmonellae (Staley *et al.,* 1969; Hampton and Rosario 1965; Bergdoll, 1967; Duncan and Strong, 1969; Keusch *et al.,* 1970; Gyles and Barnum, 1969; Smith and Halls, 1967). Enterotoxins apparently do not cause structural altera-tions in the intestinal epithelial cells (Gangarosa *et al.,* 1960), but they influence water and electrolyte secretion with diarrhea a possible result. Information on the effect of enterotoxins on intestinal function is given in the excellent reviews of Binder and Powell (1970) and Banwell and Sherr (1973).

Judging from the percent dry matter of the intestinal contents, water absorption is probably similar in the small intestine of conventional and germfree mice (Combe *et al.,* 1965) (Table VIII). Csaky (1968) found that in germfree rats there was an initial lag in water absorption from the perfused ileum, but the ultimate rate was not different in CV and GF animals. He suggested that in the small intestinal contents of GF animals a substance may be present which, when washed out, results in normal water absorption. In conventional life, intestinal bacteria may destroy the substance inhibiting water absorption. Recently, Andrieux *et al.* (1979) have noted a decrease in the apparent absorption of sodium and sodium potassium from the small intestine of GF rats.

In the large intestine sodium and water absorption are better under-stood in the GF rodent because the attention of investigators has been focused on the possible role of Cl⁻ in the development of cecal distention and chronic diarrhea. According to Asano (1967), relative anion defi-ciency, especially a low Cl⁻ level (Table VI), is an important characteristic of GF cecal contents. Feeding a chloride-yielding resin to GF animals increased chloride levels in the gut contents and considerably improved water absorption from the lower bowel (Asano, 1969a, 1969b) (Table VIII). Combe *et al.* (1965) found that the sodium concentration in the cecal contents of GF rats was quite similar to that of CV animals.

The threshold concentrations of NaCl necessary for the maintenance of solute-coupled water absorption from the cecum of germfree and con-ventional rats were investigated by Nakamura and Gordon (1973). These authors found that in the cecal wall of the GF rat there is no inherent defect in the transport of water, sodium, or chloride (Table IX); in fact, the GF cecum proved more efficient in this regard than its conventional counterpart. After replacing the cecal contents of GF and CV rats with saline *in vivo,* Loeschke and Gordon (1970) found that both sodium transport and water absorption were greater in the germfree cecum.

Osmotic permeability, on the other hand, was not different (Table X, Fig. 1).

Mucus and related substances are present at elevated levels in germfree intestinal contents (Combe et al., 1965; Csaky, 1968; Lindstedt et al., 1965; Loesche, 1968) and may be involved in the impaired water absorption from the GF lower bowel. Gordon and Nakamura (1969) found that in germfree rat cecal contents colloid osmotic pressure was 60 to 70 mm Hg higher than in blood plasma, whereas in conventional animals the values were similar. This excess of non-absorbable macromolecules in the GF cecum and colon might explain in part the impairment in water absorption.

To test the hypothesis that some of these characteristics of GF rats are due to the presence of large molecules, Loeschke et al. (1971) added non-absorbable polyethyleneglycol (PEG, mol. wt. 4000) to the drinking water of conventionally raised adult rats. Two- to five-fold increases were noted in cecal dry weight, gross surface area and net sodium-coupled water transport. Similarly, the transmural potential difference and steady state concentrations of sodium, chloride, and bicarbonate (but not of potassium) were significantly altered. These observations are compatible with the concept of enhanced active sodium transport.

Macromolecules derived from intestinal secretions and sloughed epithelial cells which are not degraded by microflora may accumulate in GF cecal contents. These large polyvalent cations may bind available anions and thereby result in a low diffusible anion concentration. Low chloride levels in the cecal contents may depress sodium and water transport because development of an appropriate electrochemical balance on both sides of the absorptive membrane presupposes that active cation transport is followed by an equivalent amount of anions (Visscher, 1957). Loss of water and electrolytes in the GF animal under such conditions may be partially compensated by an increased water intake (Gordon, 1968), increased absorption, as shown by more rapid transport of water from the cecal lumen (Loeschke and Gordon, 1970), an increase in mucosal ATPase, which is indicative of "active transport status" (Simonetta et al., 1975), and by an apparent involvement of macromolecules in stimulating mucosal ATPase activity (Loeschke et al., 1971, 1974). It is generally accepted that absorption of electrolytes, water and other nutrients is affected by the regional blood flow to the various intestinal segments. In GF rats blood flow has been found to be 20-45% less than in CV animals (Gordon, 1968). Various bioactive substances which are normally degraded by microbes have been implicated in this anomaly (Gordon and Pesti, 1971). This reduction in regional blood flow is also a probable factor in electrolyte and water imbalances.

It is generally accepted that normal intestinal digestion and absorption depend on the presence of a specific indigenous gastrointestinal

Table VIII. Germfree Mice (A) and Germfree Mice Fed Cl-Resin Diet (B)   (Mean ± S.E.[a])

	Cecal wt % body wt	% Water in cecal contents	% Water in cecal wall	Osmolarity (mOsm/L)	Na (mM)	K (mM)	Cl (mM)	CO_2 (mM)
A	10.4 ± .4	86.7 ± 1.4	75.0 ± .4	310 ± 3	56.9 ± 1.9	13.0 ± .8	3.6 ± .2	2.8 ± .3
B	5.66 ± .9	67.1 ± .9	67.2 ± .8	268 ± 6	60.7 ± 3.4	18.7 ± .9	12.7 ± .5	- - -

[a] Asano (1969a)

Table IX. Net Absorption of Water, Na$^+$ and Cl$^-$ from Ligated Cecal Pouches Containing NaCl[a]

Group	Initial NaCl conc. (mM)[b]	Absorption/hr/cm^2 mucosal surface area		
		Water (μL)	Na$^+$ (μM)	Cl$^-$ (μM)
Germfree (8 rats)	0	-5.7 ± 2.5[c]	-3.6 ± 0.4	-0.7 ± 0.1
	15	6.5 ± 1.2	0.3 ± 0.1	2.0 ± 0.3
	40	11.0 ± 1.6	1.8 ± 0.4	4.5 ± 0.9
	77	37.6 ± 4.7	5.6 ± 0.5	9.5 ± 1.5
	155	42.6 ± 5.5	6.2 ± 0.8	12.2 ± 2.2
Conventional (9 rats)	0	-7.5 ± 0.9	-3.3 ± 0.2	-0.9 ± 0.1
	40	-2.7 ± 1.2	-0.0 ± 0.1	2.9 ± 0.5
	77	10.4 ± 2.6	1.4 ± 0.5	5.1 ± 0.6
	155	19.0 ± 2.7	2.8 ± 0.7	6.9 ± 0.7

[a]Nakamura and Gordon (1973) [b]All cecal solutions were adjusted to 300 mOsm by adding mannitol, when applicable. [c]Negative numbers indicate efflux from tissue to lumen.

Table X. Osmotic Water Flow and Osmotic Water Permeability Determined in Ceca of Germfree and Conventional Rats [a,b]

	Osmotic water flow (ml/2 hr)	Osmotic water permeability	
		(μL/cm^2 x hr x mOsm)	(μL/0.1g x hr x mOsm)
Hypotonic perfusion			
Germfree	-4.6 ± .9	0.3 ± .2	6.1 ± 1.9
Conventional	-2.9 ± .8	0.6 ± .0	9.8 ± 2.5
Hypertonic perfusion			
Germfree	+3.1 ± 1.7	0.3 ± .1	4.3 ± 2.3
Conventional	+1.4 ± .7	0.3 ± 0.2	4.1 ± 2.1

[a]- = loss; + = gain of luminal fluid. Three rats per group. Means ± S.D. The osmotic permeability coefficients were calculated as PoA = Q/A. ΔmOsm, where Q is the rate of osmotic water flow, A the surface area (or dry wt) of the cecal sac, and ΔmOsm the mean osmolar concentration difference between lumen and plasma. [b]Loeschke and Gordon (1970)

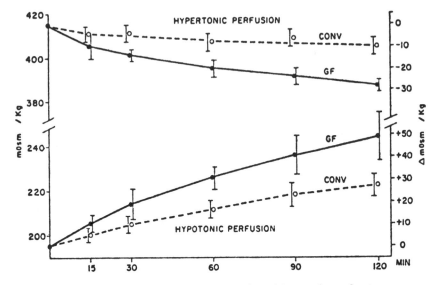

Fig. 1. Osmolality change of hypertonic and hypotonic perfusates
in ceca of germfree (GF) and conventional (CONV) rats. Mean ± SD.

microflora. The absence of this microflora disrupts the normal course of
water and electrolyte turnover and gives rise to chronic mild diarrhea. In
GF animals disturbances in water and electrolyte absorption and secretion
are, for the most part, limited to the lower bowel. Low anion levels,
increased mucopolysaccharides and elevated colloid osmotic pressure in
the lower bowel contents, along with reduced intestinal blood flow, are
involved in the development of diarrhea. To compensate for this excess
water loss, the GF animal drinks more water and develops higher mucosal
ATPase levels in the cecum to increase the absorptive efficiency of this
organ.

7. Minerals: Absorption and Utilization in Gnotobiotic Animals

7.1. Iron

Absorption of iron is regulated by the amount of total body iron and can
be modified by various intraluminal factors, such as pH and the intestinal
microflora. Excellent discussions pertaining to iron absorption are given by
Callender (1974), Wiseman (1964) and Soergel and Hofmann (1972).

The presence of the microflora influences the absorption, retention
and distribution of iron in the host animal (Table XI). Supplementation of
the diet with Aureomycin has been reported to improve iron utilization in
CV pigs (Kirchgessner, 1965), whereas supplementation with penicillin,

Table XI. Iron Content of Tissues and Total Body of Adult Germfree and Conventional Rats[a] (Mean ± S.E.M.)

	Germfree		Conventional		P values	
	Male	Female	Male	Female	Status	Sex
Liver						
mg/100 g fresh tissue	21.7±1.27	66.2±1.46	16.8±0.32	41.7±1.63	<0.005	<0.005
mg/liver	2.21±0.10	4.29±0.08	2.04±0.05	3.36±0.10	<0.005	<0.005
Spleen						
mg/100 g fresh tissue	51.7±2.90	79.8±3.19	162 ±4.2	240±12.6	<0.005	<0.005
mg/spleen	0.32±0.018	0.37±0.009	1.25±0.03	1.20±0.04	<0.005	NS
Kidney						
mg/100 g fresh tissue	8.14±0.17	9.30±0.21	10.5±0.35	15.3±0.50	<0.005	<0.005
mg/kidney	0.13±0.002	0.12±0.003	0.22±0.007	0.22±0.011	<0.005	NS
Total body						
mg/100 g fresh wt	3.20±0.09	6.47±0.10	4.37±0.08	8.15±0.12	<0.005	<0.005
mg/total body	8.52±0.20	10.4±0.16	12.5±0.27	15.8±0.32	<0.005	<0.005
Plasma iron, µg/100 ml	145±1.4	283±4.4	154±1.7	293±2.7	<0.005	<0.005
Plasma copper, µg/100 ml	134±1.3	136±1.4	141±1.2	142±1.4	<0.005	NS
Retention, µg/day/100 g body wt	66.8±13.6		102.4±5.5			
Retention, % intake	14.4±2.83		22.3±0.97			

[a]Reddy et al. (1965a) [b]Reddy et al. (1972)

Terramycin or neomycin has been found to depress iron absorption (Stern *et al.*, 1954; Forrester *et al.*, 1962). Concomitant feeding of the host with an antibiotic resistant strain of *E. coli* restored iron absorption to pretreatment values (Stern *et al.*, 1954). Iron absorption and retention in rats seem to be reduced in the absence of the microflora (Reddy *et al.*, 1972; Andrieux *et al.*, 1980). However, Geever *et al.* (1968) found that following oral administration of ^{59}Fe, the GF rat retained more of the isotope than did its CV counterpart. Plasma, spleen and kidney levels of iron were lower, and liver levels higher, in GF rats (Reddy *et al.*, 1965). In GF mice iron levels were lower in the plasma, spleen, kidney, and liver than in CV mice (Donati *et al.*, 1969). The finding that body iron was reduced by 23% in GF rats (Reddy *et al.*, 1965) indicates that absorption is depressed and/or elimination of iron is enhanced in the absence of the microflora. Not only the absence of microbes in the GF animal but the mode of diet sterilization has been shown to influence iron utilization (Andrieux *et al.*, 1979). Data for the iron content of the tissues of GF and CV rats are given in Table XI.

7.2. Copper

The mechanism of copper absorption is poorly understood, but a copper binding protein in the intestinal epithelium is believed to be involved in controlling copper absorption (Evans and LeBlanc, 1976). Data on the effect of the microflora on copper utilization are conflicting.

The absorption of copper was found by Reddy *et al.* (1965) to be reduced in the GF rat (Table XII). However, after oral administration of a dose of ^{64}Cu in the acetate form, Dowdy *et al.* (1969) found an increased absorption and reduced loss of this isotope in the feces and urine of the GF rat. This finding is supported by the observations of Andrieux *et al.* (1979, 1980) who noted a higher apparent absorption and retention of copper in GF rats. The rabbit showed no difference in copper utilization in the presence or absence of a microflora (Reddy *et al.*, 1972). GF rats had lower plasma, kidney, and spleen concentrations of copper than CV rats, whereas the copper concentration in the liver was elevated (Reddy *et al.*, 1968). Oral administration of Aureomycin to CV pigs increased copper retention (Kirchgessner, 1965).

7.3. Manganese

Reddy *et al.* (1965) found that the concentration of manganese in the liver, spleen, kidney, and in the total body were similar in GF and CV rats (Table XII). In contrast, Andrieux *et al.* (1979) found an increased absorption and retention of manganese in GF rats; there was also an

Table XII. Copper and Manganese Content of Tissues and Total Body of Adult Germfree and Conventional Rats[a] (Mean ± S.E.M.)

	Germfree		Conventional		P values	
	Male	Female	Male	Female	Status	Sex
Manganese						
Liver						
µg/100 g fresh tissue	237±11	239±9	227±8	242±12	NS	NS
µg/liver	22.3±1.40	15.2±0.81	27.4±0.88	19.7±0.93	<0.005	<0.005
Spleen						
µg/100 g fresh tissue	34.3±2.86	28.8±1.43	32.8±1.36	35.9±2.33	NS	NS
µg/spleen	0.21±0.016	0.13±0.006	0.26±0.017	0.19±0.012	<0.005	<0.005
Kidney						
µg/100 g fresh tissue	117±5	107±2	109±3	128±8	NS	NS
µg/kidney	1.80±0.09	1.34±0.03	2.23±0.08	1.79±0.11	<0.005	<0.005
Total body						
µg/100 g fresh wt	90±4	108±4	88±4	98±3	NS	<0.005
µg/total body	230±7	175±7	248±10	184±6	NS	<0.005
Copper						
Liver						
µg/100 g fresh tissue	1094±35	1374±84	832±28	891±44	<0.005	<0.005
µg/liver	102±4.9	88±6.49	101±3.4	73±4.3	NS	<0.005
Spleen						
µg/100 g fresh tissue	493±24	516±27	700±16	743±34	<0.005	NS
µg/spleen	3.01±0.15	2.41±0.12	5.34±0.43	3.89±0.15	<0.005	<0.005
Kidney						
µg/100 g fresh tissue	986±52	1147±43	1491±48	1767±70	<0.005	<0.005
µg/kidney	15.1±0.66	14.4±0.66	30.5±1.21	24.7±1.25	<0.005	<0.005
Total body						
µg/100 g fresh wt	194±5	223±7	270±6	295±12	<0.005	<0.005
µg/total body	501±8	360±8	762±22	552±17	<0.005	<0.005

[a]Reddy et al. (1965a).

effect of the mode of diet sterilization. As with iron and copper, an increase in manganese retention was observed in CV pigs treated with Aureomycin (Kirchgessner, 1968).

7.4. Zinc

Reddy *et al.* (1972) observed that zinc absorption was similar in GF and CV rats. However, Andrieux *et al.* (1980) noted increased absorption and retention of zinc in GF rats. This increase was not modified by the mode of sterilization (irradiation or steam) but was apparently decreased by including lactose in the diet. Giddens *et al.* (1972) observed esophageal and gastric parakeratosis in GF rats despite feeding a diet that contained a normal dietary zinc level. In contrast, Smith *et al.* (1972) found that GF animals had a lower zinc requirement. Alterations of the intestinal microflora caused by oral antibiotic administration appear to improve zinc retention (Kirchgessner, 1965; Pepper *et al.*, 1952).

7.5. Calcium

Gustafsson and Norman (1962) were the first to report differences in calcium metabolism between GF and CV animals. They observed that GF rats were more inclined to urinary calculus formation. Many authors have found that the absorption and retention of calcium are greater in the GF animal (Yoshida *et al.*, 1969; Wostmann, 1975; Yoshida *et al.*, 1968; Reddy *et al.*, 1969b; Reddy, 1971; Reddy, 1972; Reddy and Pollard, 1972; Garnier and Sacquet, 1969; Edwards and Boyd, 1963) (Table XIII). Reddy (1972) has attributed this increase to the enhanced Ca^{++} – ATPase, alkaline phosphatase and calcium binding protein activity which he observed in intestinal brush border preparations. The increased Ca absorption in GF rats is not altered by the mode of sterilization or by high levels of dietary lactose (Andrieux *et al.*, 1980). In the GF chick, calcium is apparently more efficiently removed from the enterocyte following absorption (Palmer and Rolls, 1981). Other factors also may play a role in the elevated calcium absorption in GF animals: increased fat absorption (Luckey, 1963; Evrard *et al.*, 1964; Yoshida *et al.*, 1968; Boyd and Edwards, 1967); elevated amino acid absorption (Kelleher and Bruckner, 1977; Phillips and Newcomb, 1968); the presence of conjugated bile acids (Kellog and Wostmann, 1969; Gustafsson *et al.*, 1957); low anion levels (Asano, 1967, 1969a, 1969b); slower intestinal transit time (Abrams and Bishop, 1967); and higher amino acid levels in the lower bowel contents (Alcock and MacIntyre, 1960).

Table XIII. Effect of Intestinal Microflora on the Absorption and Retention of Calcium, Magnesium and Phosphorus in Germfree and Conventional Rats[a]

	Calcium		Magnesium		Phosphorus	
	GF	CV	GF	CV	GF	CV
Plasma, mg/100 ml	11.4 ± .3[b]	11.0 ± .2	1.9 ± .1	2.0 ± .1	7.6 ± .2	7.5 ± 0.3
Intake, mg/day per 100 g body wt	32.2 ± .8	33.4 ± .9	4.3 ± .1[c]	4.5 ± .1	31.4 ± .8	32.5 ± 1.0
Fecal, mg/day per 100 g body wt	19.6 ± 1.3[c]	25.0 ± 1.0	1.2 ± .1[c]	2.7 ± .1	12.4 ± .8	14.6 ± .8
Absorption, mg/day 100 g body wt	12.7 ± .7[d]	8.4 ± .5	3.1 ± .1[c]	1.8 ± .2	19.0 ± .7	17.9 ± .6
Absorption, % intake	39.4 ± 2.6[d]	25.3 ± 1.6	72.0 ± 2.1[d]	39.9 ± 3.4	60.6 ± 2.0[d]	55.2 ± 1.6
Urinary, % intake	3.9 ± 0.3[d]	2.0 ± .4	27.7 ± 2.0[d]	13.0 ± 1.2	12.0 ± 1.2[d]	19.5 ± 1.2
Retention, mg/day 100 g body wt	11.3 ± .6[d]	7.8 ± .5	1.9 ± 0.1	1.2 ± 0.2	15.2 ± .6[d]	11.6 ± 0.6
Retention, % intake	35.5 ± 2.5	23.3 ± 1.6	44.2 ± 1.9	27.0 ± 3.7	48.5 ± 1.7[d]	35.7 ± 1.3

[a]Reddy (1971) [b]Mean ± S.E.M. [c]P<0.05 [d]P<0.01

7.6. Magnesium

The absence of the microflora affects magnesium utilization in a manner similar to that of calcium (Table XIII). Enhanced absorption, retention and urinary excretion of Mg have been found in GF animals (Reddy, 1971; Reddy, 1972; Yoshida *et al.*, 1969; Reddy and Pollard, 1972; Garnier and Sacquet, 1969: Andrieux *et al.*, 1980). Elevations in intestinal brush border $Mg^{++}-$ATPase and alkaline phosphatase activities were observed which were similar to those reported for calcium (Reddy, 1972).

7.7. Phosphorus

Most investigators have observed that GF animals excrete less urinary phosphorus than do CV animals (Gustafsson and Norman, 1962; Reddy *et al.*, 1969b) (Table XIII). Garnier and Sacquet (1969) found increased phosphorus absorption in GF rats but Reddy *et al.* (1969b) found no change. Yoshida *et al.* (1968, 1969) found an increase in phosphorus absorption in GF Japanese quail and rabbits. Like calcium and magnesium, phosphorus retention also was greater in GF animals (Andrieux *et al.*, 1980).

The microflora therefore appears to influence the utilization of most minerals. In general, Ca, P, Mn, Mg, and Zn absorption and retention are increased in GF animals. Iron and copper seem to be better absorbed in CV animals. Many of the differences observed appear to be the result of indirect effects of the microflora, e.g., an effect on bile acid deconjugation, saturation of polyunsaturated fatty acids, amino acid balance, etc., which affect mineral absorption. Some of the differences may be due to methods of sterilization and/or to diet composition. The influence of antibiotics on select microbial populations which affect mineral utilization requires more research.

8.   Vitamins

8.1. Thiamin

Thiamin and other B vitamins can be synthesized by the large intestinal microflora (Leong, 1937; Wostmann *et al.*, 1962; Wostmann and Knight, 1961; Pasteur, 1885). However, these nutrients do not appear to be absorbed (Wostmann and Knight, 1961). The liver of GF rats has a lower concentration of thiamin than that of CV rats fed identical diets (Wostmann *et al.*, 1962). CV rats have access, via coprophagia, to thiamin synthesized by the microflora (Yudkin, 1963; Mameesh *et al.*, 1959) but the utilization of this vitamin may be low (Barnes *et al.*, 1960). The everted small intestine of chickens absorbs thiamin more efficiently than that of CV controls (Ford and Coates, 1971). Gordon *et al.* (1960) found that thiamin disappeared faster from the digestive tract of GF animals.

8.2. Riboflavin

In the CV rat removal of the cecum depresses riboflavin absorption (Yoshida *et al.*, 1969), suggesting that alterations of the lower bowel microflora influence the utilization of dietary or endogenously synthesized riboflavin. Microbial riboflavin synthesis in the chicken has been demonstrated by Coates *et al.* (1968) and absorption of riboflavin from intestinal loops was found to be slightly greater in GF chickens (Ford and Coates, 1971).

8.3. Vitamin B_6, Nicotinic Acid, Pantothenic Acid, and Biotin

All four vitamins can be synthesized by the intestinal microflora of the CV chicken (Coates *et al.*, 1968). Synthesis of biotin (Hotzel and Barnes, 1966) and pantothenic acid (Daft *et al.*, 1963) has been shown to occur in the rat colon, and if coprophagia is permitted no dietary requirement can be demonstrated. However, the growth rates of biotin deficient GF and CV chickens were depressed equally (Asano, 1967). Higher rates of absorption of nicotinic acid, pantothenic acid and biotin were noted in isolated intestinal loops from GF chickens (Mameesh *et al.*, 1959). Recently Latymer and Coates (1981) showed that the GF chicken requires only two-thirds the amount of calcium pantothenate needed by CV birds. Alteration of the microflora by supplementation of the diet with penicillin or Aureomycin promoted growth in rats fed suboptimal quantities of vitamin B_6 (Linkswiler *et al.*, 1951; Sauberlich, 1952.) Oral administration of Aureomycin, streptomycin, or penicillin had similar effects on rats under conditions of pantothenic acid deficiency (Sauberlich, 1952; Lih and Baumann, 1951; Schendel and Johnson, 1954).

8.4. Vitamin B_{12}

Certain microorganisms, such as *Propionibacterium shermanii*, synthesize vitamin B_{12} in amounts which far exceed their own requirements and therefore supply the host and B_{12}-dependent microorganisms, e.g., *Lactobacillus lactis* (Perlmann, 1959). Dietary vitamin B_{12} also can be consumed by intestinal bacteria, resulting in B_{12} deficiency (Corcino *et al.*, 1970; Donaldson, 1962).

Coates *et al.* (1968) demonstrated that B_{12} can be synthesized by the colon flora of chickens, but that utilization of the vitamin occurred only if facilitated by coprophagia. Pseudo-vitamin B_{12} and factors A, B, and F were found in the cecal contents and feces of CV chickens reared on a purified diet with cyanocobalamin as the only source of vitamin B_{12}, but not in their GF counterparts (Coates *et al.*, 1963). Valencia *et al.* (1965) found large amounts of cobalamins in feces from CV rats, but only trace amounts in feces from GF rats. Monoassociation with a strain of Veillonella increased fecal cobalamins but did not increase tissue levels (Valencia *et al.*, 1965). The vitamin B_{12} requirement of GF chickens may be reduced because of a lower basal metabolic rate, increased intestinal transit time and other physiological anomalies observed in the GF state (Oace and Abbot, 1972). The fact that antibiotic administration has a "sparing" effect on the B_{12} requirement (Mickelson, 1956) suggests that the GF animal has a decreased requirement for this vitamin.

8.5. Folic Acid

Studies on germfree rats indicate that they have a folic acid requirement, but estimates of the requirement differ markedly (Daft *et al.*, 1963; Valencia *et al.*, 1968). The vitamin is synthesized by flora of the large intestine (Coates *et al.*, 1968; Daft *et al.*, 1963). Monoassociation of GP rats with either Aerobacter sp., Alcaligenes sp., *E. coli*, or Proteus protects the host against folic acid deficiency (Daft *et al.*, 1963). Miller and Luckey (1963) found the GF chickens monoassociated with a strain of *E. coli* had higher hemoglobin and tissue folic acid values than their GF counterparts, even though coprophagy was prevented.

8.6. Ascorbic Acid

Many intestinal flora are capable of degrading vitamin C (Kendall and Chinn, 1938; Young and James, 1942; Young and Rettger, 1943) and hence could account for the development of vitamin C deficiency under certain conditions. GF and CV guinea pigs fed a scorbutogenic diet both developed scurvy in about five weeks, but on administration of vitamin C the GF animals recovered more rapidly (Phillips *et al.*, 1959). In a more extensive study, Levenson *et al.* (1962) found that GF guinea pigs survived almost twice as long as CV animals when fed a vitamin C deficient diet.

8.7. Vitamin A

Little work related to vitamin A has been done using GF animals; however, it has been shown that when depleted of vitamin A they survive longer than CV animals (Kessner and Epstein, 1966; Rogers *et al.*, 1971). This might be due to the better fat absorption observed in the GF state, degradation of the vitamin by microflora, or the immunological stress of the conventional state.

8.8. Vitamins D and E

Growth rates of GF and CV chickens were comparable when they were fed a vitamin D or vitamin E deficient diet; however the amount of muscle ash in the GF vitamin D deficient group was considerably higher than that found in deficient CV chickens (Reyniers *et al.*, 1946). This finding may be explained by the increased mineral absorption and retention in the GF state.

8.9. Vitamin K

Vitamin K_2 is synthesized by the microflora of the large intestine in amounts which are sufficient to meet the requirements of the CV rat (Hollander *et al.*, 1976). High doses of vitamin A exacerbate vitamin K deficiency in rats, but the microflora do not seem to be involved since this phenomenon occurs with equal severity in GF and CV rats (Wostmann and Knight, 1965). GF rats given diets low in vitamin K soon developed symptoms of deficiency; association with an *E. coli* and sarcine-like micrococcus (unidentified) reversed the deficiency symptoms (Gustafsson and Norman, 1962).

9.   Digestive Enzymes

Excellent reviews on gastrointestinal enzymes in CV and GF animals have been published by Stroud (1974) and Corring *et al.* (1981).

9.1. Digestive Enzymes in Gnotobiotic Animals

Digestive enzyme activity is affected by the intestinal microflora both directly and indirectly (Corring *et al.*, 1981). Direct effects include the microbial synthesis of enzymes analogous to those of the host, resulting in increased total enzyme activity and, conversely, the splitting or inhibition of host enzymes with a resultant decrease in total enzymes (Borgstrom *et al.*, 1959). Moreover, microbial synthesis of enzymes not produced by the host (e.g., cellulase) extends the digestive capability of the host and improves nutrient utilization (Beresford *et al.*, 1971; Cranwell, 1968; Juhr, 1980). The microflora may indirectly influence host enzymatic activity through luminal pH changes, alterations of secretory and absorptive func-

Table XIV. Digestive Enzyme Activity in Germfree and Conventional Rats
(Mean ± S.D. or Range)

| | | Pancreas | Small Intestinal Mucosa | | |
			Upper	Middle	Lower
Protease[a] mEq tyrosine released in 10 min/g dry content	GF CV	6.2 ± 2.5 4.8 ± 1.4			
Amylase[a,c,e] mg starch digested in 15 min x 10^{-3}/g dry content	GF CV	630 ± 122 567 ± 48			
U/100 g body wt U/g content	GF CV	1545 ± 134 1407 ± 129			
U/day	GF CV				
Lipase[a,c] mEq fatty acids released in 20 min/g dry content	GF CV	258 ± 70 259 ± 89			
μmoles of fatty acid liberated/min/g content	GF CV	560 ± 62 545 ± 54			
Trypsin[a,c,e] % total enzyme activity	GF CV	78 ± 1 79 ± 3			
U/100 g body wt U/g content mg/day	GF CV GF CV	943 ± 66 955 ± 60			
Maltase[b] μmoles of disaccharide hydrolyzed/mg prot/60 min	GF CV		21 ± 1 11 ± .7	20 ± 1 10 ± .8	15 ± .9 6 ± .4
Invertase[b,e] μmoles of disaccharide hydrolyzed/mg prot/60 min	GF CV		4.8 ± .3 2.0 ± .1	3.0 ± .2 1.6 ± .1	.6 ± .0 .2 ± .0
U/day	GF CV				
Trehalase[b] μmoles of disaccharide hydrolyzed/mg prot/60 min	GF CV		8.8 ± .3 3.3 ± .2	4.4 ± .5 1.6 ± .2	.9 ± 1 .3 ± .0
Lactase[b] μmoles of disaccharide hydrolyzed/mg prot/60 min	GF CV		.9 ± 1 .3 ± .0	2 ± .1 .6 ± .1	.6 ± .1 .2 ± .0
Cellobiase[b] μmoles of disaccharide hydrolyzed/mg prot/60 min	GF CV		.3 ± .0 .1 ± .0	.6 ± .1 .1 ± .0	.1 ± .0 .03 ± .0
Chymotrypsin[c] U/100 g body wt U/g content[b]	GF CV	1194 ± 104 1072 ± 28			
L-Ala-L-Phenylalanine[d] Peptide hydrolase mU/mg protein	GF CV	42 ± 10 38 ± 15			
Glycyl-L-Phenylalanine[d] Peptide hydrolase mU/mg protein	GF CV	35 ± 5 27 ± 3			
L-Leu-Gly-Glycine[d] Peptide hydrolase mU/mg protein	GF CV	2.1 ± .8 2.1 ± .8			

[a]Lepkovsky et al. (1966) [b]Reddy and Wostmann (1966) [c]Reddy et al. (1969) [d]Szabo (1977)

| Small Intestinal Contents | | | Cecal Contents | Colon Contents | Feces |
Upper	Middle	Lower			
.7 ± .3	1.8 ± .5	1.5 ± .3	1.1 ± .2	.9 ± .3	.8 ± .3
.9 ± .2	2.0 ± .6	1.2 ± .3	.8 ± .3	.8 ± .1	.5 ± .1
79 ± 26	119 ± 20	16 ± 4	2.2 ± .3	1.6 ± .3	1.1 ± .1
113 ± 40	113 ± 69	112 ± 45	9.7 ± 4.7	6.7 ± 2.6	3.6 ± 3.2
	445 ± 60		10 ± 1	18 ± 3	
	192 ± 20		6.0 ± .7	1.6 ± .1	1.6 ± .8
					40-100
					40-100
11 ± 4	15 ± 4	8.7 ± 4.7	5.1 ± 1	6.2 ± 1.5	5.7 ± .7
17 ± 4	17 ± 3	9.2 ± 2.5	5.7 ± 2	4.3 ± 2.1	4.7 ± 1.3
	96 ± 11		7.46 ± .68	11 ± 2	
	96 ± 13		2.37 ± .11	1.1 ± .3	
74 ± 4	78 ± 5	78 ± 1	81 ± 2	79 ± 4	77 ± 2
74 ± 3	75 ± 3	74 ± 4	76 ± 2	77 ± 4	73 ± 3
231 ± 12			39 ± 2	32 ± 3	32 ± 3
139 ± 10			14 ± 1	9.2 ± .7	9 ± .7
					1-6 [f]
					n.d.
					12-25 [f]
					n.d.
	306 ± 9		17 ± 1 [b]	14 ± 1	
	174 ± 10		6.5 ± .6 [b]	5.0 ± .3	
	140 ± 17		172 ± 18	145 ± 11	
	417 ± 43		88 ± 23	72 ± 17	
	83 ± 14		80 ± 10	64 ± 7	
	153 ± 30		44 ± 19	20 ± 6	
	20 ± 4		30 ± 5	31 ± 4	
	75 ± 4		9.2 ± 2	3.3 ± 1.7	

[e] Borgström et al. (1959)

tions, and changes in intestinal epithelial cell renewal rates (Abrams *et al.*, 1963; Guenet *et al.*, 1970; Khoury *et al.*, 1969; Lesher *et al.*, 1964).

Absence of the microflora does not seem to influence the secretion of pancreatic trypsinogen, chymotrypsinogen, amylase, lipase, or peptide hydrolase (Levenson and Tennant, 1963; Reddy *et al.*, 1969a; Szabo, 1979) (Table XIV). On reaching the intestine, however, the activity of these enzymes may be reduced by the intestinal microflora. Digestive enzyme activity is related to the density of the microbial population, i.e., as bacterial counts increase from the small to the large intestine, host enzyme activity decreases (Reddy *et al.*, 1969a; Borgstrom *et al.*, 1959). In the GF animal this decrease does not occur and in some cases an increase has been noted (Lepkovsky *et al.*, 1966; Loesche, 1968b; Borgstrom *et al.*, 1959; Juhr, 1980). Jervis and Biggers (1964) observed a lower brush border alkaline phosphatase activity in CV than in GF mice. Succinic dehydrogenase and esterase activities have been found to be similar (Hershovic *et al.*, 1967). Peptidase and disaccharidase activity in the small intestinal epithelial cells of GF rats and piglets has been found to be increased (Szabo, 1977, 1979) (Tables XV, XVI, XVII). In the host's intestinal contents the situation is reversed, i.e., there is greater peptidase activity in the CV animal, possibly because of an increased epithelial cell desquamation rate (Szabo, 1977). The peptidases and disaccharidases (except for lactase) exhibit a progressive decrease in activity from the small intestine to the cecum and colon in CV pigs, rats, and chickens (Szabo, 1979; Lepkovsky *et al.*, 1964). GF animals, on the other hand, were found to have enzymatic activities in the lower bowel similar to values found for the small intestine (Szabo, 1979; Lepkovsky *et al.*, 1964). Chymotrypsin and elastase activities appear to be more sensitive to bacterial inactivation than is trypsin activity. In the GF rodent these enzymes are not degraded in the large intestine (Genell *et al.*, 1976). Lactase (Szabo, 1979), urease (Delluva *et al.*, 1968; Juhr, 1980) and histidine decarboxylase (Wostmann, 1966-67) are at low levels or completely lacking in GF animals and their presence is apparently dependent on microbial synthesis. Urease activity in the cecum of the CV rat has been attributed mainly to certain strains of Lactobacillus, Staphylococcus, and Actinobacillus (Table VII) (Ducluzeau *et al.*, 1966). Lipolytic activity in the pancreas of GF and CV rats is identical; however, as with many of the other enzymes, lipase activity is decreased in the lower bowel of CV animals but not in GF animals (Lepkovsky *et al.*, 1966; Szabo, 1977; Reddy *et al.*, 1969a; Juhr, 1980). In summary, degradation of many digestive enzymes is enhanced as the intestinal microbial population increases, i.e., from the small intestine to the large intestine. In the absence of the microflora the persistence of various host enzymes in the lower bowel may result in more complete substrate degradation, as indicated by increased levels of free amino acids and free fatty acids in the

Table XV. Digestive Enzyme Activity in Three-Week-Old Germfree and Conventional Piglets[a]   (Mean ± S.D.)

		Pancreas	Duodenal Mucosa	Ileal Mucosa	Cecal Contents	Large Intestinal Contents
Proteolytic Activity mU/mg protein	GF	118.8 ± 40.02			39 ± 5	37 ± 10
	CV	86.21 ± 32.82			63 ± 28	15 ± 5
Trypsin mU/mg protein	GF	n.d.[b]	n.d.	n.d.	n.d.	n.d.
	CV					
GLY – PHE mU/mg protein	GF	84 ± 17	1344 ± 54	2942 ± 1119	569 ± 194	806 ± 87
	CV	72 ± 14	330 ± 225	600 ± 82	408 ± 206	164 ± 70
ALA – PHE mU/mg protein	GF	326 ± 43	2146 ± 436	4971 ± 1274	1189 ± 237	1370 ± 38
	CV	367 ± 63	1116 ± 360	1680 ± 592	1435 ± 560	505 ± 163
LEU – GLY – GLY mU/mg protein	GF	50 ± 16	722 ± 211	1279 ± 378	350 ± 48	284 ± 85
	CV	66 ± 8	464 ± 108	664 ± 216	572 ± 282	82 ± 29
Carbohydrases						
α-Amylase U/mg protein	GF	2.7 ± .5			1225 ± 415	1.0 ± 0.1
	CV	3.0 ± .5			.07 ± .04	.03 ± .02
Lactase U/mg protein	GF		26 ± 8	14 ± 8	16 ± 9	18 ± .2
	CV		5.7 ± 2.4	3.1 ± 1.6	10 ± .2	21 ± 16
Cellobiase U/mg protein	GF		18 ± 7	10 ± 4.7	9.4 ± 5.3	12 ± 6
	CV		3.7 ± 1.5	1.5 ± .5	4.1 ± 3.5	6.4 ± 5.5
Saccharase U/mg protein	GF		5.2 ± 1.6	7.1 ± 1.8	16 ± 4	12.5 ± 4
	CV		2.8 ± 1.3	3.4 ± 1.9	3.8 ± 2.5	3.4 ± 1.8
Maltase U/mg protein	GF		74 ± 18	106 ± 63	152 ± 40	132 ± 42
	CV		12 ± 4	33 ± 18	151 ± 95	78 ± 55
Lipase U/mg protein	GF	3502 ± 765			4.5 ± 1.3	5.5 ± 2.1
	CV	5226 ± 1746			35 ± 14	95 ± 56

[a] Szabo (1979)   [b] None detectable

Table XVI.   The Protease, Amylase and Lipase Activity of the Intestinal, Cecal, and Colon and Cloacal Contents of Germfree and Conventional Chickens[a] (Mean ± S.E.M.)

	N	Proteases[b]	Amylase[b]	Lipase[b]	% Trypsin[a]
Upper Intestine					
Germfree	9	1.2 ± .2[c]	90 ± 28	5.6 ± 1.2	69
Conventional	9	1.2 ± .2	105 ± 53	6.5 ± 2.4	82
Lower Intestine					
Germfree	9	1.4 ± .2	90 ± 20	4.0 ± 1	74
Conventional	9	1.7 ± .2	104 ± 38	4.4 ± 1.1	80
Ceca					
Experiment 1					
Germfree	3	1.5 ± .1	9.0 ± 4.1	1.6 ± .5	80
Conventional	3	.2 ± .04	4.5 ± 2.3	1.4 ± .6	62
Experiment 2					
Germfree	6	1.3 ± .2	6.6 ± 5.0	.7 ± .4	
Conventional	6	1.0 ± .3	1.5 ± 1.0	2.4 ± .8	
Colon					
Experiment 1					
Germfree	3	1.0	13.7	1.7	70
Conventional	3	1.5	12.2	2.1	83
Cloaca					
Experiment 1					
Germfree	3	.9	16.7	1.8	70
Conventional	3	.6	4.9	.9	
Colon + Cloaca					
Experiment 2					
Germfree	6	1.1 ± .3	15 ± .6	1.1 ± .2	
Conventional	6	.8 ± .2	8 ± 3	1.0 ± .3	

[a]Lepkovsky (1964)   [b]Proteases expressed as mEq. of tyrosine released in 10 min. Amylase expressed as mg. starch digested in 15 min x 10^{-3}. Lipase expressed as mEq. of fatty acids released in 20 min. All enzymes incubated at 37°C and expressed as per gm. of dry solids.

cecal and fecal contents of GF rats. Certain enzymes are, on the other hand, synthesized by intestinal microorganisms and aid in the breakdown of substances otherwise unavailable to the host, e.g., urea, lactose, and cellulose.

10. Bile Acids

Bile acid composition in the intestinal lumen of CV animals is affected by the type of primary bile acids synthesized and secreted by the host animals hepatocytes, transformation of the primary bile acids by bacterial enzymes to yield secondary bile acids, reabsorption of the various transformated products, and reentry of these products into the hepatocyte

Table XVII. Intestinal Disaccharidase Activities in Germfree and Conventional Chick Given a Practical Chick Mash[a] (N =15)

	GF	CV	GF–CV
Body wt, g	228	186	42***
Total activity units in small intestine			
Maltase	16337	10592	5745***
Sucrase	3112	1605	1507***
Palatinase	755	480	275***
Lactase	44	53	-8
Total activity units in large intestine			
Maltase	3967	2834	1133
Sucrase	677	609	68
Palantinase	147	200	-53*
Lactase	7	227	-220***

[a]Siddons and Coates (1972)
*P<0.05***P<.001

where they are subjected to additional enzymatic changes. Not only do the microflora influence bile acid metabolism in the host animal, but the bile acids in turn can affect which microorganisms constitute the "climax" indigenous microflora. The influence of bile salts on the growth of microorganisms is well covered in a review by Talalay (1957) and has been re-examined by Sakai et al. (1980).

The intestinal microflora appears to influence the total bile acid pool (Table XVIII). In the GF animal total biliary acids are elevated but the amount excreted in the feces is reduced (Eyssen et al., 1976; Gustafsson et al., 1957, 1960; Kellog and Wostmann, 1969; Mott et al., 1973; Sacquet et al., 1971). In the absence of the microflora, the conjugated primary bile acids are excreted in the feces virtually unchanged (Kellog and Wostmann, 1969; Gustafsson et al., 1957). The intestinal microflora deconjugate and dehydroxylate these acids to yield a complex mixture of primary and secondary fecal bile acids. Microbial 7α- or 7β-dehydroxylation of cholic acid and chenodeoxycholic acid yields deoxycholic acid and lithocholic acid, respectively, along with numerous other metabolites (Gustafsson et al., 1966; Midvedt and Norman, 1967, 1968; Draser et al., 1966; Aries and Hill, 1970). The enzyme 7α-decarboxylase acts only on free bile acids and not on conjugated bile salts or the methyl esters (Gustafsson et al., 1968; Aries and Hill, 1970). A strain of Lactobacilli, isolated from rat feces, was capable of 7α-dehydroxylation after colonization of the intestinal tract of the GF rat (Gustafsson et al., 1968). However, as shown by Dickinson et al. (1971) some organisms with 7α-dehydroxylase activity in vitro lose this ability when they are used to colonize the GF gastrointestinal tract. In rats deoxycholic acid is rehy-

Table XVIII. Bile Acids and Sterols Found in Intestinal, Fecal and Bile Contents of Germfree and Conventional Animals (Mean ± S.D.)

	Species	Age or Body wt.	GF	CV
Total Bile Acids				
Conjugated, uM/ml	rat[l]		60	21
Conjugated, mg total	pig[e]	6 wk	964	1009
Conjugated, mg/ml	pig[f]		37 ± 3	28 ± 3
Bile and Small Intestine				
Mg/animal	pig[g]	6-7 mo	7.7	3.7
Mg/100 g body wt.			21.1	11.5
Total Bile Acids in Small Intestinal Contents				
Mg/kg body wt.	rat[h]	250g	254 ± 78	41 ± 21
Cholic acid and metabolites, mg	rat[a]	135g	18.4	14.6
Mg/total contents	mouse[k]		11.2 ± 2.4	3.3 ± 0.4
Mg/total contents	rat		4.9 ± 2.0(♂)	2.5 ± 1.4(♂)
			3.7 ± 3.8(♀)	1.2 0.1(♀)
Total Fecal Bile Acids				
Mg/day	pig[c]			413
Mg/kg body wt/day	rat[d]	3-4 mo	11 ± 2	21 ± 10
Mg/kg body wt/day	rat[d]		6.5 ± 2.2	13 ± 2
			8.3 ± 1.5	14 ± 4
Mg/animal/24 hr	mouse[j]	3 mo	1.4	
Mg/100g body wt/24 hr	mouse[g]	7 mo	2.9	4.1
Cholic acid and metabolites	rat[a]	134 g	1.9	5.1
Total Fecal Neutral Sterols				
Endogenous	rat[d]	100 d	14 ± 3	21 ± 9
Mg/kg body wt/day	rat[d]		12 ± 2	20 ± 6
Mg/kg body wt/day	rat[b]	3-4 mo	13 ± 3	19 ± 5

[a]Gustaffson et al. (1960)   [b]Kellog and Wostmann (1969)   [c]Mott et al. (1973)
[d]Gustaffson et al. (1957)   [e]Eyssen et al. (1975)   [f]Beaver and Wostmann (1974)
[g]Eyssen et al. (1976)   [h]Gustaffson et al. (1975)   [i]Gustaffson et al. (1975)
[j]Eyssen et al. (1973)   [k]Chang et al. (1975)   [l]Sacquet et al. (1971)

droxylated by the liver during enterohepatic cycling (Bergstrom et al., 1953) whereas the rabbit liver lacks this capacity. In the CV rabbit the bile acid pool is almost completely converted to deoxycholic acid after a number of enterohepatic cycles (Ekdahl and Sjovall, 1955), whereas the GF rabbit microbial dehydroxylation of the primary bile acids does not occur and therefore the main bile constituent is cholic acid (Hoffman et al., 1964). Lithocholic acid, a secondary bile acid, is not found in GF animals but can be found in the bile and feces of several CV mammalian species (Eyssen et al., 1976; Demarne et al., 1972). However, acid is not present in all CV animals, indicating that there may be differences in their indigenous microflora (Eyssen et al., 1976). Kellog and Wostmann (1969) have shown that in GF rats 50-75% of the bile acids is β-muricholic acid, an acid which constitutes only 25% of the bile acids found in CV rats. The same differences in bile acid spectrum have been observed in mice (Eyssen et al., 1973).

The mechanism(s) underlying the differences in the bile acid spectrum of GF and CV animals, illustrated by reduced amounts of β-muricholate and the appearance of hyodeoxycholate (HDC) and ω-muricholate (ω-MC) both of which are secondary bile acids in the CV state, is now better understood. HDC is a metabolite of β-muricholate (β-MC) and was presumed to be the intermediate between β-MC and ω-MC. Whether this conversion was due to microbial or hepatic activity was not clear until Madsen and Wostmann (1975) showed that oral administration of HDC to germfree rats resulted in the appearance of ω-MC, indicating that a hepatic route existed for conversion from HDC to ω-MC. Recently, Eyssen et al. (1981) showed that the conversion of β-muricholate to hyodeoxycholate can be carried out by single strains of intestinal bacteria in the absence of other intestinal microorganisms. These mechanisms explain the lower β-muricholate in CV animals because the transformed metabolites of β-MC (HDC and ω-MC) are not well absorbed. The ratio of muricholic acid to cholic acid is twice as high in male GF mice as in females, indicating that not only the microflora but also the sex of the animal influences bile acid composition (Eyssen et al., 1973). Sex differences in bile acid metabolism are also evident in GF rats, in which the female was found to have ten-fold increases in bile salt sulfates relative to the male (Eyssen and Parmentier, 1978).

The ratio of HDC to ω-MC also may be altered by inclusion of lactose in a conventional rat diet (Wostmann et al., 1977). By feeding germfree rats chenodeoxycholic acid, Gustafsson et al. (1981) demonstrated that 60-70% of this acid was converted to α- and β-muricholic acid and that the excretion of cholic acid was almost completely inhibited. Cecectomy of CV rats causes a marked reduction in the amount of hyodeoxycholic acid, a slight decrease in the total amount of bile acids, and a relative increase of trihydroxylated acids (Van Heijenoort et al., 1972). GF animals have not been studied in this context (Puhr and Kellog, 1975 unpublished). The microflora in the CV animal increase the oxidative catabolism of cholesterol, resulting in the production of bile acids (Wostmann et al., 1966, 1973, 1976 unpublished) and a decreased half life of cholic acid (Gustafsson et al., 1957).

It is apparent that the intestinal microflora alters the metabolism and excretion of bile acids dramatically. As a result of microbial activity, the bile acid pool is reduced and the fecal excretion of bile acids and neutral sterols is increased. The acceleration in bile acid turnover results in a faster rate of transformation of cholesterol to bile acids in the CV animal. The relative amount of β-muricholic acid in bile is reduced and hyodeoxycholate and ω-muricholate occur as secondary bile acids. These differences between GF and CV animals are illustrated in Tables XVIII and XIX.

11. Fats and Fatty Acids

The role of the intestinal microflora in fatty acid metabolism and thereby in prostanoid-mediated processes recently has been discussed (Bruckner, 1981a, 1981b).

A reduction or alteration in the microflora brought about by oral administration of antibiotics or by maintaining strict sanitary housing conditions generally enhances fatty acid absorption (Afifi, 1959; Supplee, 1960; Yoshida et al., 1971; Yang et al., 1963). In the absence of the microflora, enhanced fat absorption has been noted in rats, rabbits, chick-

Table XIX. Bile Acid Composition of Intestinal, Fecal and Bile Contents of

	GF or CV	Litho-cholic	Deoxy-cholic	Chenodeoxy-cholic
Mouse, 6-7 mo				
% of total bile acids:[a]				
Bladder & small intest.	GF			1.5
Bladder & small intest.	CV	tr	1.4	2.5
Cecum & colon	GF			3.4
Cecum & colon	CV	tr	6.5	2.5
Mg/mouse:[b]				
Gall bladder	GF			
Gall bladder	CV			
Rat, 200-300 g				
Relative % bile acids:				
Small intest.[c]	GF		0	17 ± 0.5
Small intest.[c]	CV		12 ± 5	2.3 ± 2.6
Bile acids[d]	GF	0	0	4.3
Bile acids[d]	CV	4.6	4.3	4.3
Small intest.[e]	GF			
Small intest.[e]	GF			
Small intest.[e]	CV			
Small intest.[e]	CV			
Pig, 6 wk[f]				
% of total bile acids:[f]				
Bladder	GF	0		21
Bladder	CV	tr		12
Fecal	CV	16		6
Dog[g]				
Bladder	GF			4.6 ± 0.6
Bladder	CV		12 ± 3	3.7 ± 1.6
Rabbit				
% of total bile acids				
Bladder	GF			1.4

[a]Eyssen et al. (1976) [b]Eyssen et al. (1973) [c]Gustaffson et al. (1975)
[g]Beaver and Wostmann (1974) [h]Hoffman et al. (1964)

ens, and mice (Evrard *et al.,* 1964, 1965; Yoshida *et al.,* 1968; Boyd and Edwards, 1967; Chang *et al.,* 1975).

The effect of the intestinal microflora on fat absorption may vary depending on the composition of the fat fed. As first shown by Boyd and Edwards (1967), the absorption of polyunsaturated fat is not as strongly influenced by the microflora as that of the saturated fatty acids. These workers indicated that the apparent digestion of corn oil in CV and GF chickens was about the same, but that of tallow was lower reduced in the CV state. Demarne *et al.* (1970a, 1970b) obtained similar results in the rat. The better absorption of saturated fat in the GF rat is attributable

GF and CV Animals   (Mean ± S.D.)

Cholic	Allo-cholic	α-Muri-cholic	ω-Muri-cholic	β-Muri cholic	Hyo-cholic	Hyodeoxy-cholic
23.6	2.3	8.6		61.7		
37.8	4.3	11.4	10.9	28.2		
54.8	5.6	3.3		32.9		
48.9	5.3	3.9	16.4	10.2		
1.8				7.7		
4.6				5.0		
58 ± 2		6.3 ± .5		18 ± 3		0
78 ± 7		4.0 ± .8		.8 ± 1.5		2 ± 4
39				39		0
54				14		9.5
				35 ± 7		
				13 ± 3		
				6.4 ± 5.0		
				0		
					74	5
					47	41
					10	64
95 ± .7						
84 ± 2	4.8					
93.8						

[d]Demarne et al. (1972) [e]Gustaffson et al. (1975) [f]Eyssen et al. (1975)

Table XX.   Fat Content, Composition, and Absorption in the GI Tract of Germfree and
             Conventional Animals   (Mean ± S.D.)

	Species	GF	CV
Fecal Fat			
Mg fecal fatty acids/g feed	rat[a]	5.3	5.2
Mg fecal fatty acids/g feed	rat[b]	1.9	2.7
		1.6	1.2
Mg fecal fat/g dry matter	pig[c]	47	262
% fat in feces	rat[d]	4.6	5.6
% fat retained	rat[d]	97	95
Mg C-18 fatty acids/24 hr/kg body wt	rat[f]	15 ± 2	14 ± 2
Cecal Fat			
% of cecal contents	chicken[e]	2.8	3.3
Fat Absorption[g,h]			
Stomach Oleic	rat	36 ± 6	26 ± 4
Triolein		45 ± 10	23 ± 4
Small intest. Oleic	rat	9 ± 1	8 ± 2
Triolein		10 ± 2	12 ± 3
Cecum Oleic	rat	.8 ± .4	4 ± 1
Triolein		.3 ± 0	4 ± 1
Colon Oleic	rat	.1 ± 0	.2 ± .1
Triolein		.1 ± 0	.3 ± .1
% Absorption Oleic	rat	84 ± 3	85 ± 3
Triolein		78 ± 5	77 ± 4

[a]Sacquet et al. (1966) [b]Evrard et al. (1964) [c]Eyssen et al. (1975) [d]Luckey (1963)
[e]Reynier (1960) [f]Eyssen et al. (1972) [g]Evrard et al. (1965) [h]% of ^{131}[I]oleic acid or
triolein remaining 6 hr after administration

primarily to the more efficient utilization of palmitic and stearic acid in
the absence of the microflora (Demarne et al., 1970a, 1970b). The
absorption of oleic acid and triolein is not affected (Boyd and Edwards,
1967; Tennant et al., 1969; Sharrer and Riedel, 1972). In the CV chicken
the bacteria which seem to be primarily responsible are *Clostridium wel-
chii* and *Streptococcus faecalis* (Cole and Boyd, 1967).

Intestinal bacteria are capable of hydrogenating unsaturated fatty
acids (Eyssen et al., 1973) and thereby may decrease intestinal fat
absorption. The microflora affect not only the absorption of fatty acids,
but the type of dietary and endogenous fatty acids which are excreted in
the feces. When conventional rats were cecectomized, fecal stearic acid
excretion was almost completely suppressed, indicating that the cecum is
the main site of biohydrogenation (Eyssen and Parmentier, 1974). The
following additional observations have been made: 95% of the fatty acids
in GF feces have an even number of carbon atoms compared to 60% in
CV feces; mono- and polyunsaturated fatty acids of endogenous origin are
50% lower in CV than in GF feces; the most prevalent fatty acid excreted
in the feces of both GF and CV rats is palmitic acid (Combe et al., 1976;
Demarne et al., 1979; Evrard et al., 1965; Eyssen et al., 1972). Data on
fat absorption in GF and CV animals are given in Tables XX-XXII.

Table XXI. The Apparent Digestibility of Fatty Acids and Total Fatty Material by Germfree and Conventional Rats[a] (Range, %)

Fatty Acid[b]	Germfree	Conventional
C_{12}	95.0 - 97.3	96.0 - 98.6
C_{14}	85.0 - 89.4	83.3 - 88.9
C_{16}	76.9 - 82.3	63.4 - 70.0
C_{18}	67.9 - 79.3	50.0 - 63.8
$C_{18:1}$	96.7 - 97.2	94.3 - 96.4
$C_{18:2}$	92.6 - 100.0	93.6 - 97.6

[a]Demarne et al. (1970) [b]Composition of fatty material: 25% coconut oil, 25% cotton oil, 50% cacao oil (13% of the diet)

12. Carbohydrates

Polysaccharide degradation to disaccharide units occurs mainly as a result of pancreatic amylase action in the upper small intestine (Auricchio et al., 1967; Fogel and Gray, 1969). The splitting of disaccharides to absorbable monosaccharide units is completed by the disaccharidases associated with the brush border glycocalix (Miller and Crane, 1961a, 1961b; Eicholz and Crane, 1965). The monosaccharides are absorbed in the proximal as well as the more distal segments of the small intestine (Gray and Ingelfinger, 1966). Factors which affect intestinal transit time, enzymatic activity, electrolyte composition, or mucosal brush border morphology and function may influence carbohydrate digestion and absorption.

Pancreatic and small intestinal carbohydrase activity is similar in GF

Table XXII. The Apparent Digestibility of Fatty Acids and Fat by Germfree and Conventional Chickens[a] (Range, %)

Fatty Acid[b]	Germfree	Conventional
C_{16}	81.3 - 85.9	75.4 - 80.8
C_{18}	77.6 - 79.6	66.9 - 67.4
$C_{18:1}$	81.9 - 89.6	86.3 - 87.6
$C_{18:2}$	88.3 - 89.9	89.0 - 88.6

[a]Boyd and Edwards (1967) [b]Corn oil at 6% of the diet

and CV animals (Larner and Gillespie, 1957; Lepkovsky *et al.*, 1964, 1966; Reddy and Wostmann, 1966; Reddy *et al.*, 1968, 1969a; Szabo, 1979). However, in CV animals the microflora may contribute to the "total enzyme pool" by supplying carbohydrases which are then available to the host, or reduce the "enzyme pool" by degradading endogenous enzymes. Gluco- and galactosidase (Hawksworth *et al.*, 1971), amylase (Griffiths and Davies, 1963) and cellulase (Beresford *et al.*, 1971; Cranwell, 1968; Juhr, 1980) are examples of carbohydrases of bacterial origin. Possible degradation of endogenous carbohydrases by the microflora is indicated by the lower disaccharidase activity observed in the CV rat and rabbit (Yoshida *et al.*, 1968; Reddy and Wostmann, 1966). However, other investigators have found no differences in disaccharidase activity between the GF and CV rat (Larner and Gillespie, 1957; Dahlquist *et al.*, 1965) or chicken (Siddons and Coates, 1972). The apparent digestibility of starch was found to be decreased in antibiotic-treated CV rabbits (Stern *et al.*, 1954) and a similar reduction has been observed by Yoshida *et al.* (1968) in GF rabbits.

Heneghan (1963) found that xylose absorption in GF rats and mice was increased twofold. Active transport of glucose was not affected. Endogenous carbohydrate-containing materials have been reported to accumulate in the cecum of the GF rat (Loesche, 1969a). In the presence of the microflora carbohydrates yield primarily volatile fatty acids and lactic acid as end products of fermentative processes (Bolton, 1965). These products are available as energy sources to the host, and they may constitute as much as 20% of the energy required by monogastric animals (Savage, 1977). A corollary to this is that microorganisms generally not considered enteropathogenic can adversely affect intestinal sugar transport when microbial fermentation is excessive, and they may contribute to the diarrhea observed in some malnourished children (Gracey *et al.*, 1975). The amounts of some products of carbohydrate digestion found in the intestinal contents of GF and CV animals are given in Table VI.

13. Proteins

Endogenous and exogenous proteins can be degraded by microbial proteases as well as by enzymes secreted into the gut. Amino acids in the intestinal lumen are used for microbial protein synthesis or degraded to urea, keto-acids, amines and NH_3, which may be absorbed by the host (Yang *et al.*, 1972). Nitrogen fixation has been reported to occur in the intestinal tract of humans, pigs, and guinea pigs (Bergersen and Hipsley, 1970).

Many investigators have analyzed the contents of the intestine of GF and CV animals in attempts to determine the effect of the microflora on the host's protein status (Combe *et al.*, 1965; Salter and Coates, 1970, 1971; Lindstedt *et al.*, 1965; Miller, 1967; Levenson and Tennant, 1963; Evrard *et al.*, 1964). The results are relatively consistent in indicating that there are no differences in protein digestion in the small intestine of GF and CV animals but that there are significant differences in the lower bowel. The fate of nonabsorbed and endogenous nitrogen is greatly modified by the presence of a microflora. In the cecal contents urea, amino acids, mucoprotein and peptide concentrations were higher in GF than in CV rats and chickens (Combe and Sacquet, 1966; Combe *et al.*, 1965; Salter and Coates, 1970, 1971).

Most of the nitrogen-containing material in the GF rat cecum is in a soluble form, while insoluble fractions predominated in CV rats (Table XXIII). The same differences were noted by Yamanaka *et al.* (1980). They reported that while the amount of total nitrogen in the intestinal tract of GF and CV mice fed amino acid or purified whole egg protein diets was similar, there were significant alterations in the ratio of protein nitrogen to total nitrogen and of water insoluble nitrogen to protein nitrogen, the CV animals having higher values. Free amino acids (AA) were elevated in the cecal contents of GF animals. Yoshida *et al.* (1971) found that the microflora improved protein quality in fecal samples derived from CV rabbits. This improvement was due to an increase in the ratio of essential to total AA and in amino acid synthesis from nonprotein nitrogen sources, e.g., urea and NH_3 (Tables VI and VII). The true digestibility of nitrogen in whole egg diets fed at 15% of the diet, as well as net protein utilization and biological value, were found to be higher in CV than in GF rats. However, when the animals were fed the same protein at 2.5% of the diet the reverse results were obtained (Yamanaka and Nomura, 1980).

Protein and AA digestibility were investigated in GF and CV chickens by Salter and Fulford (1974). Their results indicated that the microflora participate in the splitting of dietary and endogenous proteins and that the soluble protein fractions are more susceptible to bacterial degradation than the insoluble ones. Enzymes and mucoproteins were found in the soluble fraction, whereas epithelial cells were the main constituents of the insoluble fraction. Threonine, serine, and glucosamine were degraded by the microflora, while methionine, isoleucine, leucine, and phenylalanine were synthesized. Utilization of the synthesized AA is facilitated by coprophagia. No differences were observed between GF and CV rats in the active transport of alanine or in passive absorption of glutamate in the ileum, cecum, or colon (Kelleher and Bruckner, 1977). However, others have noted an increased absorption of some AA in GF rats (Phillips and Newcomb, 1965). Peptide absorption has not been studied in GF animals.

The microbial degradation of nitrogen compounds in the gut lumen results in markedly elevated NH_3 levels in the portal blood of CV pigs (Warren and Newton, 1959) and chickens (Salter, 1973) relative to those

Table XXIII.   Percentage Distribution of N in the Cecal Contents of Germfree (GF), Conventional (CV) and Traditional (TR) Rats[a]

Rats	Soluble 80% ethanol	Soluble 10% TCA	Insoluble	Total N mg/100 g
GF	68	21	11	500
CV	20	32	48	915
TR	23	20	57	685

[a]Combe and Pion (1966)

of GF animals. This results in a lower apparent digestion of nitrogen in GF animals (Yamanaka *et al.*, 1973) (Table XXIV). The complexity of these events is illustrated by the observation that when GF animals were introduced to various microbial associates, staphylococcus increased nitrogen retention in the host whereas Bacteroides sp., *E. coli,* Lactobacillus, *Staphylococcus epidimis* and *Streptococcus faecalis* had no effect (Yamanaka *et al.*, (1973). Contrary to Wostmann (1959), Yamanaka *et al.* (1972) indicated that the "biological value" of dietary protein may be lower for GF mice. This difference may be due to the fact that different levels of protein were fed (Yamanaka and Nomura, 1980).

With one exception (Harmon *et al.*, 1968) it has been found that GF rats excrete more nitrogen in the feces and less in the urine than CV rats do (Reddy *et al.*, 1969a; Luckey, 1963; Evrard *et al.*, 1964; Hoskins and Zamcheck, 1968) (Table XXIV). The increased level of nitrogen in the lower bowel of GF animals apparently represents undegraded endogenous compounds rather than dietary protein (Loesche, 1968b; Combe, 1971; Combe *et al.*, 1967).

Table XXIV.   Daily Nitrogen Balance (mg N) of Conventional (CV) and Germfree (GF) Rats During an 8-day Period[a]

Intake		Urinary N		Fecal N		P for
K cal	N	CV	GF	CV	GF	fecal N
0	0	112	121	9	21	.05
61	416	268	214	53	88	.05
60	640	455	421	60	130	.001
60	1280	865	769	81	180	.01

[a]Levenson and Tennant (1963)

14. Acknowledgments

The authors express their gratitude to Helmut A. Gordon, George E. Mitchell, Jr., and Ferenc Kovacs for their advice and encouragement, and to Ann Chow, Susanna Cherian, Colleen Marquis, Nancy King, and Beverly Bruckner for their help in typing, editing, and assembling information.

References

Abrams, G.D., and Bishop, J.E., 1967, Effect of the normal microbial flora on gastrointestinal motility, *Proc. Soc. Exp. Biol. Med.* **126**:301.

Abrams, G.D., Bauer, H., and Sprinz, H., 1963, Influence of the normal flora on mucosal morphology and cellular renewal in the ileum. A comparison of germfree and conventional mice, *Lab. Invest.* **12**:355.

Afifi, M., 1959, Uber den einfluss antibiotischer wirkstoffe in verschiedenen posierungen auf kukenwachstrun und futterverdaulichkeit, *Arch. Kleintierzucht* **23**:404.

Alcock, N., and MacIntyre, I., 1960, Interrelation of calcium and magnesium absorption, *Biochem. J.* **76**:19.

Andrieux, C., Gueguen, L., and Sacquet, E., 1979, Influence du mode sterilisation des aliments sur l'absorption des mineraux chez le rat axenique et holoxenique, *Ann. Nutr. Aliment.* **33**:1257.

Andrieux, C., Gueguen, L., and Sacquet, E., 1980, Effects of lactose and mode of sterilization of a lactose diet on mineral metabolism in germfree and conventional rats, *Reprod. Nutr. Develop.* **20**:119.

Aries, V., and Hill, M.J., 1970, Degradation of steroids by intestinal bacteria. II. Enzymes catalyzing the oxido-reduction of the 3α, 7α and 12α hydroxy group in cholic acid and the dehydroxylation of 7-hydroxyl groups, *Biochim. Biophys. Acta* **202**:535.

Asano, T., 1967, Inorganic ions in cecal content of gnotobiotic rats, *Proc. Soc. Exp. Biol. Med.* **124**:424.

Asano, T., 1969a, Anion concentration in cecal content of germfree and conventional mice, *Proc. Soc. Exp. Biol. Med.* **131**:1201.

Asano, T., 1969b, Modification of cecal size in germfree rats by long-term feeding of anion exchange resins, *Am. J. Physiol.* **217**:911.

Auricchio, S., Della Pietra, D., and Vegnente, A., 1967, Studies on the intestinal digestion of starch in man. II. Intestinal hydrolysis of amylopectin in infants and children, *Pediatrics* **39**:853.

Banwell, J.G., and Sherr, H., 1973, Effect of bacterial enterotoxins on the gastrointestinal tract, *Gastroenterology* **65**:467.

Barnes, R.H., Kwong, E., Delany, K., and Fiala, G., 1960, The mechanism of the thiamine-sparing effect of penicillin in rats, *J. Nutr.* **71**:149.

Beaver, M., and Wostmann, B.S., 1962, Histamine and 5-hydroxytryptamine in the intestinal tract of germfree animals, animals harboring one microbial species and conventional animals, *Br. J. Pharmacol.* **19**:385.

Bekkum, D.W. van, 1966, The germfree animal in research, in: *The Germfree Animal in Research* (M.E. Coates, ed.), p. 253, Academic Press, New York.

Beresford, C.H., Neale, R.J., and Brooks, O.G., 1971, Iron absorption and pyrexia, *Lancet* **1**:568.

Berg, R.D., 1980, Mechanisms confining indigenous bacteria to the gastrointestinal tract, *Am. J. Clin. Nutr.* **33**:2472.

Bergdoll, S., 1967, in: *Biochemistry of Some Foodborne Microbial Toxins* (R.I. Mateles and G.N. Wogan, eds.), p. 1, M.I.T. Press, Cambridge, Mass.

Bergersen, F.J., and Hipsley, E.H., 1970, The presence of N_2-fixing bacteria in the intestines of man and animals, *J. Gen. Microbiol.* **60**:61.

Bergström, S., Rottenberg, M., and Sjövall, J., 1953, Uber den einfluss staffwechsel der cholsäure and desoxycholsaüre in der ratte: X. Mitteil über steroide and gallensaüren, *Hoppe-Seylers Z. Physiol. Chem.* **295**:278.

Binder, H.J., and Powell, D.W., 1970, Bacterial enterotoxins and diarrhea, *Am. J. Clin. Nutr.* **23**:1582.

Bolton, W., 1965, Digestion in the crop of the fowl, *Br. Poult. Sci.* **6**:97.

Borgström, B., Dahlquist, A., Gustafsson, E., Lundh, G., and Malmquist, J., 1959, Trypsin invertase and amylase content of feces of germfree rats, *Proc. Soc. Exp. Biol. Med.* **102**:154.

Bornside, G.H., Donovan, W.E., and Myers, M.B., 1976, Intracolonic tensions of oxygen and carbon dioxide in germfree, conventional and gnotobiotic rats, *Proc. Soc. Exp. Biol. Med.* **151**:437.

Boyd, F.M., and Edwards, H.M., 1967, Fat absorption in germfree chicks, *Poult. Sci.* **46**:1481.

Bruckner, G., 1981a, Inhibition of metarteriole "Norepinephine refractoriness" by Na salicylate in antibiotic treated rats, in: *Recent Advances in Germfree Research* (S. Sasaki, A. Ozawa and K. Hashimoto, eds.), p. 745, Tokai University Press.

Bruckner, G., 1981b, Prostaglandin and fatty acid status of germfree rats: possible cardiovascular interactions, Proceedings, 19th Annual Meeting of the Association for Gnotobiotics, p. 21, Roswell Park Memorial Institute, Buffalo, New York.

Bruckner, G., Tucker, R.E., Grunewald, K., and Mitchell, G.E. Jr., 1984, Essential fatty acid status and characteristics associated with colostrum deprived gnotobiotic and conventional lambs. Growth, organ development, cell membrane integrity and parameters associated with lower bowel function, *J. An. Sci.* (in press).

Callender, S.T., 1974, in: *Biomembranes, 4B* (D.H. Smyth, ed.), p. 76, Plenum Press, New York.

Celesk, R.A., Asano, T., and Wagner, M., 1976, The size, pH, and redox potential of the cecum in mice associated with various microbial floras, *Proc. Soc. Exp. Biol. Med.* **151**:260.

Chang, L., Pleasants, J., and Wostmann, B.S., 1975, Intestinal metabolism of lipids in GF C3H mice fed chemically defined diets, Proceedings, 5th International Symposium on Gnotobiology, Karolinska Institute, Stockholm, Sweden (abstract).

Cherian, S., Bruckner, G., Miniats, O.P., Volk, K., and Jackson, S., 1979, Characteristics of lower bowel contents in germfree and conventional piglets, in: *Clinical and Experimental Gnotobiotics* (T. Fliedner, H. Heit, D. Niethammer, and H. Pflieger, eds.), p. 117, Gustav Fischer Verlag, Stuttgart and New York.

Coates, M.E., Gregory, M.E., Poster, J.W.G., and Williams, A.P., 1963, Vitamin B_{12} and its analogues in the gut contents of germfree and conventional chicks, *Proc. Nutr. Soc.* **22**:XXVII.

Coates, M.E., Ford, J.E., and Harrison, G.F., 1968, Intestinal synthesis of vitamins of the B complex in chicks, *Br. J. Nutr.* **22**:493.

Cole, J.R. Jr., and Boyd, F.M., 1967, Fat absorption from the small intestine of gnotobiotic chicks, *Appl. Microbiol.* **15**:1229.

Cole, J.S., and Wiseman, R., 1971, Levels of uric acid in the intestinal tracts of germfree, gnotobiotic and conventional mice, *Metabolism* **20**:278.

Combe, E., 1971, Etude de l'azote endogene dans le contenu digestif des rats holoxeniques et axeniques a l'aide d'acides amines, *C. R. Soc. Biol. (Paris)* **165**:289.

Combe, E., et Pion, R., 1966, Note sur la composition en acides amines du contenu de caecum des rats axeniques et des rats temoins, *Ann. Biol. Anim. Biochim. Biophys.* **6**:255.

Combe, E., and Sacquet, E., 1966, Influence de l'etat axenique sur divers composis azotes contenus dans le caecum de rats albinos recevant des quantites variable de proteins, *C. R. Acad. Sci. (Paris)* **262**:685.

Combe, E., Penot, E., Charlier, H., and Sacquet, E., 1965, Metabolisme de rat "germfree" Teneur des contenus digestifs en certains composis azotes, en sodium, en potassium; teneurs de quelques tissues en acides nucleiques, *Ann. Biol. Anim. Biochim. Biophys.* **5**:189.

Combe, E., Arnal, M., and Sacquet, E., 1967, Note sur l'etude de l'origine des composis azotes dans le caecum des rats axeniques a l'aide d'acides amines, *C. R. Soc. Biol. (Paris)* **161**:1076.

Combe, E., Demarne, Y., Gueguen, L., Ivorec-Szylit, O., Meslin, J.C., and Sacquet, E., 1976, Some aspects of the relationships between gastrointestinal flora and host nutrition, *World Rev. Nutr. Diet.* **24**:1.

Corcino, J.J., Waxman, S., and Herbert, V., 1970, Absorption and malabsorption of vitamin B_{12}, *Am. J. Med.* **48**:562.

Corring, T., Juste, C., and Simones-Nunes, C., 1981, Digestive enzymes in the germfree animal, *Reprod. Nutr. Develop.* **21**:355.

Cranwell, D., 1968, Microbial fermentation in the alimentary tract of the pig, *Nutr. Abstr. Rev.* **38**:721.

Csaky, T.Z., 1968, Intestinal water permeability regulation involving the microbial flora, in: *The Germfree Animal in Research* (M.E. Coates, ed.), p. 151, Academic Press, New York.

Daft, F.S., McDaniel, E.G., Harman, L.G., Romine, M.K., and Hegner, J.R., 1963, Role of coprophagy in utilization of B-vitamines synthesized by intestinal bacteria, *Fed. Proc.* **22**:129.

Dahlquist, A., Bull, B., and Gustafsson, B.E., 1965, Rat intestinal 6-bromo, 2-naphthyl glycosidase and dissacharidases activities: I. Enzymatic properties and distribution in the digestive tract of conventional and germfree animals, *Arch. Biochem. Biophys.* **190**:150.

Das, M., and Radhakrishnan, A.N., 1976, Role of peptidases and peptide transport in the intestinal absorption of proteins, *Wld. Rev. Nutr. Diet.* **24**:58.

Delluva, A.M., Markley, K., and Daview, R.E., 1968, The absence of gastric urease in germfree animals, *Biochim. Biophys. Acta* **151**:646.

Demarne, Y., Sacquet, E., Flanzy, J., Garnier, H., and Francois, A.C., 1970a, Influence de la flore intestinale sur l'utilisation digestive des acides gras chez le rat, *Ann. Biol. Anim. Biochim. Biophys.* **10**:175.

Demarne, Y., Sacquet, E., Flanzy, J., Garnier, H., and Francois, A.C., 1970b, Influence de la flore intestinale sur l'utilisation digestive des acides gras chez le rat, *Ann. Biol. Anim. Biochim. Biophys.* **10**:369.

Demarne, Y., Sacquet, E., and Garnier, H., 1972, Le flore gastro-intestinale et la digestion des matieres grasses chez le monogastrique, *Ann. Biol. Anim. Biochim. Biophys.* **12**:509.

Demarne, Y., Sacquet, E., Lecourtier, M.J., and Flanzy, J., 1979, Comparative study of endogenous fecal fatty acids in germfree and conventional rats, *Am. J. Clin. Nutr.* **32**:2027.

Dickinson, A.B., Gustafsson, B.E., and Norman, A., 1971, Determination of bile acid conversion potencies of intestinal bacteria by screening *in vitro* and subsequent establishment in germfree rats, *Acta Pathol. Microbiol. Scand. Sect. B.* **79**:691.

Donaldson, R.M. Jr., 1962, Malabsorption of Co^{60}-labeled cyanocobalamin in rats with intestinal diverticula: I. Evaluation of possible mechanism, *Gastroenterology* **43**:271.

Donati, R.M., McLaughlin, M.M., Levri, E.A., Berman, A.R., and Stromberg, L.V.R., 1969, The response of iron metabolism to the microbial flora studies on germfree mice, *Proc. Soc. Exp. Biol. Med.* **130**:920.

Dowdy, R.P., Herman, Y.F., and Sauberlich, H.E., 1969, Effect of germfree status on ^{64}Cu excretion by the rat, *Proc. Soc. Exp. Biol. Med.* **130**:1294.

Draser, B.S., Hill, M.J., and Shiner, M., 1966, The deconjugation of bile salts by human intestinal bacteria, *Lancet* **1**:1237.

Ducluzeau, R., Raibaud, P., Dickinson, A.B., Sacquet, E., and Mocquot, G., 1966, Hydrolyse de l'uree *in vitro* et *in vivo* dans le caecum des rats gnotobiotiques par differentes souches bacteriennes isolees du tube digestif de rat conventionnels, *C. R. Acad. Sci.* **262**:944.

Duncan, C.L., and Strong, D.H., 1969, Ileal loop fluid accumulation and production of diarrhea in rabbits by cell-free products of *Clostridium perfringens*, *J. Bacteriol.* **100**:86.

Dupont, J.R., Jervis, H.R., and Sprinz, H., 1965, Auerbach's plexus of the rat caecum in relation to the germfree state, *Comp. Neurol.* **125**:11.

Edwards, H.M. Jr., and Boyd, F.M., 1963, Effect of germfree environment on [47]Ca metabolism, *Poult. Sci.* **42**:1030.

Eichholz, A., and Crane, R.K., 1965, Studies on the organization of the brush border in intestinal epithelial cells, *J. Cell. Biol.* **26**:687.

Ekdahl, P.H., and Sjövall, J., 1955, Metabolism of deoxycholic acid in the rabbit, *Acta Physiol. Scand.* **34**:1.

Evans, G.W., and LeBlanc, F.N., 1976, Copper-binding protein in rat intestine: amino acid composition and function, *Nutr. Rep. Int.* **14**:281.

Evrard, E., Hoet, P.P., Eyssen, H., Charlier, H., and Sacquet, E., 1964, Fecal lipids in germfree and conventional rats, *Br. J. Exp. Path.* **45**:409.

Evrard, E., Sacquet, E., Raibaud, P., Charlier, H., Dickinson, A., Eyssen, H., and Huet, P.P., 1965, Studies on conventional and gnotobiotic rats: effect of intestinal bacteria on fecal lipids and fecal sterols, *Ernährungsforschung* **10**:257.

Eyssen, H., and Parmentier, G., 1974, Biohydrogenation of sterols and fatty acids by the intestinal microflora, *Am. J. Clin. Nutr.* **27**:1329.

Eyssen, H.J., and Parmentier, G.G., 1979, Influence of the microflora of the rat on the metabolism of fatty acids, sterols and bile salts in the intestinal tract, in: *Clinical and Experimental Gnotobiotics* (T. Fliedner, H. Heit, D. Niethammer, and H. Pflieger, eds.), p. 39, Gustav Fischer Verlag, Stuttgart and New York.

Eyssen, H., Piessens-Denef, M., and Parmentier, G., 1972, Role of the cecum in maintaining Δ^5 steroid and fatty acid-reducing activity of the rat intestinal microflora, *J. Nutr.* **102**:1501.

Eyssen, H., DePauw, G., and DeSomer, P., 1973, Biohydrogenation of long-chain fatty acids by intestinal microorganisms, in: *Germfree Research* (J.B. Heneghan, ed.), p. 277, Academic Press, New York.

Eyssen, H., Parmentier, G., and Mertens, J.A., 1976, Sulfated bile acids in germfree and conventional mice, *Eur. J. Biochem.* **66**:507.

Eyssen, H., Huijghebaert, S., DePauw, G., and Stragier, J., 1981, Studies in gnotobiotic rats: Metabolism of bile salts by strictly anaerobic intestinal microorganisms, 7th International Symposium on Gnotobiology, Tokyo, Japan, p. 64 (abstract).

Eyssen, H., Parmentier, G., Martens, J., and DeSomer, P., 1973, The bile acids of the mouse: effect of microflora, age and sex, in: *Germfree Research* (J.B. Heneghan, ed.), p. 271, Academic Press, New York.

Fogel, M.R., and Gray, G.M., 1969, Human intestinal β-galactosidases and lactase deficiency, *Clin. Res.* **17**:111.

Ford, D.J., and Coates, M.E., 1971, Absorption of glucose and vitamines of the B complex by germfree and conventional chicks, *Proc. Nutr. Soc.* **30**:10A.

Forrester, R.H., Conrad, M.E. Jr., and Crosby, W.H., 1962, Measurement of total body iron 59 in animals using whole body liquid scintillation detectors, *Proc. Soc. Exp. Biol. Med.* **11**:115.

Fry, R.J.M., Kisieleski, W.E., Brennan, P.C., Fritz, T.E., and Staffeldt, E.F., 1966, The lack of effects of monocontaminants on the intestinal mucosa of germfree mice, *Gen. Radiat. Biol.* **72**:25.

Galjaard, H., Meer-Fieggen, W. van der, and Giesen, J., 1972, Feedback control by functional villus cells on cell proliferation and maturation in intestinal epithelium, *Exp. Cell Res.* **73**:197.

Gangarosa, E.J., Beisel, W.R., Benyajati, C., Sprinz, H., and Piyaratin, P., 1960, The nature of the gastrointestinal lesion in Asiatic cholera and its relation to pathogenesis: a biopsy study, *Am. J. Trop. Med. Hyg.* **9**:125.

Garnier, H., and Sacquet, E., 1969, Absorption apparente et retention du sodium, du potassium, du calcium et du phosphase chez le rat axenique et chez le rat holoxenique, *C. R. Acad. Sci.* **269**:379.

Geever, E.F., Kan, D., and Levenson, M., 1968, Effect of bacterial flora on iron absorption in the rat, *Gastroenterology* **55**:690.

Genell, S., Gustafsson, B.E., and Ohlsson, K., 1976, Quantitation of active pancreatic endopeptidases in intestinal contents of germfree and conventional rats, *Scand. J. Gastroenterology* **11**:757.

Giddens, W.E. Jr., Pleasants, J.R., Wostmann, B.S., and Whitehair, C.K., 1972, Iron and zinc absorption and metabolism in germfree rats, *Nutr. Rev.* **30**:148.

Gordon, H.A., 1967, A substance acting on smooth muscle in intestinal contents of germfree animals, *Ann. N.Y. Acad. Sci.* **147**:83.

Gordon, H.A., 1968, Is the germfree animal normal? A review of its anomalies in young and old age, in: *The Germfree Animal in Research* (M.E. Coates, ed.), p. 127, Academic Press, New York.

Gordon, H.A., and Bruckner-Kardoss, E., 1958-1959, The distribution of reticulo-endothelial elements in the intestinal mucosa and submucosa of germfree monocontaminated and conventional chicks orally treated with penicillin, *Antibiotics Annual* p. 1012.

Gordon, H.A., and Bruckner-Kardoss, E., 1961a, Effect of normal flora on intestinal surface area, *Am. J. Physiol.* **201**:175.

Gordon, H.A., and Bruckner-Kardoss, E., 1961b, Effects of the normal microbial flora on various tissue elements of the small intestine, *Acta Anat.* **44**:210.

Gordon, H.A., and Nakamura, S., 1969, Elevated levels of colloid osmotic pressure in cecal contents of germfree rats, in: Proceedings, 8th Annual Meeting Assoc. Gnotobiotics, Oak Ridge, Tenn. (abstract).

Gordon, H.A., and Pesti, L., 1971, The gnotobiotic animal as a tool in the study of host microbial relationships, *Bact. Rev.* **35**:390.

Gordon, H.A., and Wostmann, B.S., 1973, Chronic mild diarrhea in germfree rodents: a model portraying host-flora synergism, in: *Germfree Research* (J.B. Heneghan, ed.), p. 593, Academic Press, New York.

Gordon, H.A., Bruckner-Kardoss, E., and Kan, D., 1961, Effects of normal flora on structural and absorptive characteristics of the intestine, in: Proceedings of the 5th International Congress of Nutrition, Washington, D.C., Sept. 1-7, 1960, *Fed. Proc.* Suppl. 7, p. 21.

Gordon, H.A., Bruckner-Kardoss, E., Staley, T.E., Wagner, M., and Wostmann, B.S., 1966, Characteristics of the germfree rat, *Acta Anat.* **64**:367.

Gordon, J.H., and Dubos, R., 1970, The anaerobic bacterial flora of the mouse cecum, *J. Exp. Med.* **132**:251.

Gracey, M., Burke, V., Thomas, J.A., and Stone, D.E., 1975, Effect of microorganisms isolated from the upper gut of malnourished children on intestinal sugar absorption *in vivo*, *Am. J. Clin. Nutr.* **28**:841.

Gray, G.M., and Ingelfinger, F.J., 1966, Intestinal absorption of sucrose in man: interrelation of hydrolysis and monosaccharide product absorption, *J. Clin. Invest.* **45**:388.

Griffiths, M., and Davies, D., 1963, The role of the soft pellets in the production of lactic acid in the rabbit stomach, *J. Nutr.* **80**:171.

Guenet, J.L., Sacquet, E., Gueneau, G., and Meslin, J.C., 1970, Action de la microflore totale du rat sur l'activite mitotique des crypts de lieberkühm, *C. R. Acad. Sci.* **270**:3087.

Gustafsson, B.E., 1948, Germfree rearing of rat, general technique, *Acta Pathol. Microbiol. Scand. Suppl.* **73**:1.

Gustafsson, B.E., and Maunsbach, A.B., 1971, Ultrastructure of the enlarged caecum in germfree rats, *Z. Zellforsch. Mikrosk. Anat.* **120**:555.

Gustafsson, B.E., and Norman, A., 1962, Urinary calculi in germfree rats, *J. Exp. Med.* **116**:273.

Gustafsson, B.E., Bergström, S., Lindstedt, S., and Norman, A., 1957, Turnover and nature of fecal bile acids in germfree and infected rats fed cholic acid 24-^{14} bile acids and steroids, *Proc. Soc. Exp. Biol. Med.* **94**:467.

Gustafsson, B.E., Norman, A., and Sjövall, J., 1960, Influence of *E. coli* infection on turnover and metabolism of cholic acid in germfree rats, *Arch. Biochem. Biophys.* **91**:93.

Gustafsson, B.E., Midvedt, T., and Norman, A., 1966, Isolated fecal microorganisms capable of 7α-dehydroxylating bile acids, *J. Exp. Med.* **123**:413.

Gustafsson, B.E., Midvedt, T., and Norman, A., 1968, Metabolism of cholic acid in germfree animals after the establishment in the intestinal tract of deconjugating and 7α-dehydroxylating bacteria, *Acta Pathol. Microbiol. Scand.* **72**:433.

Gustafsson, B.E., Einarsson, K., and Gustafsson, J., 1975, Influence of cholesterol feeding on liver microsomal metabolism of steroids and bile acids in conventional and germfree rats, *J. Biol. Chem.* **250**:8496.

Gustafsson, B.E., Angelin, B., Björkhem, I., Einarsson, K., and Gustafsson, J., 1981, Effects of feeding chenodeoxycholic acid on metabolism of cholesterol and bile acids in germfree rats, *Lipids* **16**:228.

Gyles, C.L., and Barnum, D.A., 1969, A heat-labile enterotoxin from strains of *Escherichia coli* enteropathogenic for pigs, *J. Infect. Dis.* **120**:419.

Hampton, J.C., and Rosario, B., 1965, The attachment of microorganisms to epithelial cells in the distal ileum of the mouse, *Lab. Invest.* **14**:1464.

Harmon, B.G., Becker, D.E., Jensen, A.H., and Baker, D.H., 1968, Influence of microbiota on metabolic fecal nitrogen in rats, *J. Nutr.* **96**:391.

Hawksworth, G., Drasar, B.S., and Hill, M.J., 1971, Intestinal bacteria and the hydrolysis of glycosidic bonds, *J. Med. Microbiol.* **4**:451.

Heneghan, J.B., 1963, Influence of microbial flora on xylose absorption in rats and mice, *Am. J. Physiol.* **205**:417.

Heneghan, J.B., 1979, Enterocyte kinetics, mucosal surface area and mucus in gnotobiotes, in: *Clinical and Experimental Gnotobiotics* (T. Fliedner, H. Heit, D. Niethammer, and H. Pflieger, eds.), p. 19, Gustav Fischer Verlag, Stuttgart and New York.

Heneghan, J.B., and Gordon, H.A., 1974, Characteristics of the gastrointestinal tract of germfree beagles, Proceedings, 11th Annual Meeting for the Assoc. of Gnotobiotics, Guelph, Ontario, Canada (abstract).

Heneghan, J.B., Gordon, H.A., and Miniats, O.P., 1979, Intestinal mucosal surface area and goblet cells in germfree and conventional piglets, in: *Clinical and Experimental Gnotobiotics* (T. Fliedner, H. Heit, D. Niethammer, and H. Pflieger, eds.), p. 107, Gustav Fischer Verlag, Stuttgart and New York.

Hershovic, T., Katz, J., Floch, M.H., Spencer, R.P., and Spiro, H.M., 1967, Small intestinal absorption and morphology in germfree, monocontaminated and conventionalized mice, *Gastroenterology* **52**:1136.

Hoffman, A.F., Mosbach, E.H., and Sweeley, C.C., 1964, Bile acid composition of bile from germfree rabbits, *Biochim. Biophys. Acta* **176**:204.

Hollander, D.K., Muralidhara, S., and Rim, E., 1976, Colonic absorption of bacterially synthesized vitamin K₂ in the rat, *Am. J. Physiol.* **230**:251.

Hoskins, L.C., 1968, Bacterial degradation of gastrointestinal mucus: II. Bacterial origin of fecal A B H (O) blood group antigen-destroying enzymes, *Gastroenterology* **54**:218.

Hoskins, L.C., and Zamcheck, N., 1968, Bacterial degradation of gastrointestinal mucus: I. Comparison of mucus constituents in the stool of germfree and conventional rats, *Gastroenterology* **54**:210.

Hotzel, D., and Barnes, R.H., 1966, Contribution of the intestinal microflora to the nutrition of the rat, *Vitam. Horm.* **24**:115.

Hudson, J.A., and Luckey, T.D., 1964, Bacteria induced morphologic changes, *Proc. Soc. Exp. Biol. Med.* **116**:628.

Itoh, K., and Mitsuoka, T., 1980, Production of gnotobiotic mice with normal physiological functions: I. Selection of useful bacteria from feces of conventional mice *Z. Versuchstierkd.* **22**:173.

Jervis, H.R., and Biggers, D.C., 1964, Mucosal enzymes in the cecum of conventional and germfree mice, *Anat. Rec.* **148**:591.

Juhr, N.C., 1980, Intestinale enzymaktivität bei keimfreien und konventionelle ratten und maüsen, *Z. Versuchstierkd.* **22**:197.

Kelleher, J., and Bruckner, G., 1977, Sodium and microbial effects on intestinal absorption of amino acids, Proceedings of the Annual Meeting for Gnotobiotics at Saranac Lake, N.Y. (abstract).

Kellog, T.F., and Wostmann, B.S., 1969, Fecal neutral steroids and bile acids from germfree rats, *J. Lipid. Res.* **10**:495.

Kendall, A.I., and Chinn, H., 1938, Decomposition of ascorbic acid by certain bacteria; studies in bacterial metabolism, *J. Infect. Dis.* **62**:330.

Kenworthy, R., and Allen, W.D., 1966, Influence of diet and bacteria on small intestinal morphology, with special reference to early weaning and *Escherichia coli.*, *J. Comp. Pathol.* **76**:291.

Kessner, D.M., and Epstein, F.H., 1966, Effect of magnesium deficiency on gastrointestinal transfer of calcium, *Proc. Soc. Exp. Biol. Med.* **122**:721.

Keusch, G.T., Mata, L.J., and Grady, G.F., 1970, Malabsorption with cholera toxin, *Clin. Res.* **18**:442.

Khoury, K.A., Floch, M.H., and Hersh, T., 1969, Small intestinal mucosal cell proliferation and bacterial flora in the conventionalization of the germfree mouse, *J. Exp. Med.* **130**:659.

Kirchgessner, M., 1965, De l'approvisionnement en oligo-elements et de leur disponibilite, Proceedings of the European Association for Animal Production, Noordirigk.

Komai, M., and Kimura, S., 1980, Gastrointestinal responses to graded levels of cellulose feeding in conventional and germfree mice, *J. Nutr. Sci. Vitaminol.* **26**:389.

Knight, P.L. Jr., and Wostmann, B.S., 1964, Influence of *Salmonella typhimurium* on ileum and spleen morphology of germfree rats, *Proc. Ind. Acad. Sci.* **72**:78.

Larner, J., and Gillespie, R.E., 1957, Gastrointestinal digestion of starch, *J. Biol. Chem.* **225**:279.

Latymer, E.A., and Coates, M.E., 1981, The influence of microorganisms and of stress on the chick's requirement for pantothenic acid, *Br. J. Nutr.* **45**:441.

Lee, A., Gordon, J., Lee, C.J., and Dubos, R., 1971, The mouse intestinal microflora with emphasis on the strict anaerobes, *J. Exp. Med.* **133**:339.

Leong, P.C., 1937, Vitamin B₁ in animal organisms; quantitative study of metabolism of B₁ in rats, *Biochem J.* **31**:373.

Lepkovsky, S., Wagner, M., Furuta, F., Ozone, K., and Koike, T., 1964, The proteases, amylase and lipase of the intestinal contents of germfree and conventional chickens, *Poult. Sci.* **43**:722.

Lepkovsky, S., Furuta, F., Ozone, K., Koike, T., and Wagner, M., 1966, The proteases, amylase and lipase of the pancreas and intestinal contents of germfree and conventional rats, *Br. J. Nutr.* **20**:257.

Lesher, S., Walburg, H.E. Jr., and Sacher, G.A. Jr., 1964, Generation cycle in the duodenal crypt cells of germfree and conventional mice, *Nature* (London) **202**:884.

Levenson, S.M., and Tennant, B., 1963, Contributions of intestinal microflora to the nutrition of the host animal: some metabolic and nutritional studies with germfree animals, *Fed. Proc.* **22**:109.

Levenson, S.M., Tennant, B., Geever, E., Laundy, L.T.R., and Daft, F., 1962, Influence of microorganisms on scurvy, *Arch. Intern. Med.* **110**:693.

Lih, H., and Baumann, C.A., 1951, Effects of certain antibiotics on growth of rats fed diets limiting in thiamine, riboflavin or pantothenic acid, *J. Nutr.* **45**:143.

Lindstedt, G., Lindstedt, S., and Gustafsson, B.E., 1965, Mucus in intestinal contents of germfree rats, *J. Exp. Med.* **121**:201.

Linkswiler, H., Baumann, C.A., and Snell, E.E., 1951, Effect of aureomycin on response of rats to various forms of vitamin B_6, *J. Nutr.* **43**:565.

Loesche, W.J., 1968a, Protein and carbohydrate composition of cecal contents of gnotobiotic rats and mice, *Proc. Soc. Exp. Biol. Med.* **128**:195.

Loesche, W.J., 1968b, Accumulation of endogenous protein in the cecum of the germfree rat, *Proc. Soc. Exp. Biol. Med.* **129**:380.

Loesche, W.J., 1969a, Accumulation of endogenous carbohydrate containing compounds in the caecum of the germfree rat, *Proc. Soc. Exp. Biol. Med.* **131**:387.

Loesche, W.J., 1969b, Effect of bacterial contamination on cecal size and cecal contents of gnotobiotic rodents, *J. Bact.* **99**:520.

Loeschke, K., and Gordon, H.A., 1970, Water movement across the cecal wall of the germfree rat, *Proc. Soc. Exp. Biol. Med.* **133**:1217.

Loeschke, K., Ulrich, E., and Halbach, R., 1971, Macromolecules induce increased sodium transport in rat cecum, in: Proceedings of the 9th International Congress of the Union of Physiological Sciences, Munich, p. 351 (abstract).

Loeschke, K., Uhlich, E., and Kuane, R., 1974, Stimulation of sodium transport and Na-K ATPase activities in the hypertrophing rat cecum, *Plügus Arch.* **346**:233.

Luckey, T.D., 1963, *Germ-free Life and Gnotobiology* (T.D. Luckey, ed.), p. 252, Academic Press, New York.

Madsen, D., and Wostmann, B.S., 1975, Conversion of hyodeoxycholate to ω-muricholate by the conventional rat: implications of this hepatic potential for control of cholesterol pools, in: Proceedings, 5th International Symposium on Gnotobiology, Karolinska Institute, Stockholm, p. 24 (abstract).

Mameesh, M.S., Webb, R.E., Norton, H.V., and Johnson, B.C., 1959, The role of coprophagy in the availability of vitamins synthesized in the intestinal tract with antibiotic feeding, *J. Nutr.* **69**:81.

Matsuzawa, T., and Wilson, T., 1965, The intestinal mucosa of germfree mice after whole body x-irradiation into 3 kiloroentgens, *Rad. Res.* **25**:15.

Meslin, J.C., 1971, Action de la microflore totale du rat sur l'epithelium intestinal, estimation de la surface absorbante, *Ann. Biol. Anim. Biochim. Biophys.* **11**:334.

Meslin, J.C., Sacquet, E., and Guenet, J.L., 1973, Action de la flore bacterienne sur la morphologie et la surface de la muqueuse de l'intestin grele du rat, *Ann. Biol. Anim. Biochim. Biophys.* **13**:203.

Meslin, J.C., Sacquet, E., and Raibaud, P., 1974, Action d'une flore microbienne que ne deconjugue pas les sels biliares sur la morphologie et le renouvellement cellulaire de la muqueuse de l'intestin quele du rat, *Ann. Biol. Anim. Biochim. Biophys.* **14**:709.

Meslin, J.C., Sacquet, E., and Riottot, M., 1981, Effect of various modifications in the diet on ileal epithelium renewal in germfree and conventional rats, *Reprod. Nutr. Develop.* **21**:651.

Metchnikoff, E., 1903, Les microbes intestinaux, *Bull. Inst. Pasteur* **1**:265.

Mickelsen, O., 1956, Intestinal synthesis of vitamins in the nonruminant, *Vitam. Horm.* **14**:1.

Midvedt, T., and Norman, A., 1967, Bile acid transformations by microbial strains belonging to genera found in intestinal contents, *Acta Pathol. Microbiol. Scand.* **71**:629.

Midvedt, T., and Norman, A., 1968, Parameters in 7-dehydroxylation of bile acids by anaerobic lactobacilli, *Acta. Pathol. Microbiol. Scand.* **72**:313.

Miller, D., and Crane, R.K., 1961a, The digestive function of the epithelium of the small intestine: I. An intracellular locus of disaccharide and sugar phosphate ester hydrolysis, *Biochim. Biophys. Acta* **52**:281.

Miller, D., and Crane, R.K., 1961b, The digestive function of the epithelium of the small intestine: II. Localization of the dissacharide hydrolysis in the isolated brush border portion of the intestine, *Biochim. Biophys. Acta* **52**:293.

Miller, H.T., and Luckey, T.D., 1963, Intestinal synthesis of folic acid in monoflora chicks, *J. Nutr.* **80**:236.

Miller, W.S., 1967, Protein utilization in germfree and conventional chicks given a purified diet, *Proc. Nutr. Soc.* **26**:X.

Miniats, O.P., and Valli, E.P., 1973, The gastrointestinal tract of gnotobiotic pigs, in: *Germfree Research* (J.B. Heneghan, ed.), p. 575, Academic Press, New York.

Mott, G.E., Moore, R.W., Redmond, H.E., and Reiser, R., 1973, Lowering of serum cholesterol by intestinal bacteria in cholesterol fed piglets, *Lipids* **8**:428.

Nakamura, S., and Gordon, H.A., 1973, Threshold levels of NaCl upholding solute-coupled water transport in the cecum of germfree and conventional rats, *Proc. Soc. Exp. Biol. Med.* **142**:1336.

Nakao, K., and Levenson, S.M., 1967, Atypical mitochondrial morphology of the intestinal absorptive cells of the germfree rat, *Experientia* **23**:494.

Nencki, M., 1886, Bemerkung su einer bemerkung Pasteur's, *Arch. Exp. Pathol. Pharmacol.* **20**:385.

Oace, S.M., and Abbot, J.M., 1972, Methylmalonate, formiminoglutamate and aminoimidazolecarboxamide excretion of vitamin B_{12}-deficient germfree and conventional rats, *J. Nutr.* **102**:17.

Palmer, M.E., and Rolls, B.A., 1981, The absorption and secretion of calcium in the gastrointestinal tract of germfree and conventional chicks, *Br. J. Nutr.* **46**:549.

Pasteur, L., 1885, Observations relatives a la note precedente de M. Ducraux, *C. R. Acad. Sci. (Paris)* **100**:68.

Pepper, W.F., Slinger, S.J., and Motzok, I., 1952, Effect of aureomycin on the manganese requirement of chicks fed varying levels of salt and phosphorus, *Poult. Sci.* **31**:1054.

Perlman, D., 1959, in: *Advances in Applied Microbiology*, Vol. 1 (W. Umbreit, ed.), p. 87, Academic Press, New York.

Phillips, A.W., and Newcomb, H.R., 1965, Gnotobiotic methodology, *Bacteriol. Proc.*, p. 64 (abstract).

Phillips, A.W., Newcomb, H.R., Smith, J.E., and LaChapelle, R., 1961, Serotonin in the small intestine of conventional and germfree chicks, *Nature (London)* **192**:380.

Phillips, B.P., Wolfe, P.A., and Gordon, H.A., 1959, Studies on rearing the guinea pig germfree, *Ann. N.Y. Acad. Sci.* **78**:183.

Pleasants, J., Bruckner-Kardoss, V., Beaver, M., and Wostmann, B.S., 1981, Effect of fiber on physiological and metabolic parameters of germfree mice fed chemically defined diet, Proceedings 19th Annual Meeting of the Association for Gnotobiotics, Buffalo, New York. p. 19 (abstract).

Raibaud, P., Dickinson, A.B., Sacquet, E., Charlier, H., and Mocquot, G., 1966, Le microflore du tube digestif du rat: IV. Implementation controlée chez le rat gnotobiotique de differents genres microbiens isolés du rat conventionnel, *Ann. Inst. Pasteur (Paris)* **111**:193.

Ranken, R., Wilson, R., Bealmer, P.M., 1971, Increased turnover of intestinal mucosal cells of germfree mice induced by cholic acid, *Proc. Soc. Exp. Biol. Med.* **138**:270.

Reddy, B.S., 1971, Calcium and magnesium absorption: role of intestinal microflora, *Fed. Proc.* **30**:1815.

Reddy, B.S., 1972, Studies on the mechanism of calcium and magnesium absorption in germfree rats, *Arch. Biochem. Biophys.* **149**:15.

Reddy, B.S., and Pollard, M., 1972, Effect of intestinal microflora on age associated changes in hepatic xanthine oxidase and in bone mineral composition in rats, *J. Nutr.* **102**:299.

Reddy, B.S., and Wostmann, B.S., 1966, Intestinal disaccharidase activities in the growing germfree and conventional rat, *Arch. Biochem. Biophys.* **113**:609.

Reddy, B.S., Wostmann, B.S., and Pleasants, J.R., 1965, Iron, copper, and manganese in germfree and conventional rats, *J. Nutr.* **86**:159.

Reddy, B.C., Pleasants, J.R., and Wostmann, B.S., 1968, Effect of dietary carbohydrates on intestinal disaccharidases in germfree and conventional rats, *J. Nutr.* **95**:413.

Reddy, B.S., Pleasants, J.R., and Wostmann, B.S., 1969a, Pancreatic enzymes in germfree and conventional rats fed chemically defined water-soluble diet free from natural substrates, *J. Nutr.* **97**:327.

Reddy, B.S., Pleasants, J.R., and Wostmann, B.S., 1969b, Effect of intestinal microflora on calcium, phosphorus and magnesium metabolism in rats, *J. Nutr.* **99**:353.

Reddy, B.S., Wostmann, B.S., and Pleasants, J.R., 1969c, Protein metabolism in germfree rats fed chemically defined, water soluble diet and semisynthetic diet, in: *Germfree Biology* (E.A. Mirand and N. Black, eds.), p. 301, Plenum Press, New York.

Reddy, B.S., Pleasants, J.R., and Wostmann, B.S., 1972, Effect of intestinal microflora on iron and zinc metabolism and on activities of metalloenzymes in rats, *J. Nutr.* **102**:101.

Reyniers, J.A., Trexler, P.C., and Ervin, R.F., 1946, Rearing germfree albino rats, in: *Lobund Reports No. 1* (J.A. Reyniers, ed.), p. 1, University of Notre Dame Press, Notre Dame, Ind.

Reyniers, J.A., Wagner, M., Luckey, T.D., and Gordon, H.A., 1960, Survey of germfree animals: the White Wyandotte Bantam and White Leghorn chicken, in: *Lobund Reports No. 3* (J.A. Reyniers, ed.), p. 1, University of Notre Dame Press, Notre Dame, Ind.

Reyniers, J.A., Trexler, P.C., Ervin, R.F., Wagner, M., Luckey, T.D., and Gordon, H.A., 1969, A complete life cycle in the germfree Bantam chicken, *Nature (London)* **163**:687.

Rogers, W.E. Jr., Bieri, J.G., and McDaniel, E.G., 1971, Vitamin A deficiency in the germfree state, *Fed. Proc.* **30**:1773.

Sacquet, E., Charlier, H., Raibaud, P., Dickinson, A.B., Evrard, E., and Eyssen, H., 1966, Etiologie bacterienne de la stéatorrhée observie chez le rat porteur d'un cul-de-sac intestinal, *C. R. Acad. Sci. (Paris)* **262**:786.

Sacquet, E., Guenet, J.L., Garnier, H., and Meslin, J.C., 1971, Influence de deux modifications chirurgicales de l'intestin grele sur le renouvellement des enterocytes du rat: la formation d'une anse aveugle chez le rat holoxenique, l'ablation de caecum chez le rat axenique, *C. R. Acad. Sci.* **272**:841.

Sacquet, E., Guenet, J.L., Raibaud, P., and Meslin, J.C., 1972, Cecal reduction in gnototoxenic rats. Effects of cecectomy on gastrointestinal transit and renewal of small intestine epithelium, Proceedings, 4th International Symposium on Germfree Research, New Orleans, Louisiana, p. 89 (abstract).

Sacquet, E., Lachkar, M., Mathis, C., and Raibaud, P., 1973, Cecal reduction in "gnototoxenic" rats, in: *Germfree Research*: Biological Effects of Gnotobiotic Environments, (J.B. Heneghan, ed.), p. 545, Academic Press, New York.

Sakai, K., Makino, T., Kawai, Y., and Mutai, M., 1980, Intestinal microflora and bile acids: effect of bile acids on the distribution of microflora and bile acid in the digestive tract of the rat, *Microbiol. Immunol.* **24**:187.

Salter, D.N., 1973, The influence of gut microorganisms on utilization of dietary protein, *Proc. Nutr. Soc.* **32**:65.

Salter, D.N., and Coates, M.E., 1970, The influence of the gut microflora on the digestion of protein in chicks, Proceedings, 8th International Congress of Nutrition, Prague, p. 425.

Salter, D.N., and Coates, M.E., 1971, The influence of the microflora of the alimentary tract on protein digestion in the chick, *Br. J. Nutr.* **26**:55.

Salter, D.N., and Fulford, R.J., 1974, The influence of the gut microflora on the digestion of dietary and endogenous proteins: studies of the amino acid composition of the excreta of germfree and conventional chicks, *Br. J. Nutr.* **32**:625.

Sauberlich, H.E., 1952, Effect of aureomycin and penicillin upon vitamin requirements of rat, *J. Nutr.* **46**:99.

Savage, D., 1977, Microbial ecology of the gastrointestinal tract, *Ann. Rev. Microbiol.* **31**:107.

Savage, D.C., 1979, Characterization of the anaerobic flora in man and animals, in: *Clinical and Experimental Gnotobiotics* (T. Fliedner, H. Heit, D. Niethammer, and H. Pflieger, eds.), p. 163, Gustav Fischer Verlag, Stuttgart and New York.

Savage, D.C., and McAllister, J.S., 1971, Cecal enlargement and microbial flora in suckling mice given antibacterial drugs, *Infect. Immun.* **3**:342.

Savage, D.C., Dubos, R.J., and Schaedler, R.W., 1968, The gastrointestinal epithelium and its autochtonous bacterial flora, *J. Exp. Med.* **127**:67.

Schendel, H.E., and Johnson, B.C., 1954, Studies of antibiotics in weanling rats administered suboptimum levels of certain B vitamins orally and parenterally, *J. Nutr.* **54**:461.

Schottelius, M., 1902, Die bedentung der darmbakterien fur die ernährung II, *Arch. Hyg.* **42**:48.

Sharrer, E., and Riedel, G., 1972, Resorptionsstudien bei keimfreien küken, *Z. Tierphysiol. Tierernahr. Futtermittelkdc.* **30**:264.

Siddons, R.C., and Coates, M.E., 1972, The influence of the intestinal microflora on disaccharidase activities in the chick, *Br. J. Nutr.* **27**:101.

Simonetta, M., Faelli, A., Cremaschi, D., and Gordon, H.A., 1975, Electrical resistance and ATPase levels in the cecal wall of germfree and conventional rats, *Proc. Soc. Exp. Biol. Med.* **150**:541.

Skelly, B.J., Trexler, P.C., and Tanami, J., 1962, Effect of a *Clostridium* species upon cecal size of gnotobiotic mice, *Proc. Soc. Exp. Biol. Med.* **110**:455.

Smith, H.W., and Halls, S., 1967, Studies on *Escherichia coli* enterotoxin, *J. Pathol. Bacteriol.* **93**:531.

Smith, J.C. Jr., McDaniel, E.C., McBean, L.D., Doft, F.S., and Halsted, J.A., 1972, Effect of microorganisms upon zinc metabolism using germfree and conventional rats, *J. Nutr.* **102**:711.

Soergel, K.H., and Hofmann, A.F., 1972, Altered regulatory mechanisms in disease, in: *Pathophysiology* (E. D. Frolich, ed.), Second Edition, p. 515, J.B. Lippincott, Philadelphia.

Speck, M.L., 1976, Interactions between lactobacilli and man, *J. Dairy Sci.* **59**:338.

Sprinz, H., 1962, Morphological response of intestinal mucosa to enteric bacteria and its implication for sprue and Asiatic cholera, *Fed. Proc.* **21**:57.

Sprinz, H., Kundel, D.W., Dammin, G.J., Horowitz, R.E., Schneider, H., and Formal, S.B., 1961, The response of the germfree guinea pig to oral bacterial challenge with *Escherichia coli* and *Shigella flexerne, Am. J. Pathol.* **39**:681.

Staley, T.E., Wynn Jones, E., and Corley, L.D., 1969, Attachment and penetration of *Escherichia coli* into intestinal epithelium of the ileum in newborn pigs, *Am. J. Pathol.* **56**:371.

Staley, T.E., Corley, L.D., and Jones, E.W., 1970a, Early pathogenesis of colitis in neonatal pigs monocontaminated with *Escherichia coli* fine structural changes in the colonic epithelium, *Am. J. Dig. Dis.* **15**:923.

Staley, T.E., Corley, L.D., and Jones, E.W., 1970b, Early pathogenesis of colitis in neonatal pigs monocontaminated with *Escherichia coli* fine structural changes in the circulatory compartments of the *lamina propria* and submucosa, *Am. J. Dig. Dis.* **15**:937.

Stern, P., Kosak, R., and Misirliga, E., 1954, Beitrag zur frage der eisenresorption, *Experientia* **10**:227.

Stevens, N.C., and Heneghan, J.B., 1976, Intestinal surface area in conventional and germfree rats and dogs, Proceedings, Association for Gnotobiotics Annual Meeting Lobund Laboratories, University of Notre Dame, p. 16 (abstract).

Stroud, R.M., 1974, A family of protein cutting proteins, *Sci. Am.* **231**:24.

Stryer, L., 1975, in: *Biochemistry* (L. Stryer, ed.), p. 153, W.H. Freeman, San Francisco.

Sugiyama, T., and Stewart, W.F., 1975, Mytotic indices of intestinal epithelium in germfree and conventional rats enterally exposed to germfree rat cecal contents, Proceedings, 5th International Symposium on Gnotobiology, Karolinska Institute, Stockholm, p. 23 (abstract).

Supplee, N., 1960, The effect of antibiotic on the response of poults to dietary corn oil, *Poult. Sci.* **39**:227.

Syed, S.A., Abrams, G.D., and Freter, R., 1970, Efficiency of various intestinal bacteria in assuming normal function of enteric flora after association with germfree mice, *Infect. Immun.* **2**:376.

Szabo, J., 1977, Peptide hydrolase activities in the intestine of germfree and conventional rats, Proceedings, Annual Meeting, Association for Gnotobiotics, Saranac Lake, New York (abstract).

Szabo, J., 1979, Protein, carbohydrate and fat degrading enzymes in the intestine of germfree and conventional piglets, in: *Clinical and Experimental Gnotobiotics* (T. Fliedner, H. Heit, D. Niethammer, and H. Pflieger, eds.), p. 125, Gustav Fischer Verlag, Stuttgart and New York.

Talalay, P., 1957, Enzymatic mechanisms in steroid metabolism, *Physiol. Rev.* **37**:362.

Tennant, B., Reina-Guerra, M., and Marrold, D., 1971, Influence of micro-organisms on intestinal absorption, *Ann. N.Y. Acad. Sci.* **176**:262.

Tennant, B., Reina-Guerra, M., Marrold, D., and Goldman, M., 1969, Influence of microorganisms on intestinal absorption of oleic acid: I. Absorption by germfree and conventionalized rats, *J. Nutr.* **92**:389.

Valencia, R., Sacquet, E., Raibaud, P., Han, N'G.C., and Charlier, H., 1965, Role de la microflore intestinale (flore autochrone, monoflore de Veillonella, absence de flore) chez le rat subcarence en vitamin B$_{12}$, *C. R. Hebd. Seances Acad. Sci.* **260**:6439.

Van Heijenoort, Y., Sacquet, E., Raibaud, P., Demarne, Y., and Mathis, C., 1972, Effect of the microbial flora on the bile acid content of the rat small intestine, *C. R. Acad. Sci.* **275**:271.

Visscher, M.B., 1957, *Metabolic Aspects of Transport Across Cell Membranes* (Q.R. Murphy, ed.), p. 57, Univ. of Wisconsin Press, Madison.

Warren, K.S., and Newton, W.L., 1959, Portal and peripheral blood ammonia concentration in germfree and conventional guinea pigs, *Am. J. Physiol.* **197**:717.

Waxler, G.L., and Drees, D.T., 1972, Comparison of body weights, organ weights and histological features of selected organs of gnotobiotic, conventional and isolator-reared contaminated pigs, *Can. J. Comp. Med.* **36**:265.

Wells, C.L., and Babish, E., 1980, Gastrointestinal ecology and histology of rats monoassociated with anaerobic bacteria, *Appl. Environ. Microbiol.* 265.

Wiseman, G., 1964, *Absorption from the Intestine,* (G. Wiseman, ed.), p. 246, Academic Press, New York.

Wostmann, B.S., 1959, Nutrition of the germfree animal, *Ann. N.Y. Acad. Sci.* **78**:175.

Wostmann, B.S., 1966-67, Histidine decarboxylase in the adult rat, *Proc. Indiana Acad. Sci.* **76**:193.

Wostmann, B.S., 1975, Nutrition and metabolism of the germfree mammal, *World Rev. Nutr. Diet.* **22**:40.

Wostmann, B.S., and Bruckner-Kardoss, E., 1959, Development of cecal distension in germfree baby rats, *Am. J. Physiol.* **197**:1345.

Wostmann, B.S., and Bruckner-Kardoss, E., 1966, Oxidation-reduction potentials in cecal contents of germfree and conventional rats, *Proc. Soc. Exp. Biol. Med.* **121**:1111.

Wostmann, B.S., and Knight, P.L., 1961, Synthesis of thiamine in the digestive tract of the rat, *J. Nutr.* **74**:103.

Wostmann, B.S., and Knight, P.L., 1965, Antagonism between vitamins A and K in the germfree rat, *J. Nutr.* **87**:155.

Wostmann, B.S., Knight, P.L., and Kan, D.F., 1962, Thiamine in germfree and conventional animals: effect of the intestinal microflora on thiamine metabolism of the rat, *Ann. Sci.* **98**:561.

Wostmann, B.S., Wiech, N.L., and Kung, E., 1966, Catabolism and elimination of cholesterol in germfree rats, *J. Lipid Res.* **7**:77.

Wostmann, B.S., Reddy, B.S., Bruckner-Kardoss, E., Gordon, H.A., and Singh, B., 1973, Causes and possible consequences of cecal enlargement in germfree rats, in: *Germfree Research* (J.B. Heneghan, ed.), pp. 261-270, Academic Press, New York.

Wostmann, B.S., Bruckner-Kardoss, E., Beaver, M., Chang, L., and Madsen, D., 1976, Effect of dietary lactose at levels comparable to human consumption on cholesterol and bile acid metabolism of conventional and germfree rats, *J. Nutr.* **106**:1782.

Wostmann, B.S., Beaver, M., Chang, L., and Madsen, D., 1977, Effect of autoclaving of a lactose-containing diet on cholesterol and bile acid metabolism of conventional and germfree rats, *Am. J. Clin. Nutr.* **30**:1999.

Yamanaka, M., and Nomura, T., 1980, Relation between protein contents of diets and biological values in germfree rats, *J. Nutr. Sci. Vitaminol.* **26**:323.

Yamanaka, M., Iwai, H., Saito, M., Yamauchi, C., and Nomura, T., 1972, Influence of intestinal microbes on digestion and absorption of nutrients in diet and nitrogen retention in germfree, gnotobiotic and conventional mice: I. Protein and fat digestion and nitrogen retention in germfree and conventional mice, *Japn. J. Zootech. Sci.* **43**:272.

Yamanaka, M., Iawi, H., Saito, M., Yamauchi, C., and Nomura, T., 1973, Influence of intestinal microbes on digestion and absorption of nutrients in diet and nitrogen retention in germfree, gnotobiotic and conventional mice. II. Protein and fat digestion and nitrogen retention in monocontaminated gnotobiotic mice, *Japn. J. Zootech. Sci.* **44**:380.

Yamanaka, M., Nomura, T., Tokioka, J., and Kametaka, M., 1980, A comparison of the gastrointestinal tract in germfree and conventional mice fed an amino acid mixture or purified whole egg protein, *J. Nutr. Sci. Vitaminol.* **26**:435.

Yang, M.G., Bergen, W.G., Sculthorpe, A.E., and Mickelsen, O., 1972, Utilization of ^{14}C-labeled *E. coli* by the rat cecum and after force feeding, *Proc. Soc. Exp. Biol. Med.* **139**:1312.

Yoshida, T., 1972, Studies on nutritional role of intestinal microflora by using germfree animal, *J. Japn. Soc. Food Nutr.* **25**:501.

Yoshida, T., Pleasants, J.R., and Reddy, B.S., 1968, Efficiency of digestion in germfree and conventional rabbits, *Br. J. Nutr.* **22**:723.

Yoshida, T., Hirano, K., Kamizono, M., and Kuramasu, S., 1969, Nutrition of germfree Japanese quail: II. On the ratio of retention of N, Ca, Mg, and P during growth, *Nutr. Food* **22**:213.

Yoshida, T. Pleasants, J.R., Reddy, B.S., Wostmann, B.S., 1971, Amino acid composition of cecal contents and feces in germfree and conventional rabbits, *J. Nutr.* **101**:1423.

Young, R.J., Garrett, R.L., and Griffiths, M., 1963, Factors affecting the absorbability of fatty acid mixtures high in saturated fatty acids, *Poult. Sci.* **42**:1146.

Young, R.M., and James, L.H., 1942, Action of intestinal microorganisms on ascorbic acid, *J. Bact.* **44**:75.

Young, R.M., and Rettger, L.F., 1943, Decomposition of vitamin C by bacteria, *J. Bact.* **46**:351.

Yudkin, J., 1963, Availability of microbially synthesized thiamine in the rat, *J. Nutr.* **81**:63.

Index

Ascorbate
  anti-allergic properties, 33
    effects on leukotrienes, 36
    effects on prostaglandins, 35
    inhibition of degranulation, 36
    neutralization of histamine effects, 35
  anti-inflammatory properties
    antioxidant effect, 36,37
    effect on extracellular matrix, 37
    other mechanisms, 40
  immunostimulatory effects, 20
    leukocyte motility, 20,26
    lymphocyte proliferation and migration, 32
    neutrophil motility, 21,29
    phagocyte antimicrobial activity, 29
  mechanism of lymphocyte proliferative effect
    antioxidant mechanisms, 33
    cyclic nucleotides, 32

Brush border membrane, 234
  distribution of enzymes, 238
  function of enzymes, 237
  morphology of the enterocyte, 234
  structure, 235

Cancer
  and dietary vitamin A, 47
    Buffalo, New York, study, 53
    English study, 56
    Japanese study, 58
    Norwegian study, 52
    Singapore study, 53
    Western Electric study, 58
  summary of studies, 60,64

Diabetes
  fiber in management of, 172

Fiber
  clinical implications, 169
    in colon cancer, 19
    in constipation, 191
    in Crohn's disease, 193
    in diabetes, 171
    in diverticular disease, 192
    in dumping syndrome, 184
    in hyperlipidemia, 185
    in other diseases, 193
    in weight control, 190
  in different foods, 179
  metabolic effects
    reduced absorption, 177
    reduced endocrine response, 176
    reduced gastric emptying time, 178
    reduced glucose response, 175
    reduced transit time, 178
  possible adverse effects, 194
  purified fiber, 187

Gnotobiotic animals
  bile acids in, 308
  characteristics of intestinal contents, 282
    biochemical characteristics, 285
    dry weight, 283
    osmolality, 284
    pH, 285
    redox potential, 285
    relative viscosity, 284
  carbohydrate metabolism in, 315
  digestive enzymes in, 303
  fat metabolism in, 312

Gnotobiotic animals (*cont.*)
  intestinal function in
    mineral absorption, 294
    vitamin synthesis and absorption, 300
    water and electrolyte transport, 287
  intestinal morphology of, 274
    cecal reduction, 277
    histology, 275
    length and weight, 274
    mitotic index, 276
    surface area, 274
  nutrient absorption in, 271
  protein metabolism in, 316

Homocystinuria
  and methionine metabolism, 15

Isomaltose
  isomaltase deficiency, 234
  malabsorption, 233

Keshan disease
  clinical manifestations, 207
    acute type, 207
    chronic type, 207
    insidious type, 209
    subacute type, 208
  epidemiology, 218
  etiology, 211
    infection, 211
    intoxication, 212
    nutrient status, 212
  food selenium and, 215
  history and epidemiology, 205
  morphological observations, 209
  selenium status and, 212
  treatment, 211

Metabolic bone disease
  and aluminum loading, 84
  biochemical findings, 70
  and circulating inhibitor, 86
  general description, 68
  in infants, 77
  and TPN, 77
  and trace elements, 84
  and vitamin D, 78

Methionine
  decarboxylation, 9
  formation of 3-methylthiopropionate from,
    10

Methionine (*cont.*)
  importance of transamination in
    catabolism, 7,13
  metabolism, early studies on, 1
  oxidation of methyl group, 3
    importance of choline, betaine and
      sarcosine, 3
    importance of S-adenosylmethionine, 6

3-Methylthiopropionate
  formation, 10
  metabolism, 12

Milk
  citrate in, 141
  picolinic acid in, 144
  prostaglandins in, 146
  zinc binding ligands in, 147

Protein synthesis, 112
  during development, 116

Protein turnover in man
  comparison with animals, 112
  effect of nutrition, 119
    dietary protein, 126
    diet composition, 123
    food intake, 120
    non-protein energy, 123
    undernutrition, 131
  end-product method techniques
    metabolic pool, 106
    model, 105
    steady state, 108
    urinary products, 106
  methods of study, 95,110
    end-product methods, 104
    general considerations, 95
    precursor methods, 97
  precursor method techniques
    choice of label and amino acid, 98
    CO_2 production, 100
    infusion, 99
    model, 97
    the precursor pool, 102
    the steady state, 104

Selenium
  biological role, 226
  concentration in foods, 215
  in prevention of Keshan disease, 220,224
  status in Keshan disease, 218

Sucrase-isomaltase deficiency
  biochemical background, 239
  diagnosis, 245,254
    disaccharidase assay, 251
    hydrogen breath test, 250
    tolerance test, 248
  genetic conditions, 251
  incidence, 251
  pathophysiology, 241
  symptoms, 243
  treatment, 259

Sucrose
  malabsorption, 233
  oral tolerance test, 248
  sucrase deficiency, 233

Total parenteral nutrition
  aluminum loading, 84
  and hypercalciuria, 85
  and metabolic bone disease, 77
  and serum vitamin D levels, 82
  and trace elements, 84

Vitamin A
  blood levels, 50
  and carcinogenesis, 47
    animal studies, 48
    epidemiological studies, 51

Vitamin D
  in metabolic bone disease, 78

Zinc
  absorption, 159
  binding ligands and complexes, 139
    casein, 143
    citrate, 141
    in bile and pancreatic fluid, 156
    in duodenum, 157
    in infant nutrition, 147
    in milk, 140
    picolinic acid, 144
    prostaglandins, 146
  genetic abnormalities
    acrodermatitis enteropathica, 148
    adema disease, 151
    Alaskan malamute chondrodysplasia,
        152
    lethal milk, 150
  homeostasis
    theories, 154